The Role of Mineral Elements in the Crop Growth and Production

The Role of Mineral Elements in the Crop Growth and Production

Editors

Gang Li
Dong-Xing Guan
Daniel Menezes-Blackburn

MDPI

Basel • Beijing • Wuhan • Barcelona • Belgrade • Novi Sad • Cluj • Manchester

Editors
Gang Li
Institute of Urban Environment
Chinese Academy of Sciences
Xiamen
China

Dong-Xing Guan
College of Environmental and Resource Sciences
Zhejiang University
Hangzhou
China

Daniel Menezes-Blackburn
Department of Soils, Water and Agricultural Engineering
Sultan Qaboos University
Muscat
Oman

Editorial Office
MDPI
St. Alban-Anlage 66
4052 Basel, Switzerland

This is a reprint of articles from the Special Issue published online in the open access journal *Agronomy* (ISSN 2073-4395) (available at: www.mdpi.com/journal/agronomy/special_issues/mineral_agriculture).

For citation purposes, cite each article independently as indicated on the article page online and as indicated below:

Lastname, A.A.; Lastname, B.B. Article Title. *Journal Name* **Year**, *Volume Number*, Page Range.

ISBN 978-3-7258-0318-7 (Hbk)
ISBN 978-3-7258-0317-0 (PDF)
doi.org/10.3390/books978-3-7258-0317-0

Contents

About the Editors

Gang Li

Dr. Gang Li has been a Professor of environmental soil chemistry at the Institute of Urban Environment, Chinese Academy of Sciences, since 2014. He obtained his PhD degree in Environmental Science in 2013 from the Chinese Academy of Sciences. His research interests focus on the biogeochemistry of heavy metals and nutrients in agricultural and urban soils. Dr. Li is a member of the International Exchange and Cooperation Committee of the Soil Science Society of China, a council member of the 10th Surface and Biogeochemistry Committee of the Chinese Society for Mineralogy, Petrology, and Geochemistry, and a council member of the Soil and Fertilizer Society of Zhejiang Province. He has published over 100 papers in international peer-reviewed journals with citations of more than 4000 times (Web of Science) and an H-index of 34.

Dong-Xing Guan

Dr. Dong-Xing Guan is an Associate Professor in Environmental and Soil Biogeochemistry. Dong-Xing obtained his Ph.D. in Environmental Science from Nanjing University, China, in 2016. Prior to working at Zhejiang University, he held positions in Environmental Geochemistry, first as a postdoc with Nanjing University and then as an Associate Professor with Tianjin University. Dr. Guan's research interests are oriented around how contaminants and nutrients cycle in the environment. His specific interests and projects focus on the following: 1) biogeochemical and interfacial processes of trace elements in the water–soil–biota continuum; 2) bioavailability characteristics and ecological remediation of chemical contaminants in soils and sediments. Up to now, he has authored more than 60 peer-reviewed publications in well-reputed journals such as *Environmental Science and Technology* and *Water Research* and contributed to one book chapter and nine licensed patents and software copyrights.

Daniel Menezes-Blackburn

Dr. Daniel Blackburn has been an Associate Professor of Soil Microbiology at SWAE-CAMS-SQU since 2017. Previously, Daniel worked as a Senior Research Associate at the Lancaster Environment Centre at Lancaster University (UK), where he studied the use of phytases and organic acids for the mobilization of soil organic phosphorus plant nutrition. Before that, he was a Humboldt Fellow at the Max Rubner-Institut Karlsruhe (Germany), working with the phytase biochemistry of soil bacteria. His PhD was performed both at the La Frontera University (Chile) and the University of Naples (Italy), where he developed biotechnologies using phytase enzymes for the sustainable plant use of organic phosphorus from soils and manures. Daniel is currently teaching Soil Microbial Ecology, Plant–Soil environment, and Waste Management at SWAE, and his research projects cover the topics of soil salinity management and the isolation and study of extremophiles from Omani environments. Daniel is part of the editorial teams of the following journals: *Journal of Soil Science and Plant Nutrition*, *Critical Reviews in Environmental Science and Technology*, *Soil & Environmental Health*, and *Journal of Environmental Quality*.

Preface

With the global population on the rise, sustainably meeting the burgeoning food demand is a critical challenge. This challenge calls for optimizing plant mineral nutrition to enhance productivity, resource efficiency, and environmental stewardship. This Reprint provides insights from a Special Issue of *Agronomy*, highlighting the role of mineral elements in crop growth and production. It emphasizes emerging management strategies designed to balance mineral element supply with crop needs. The key to sustainable agriculture lies in enhancing nutrient acquisition, customizing nutrient application, and integrating a comprehensive plant-soil perspective. This approach is fundamental for leveraging mineral nutrition for sustainable agricultural practices, ensuring food security, and promoting environmental sustainability.

The individual papers in this Reprint offer reports on the latest research advancements or comprehensive reviews of existing knowledge. Drawing from a diverse range of disciplines, these articles together aim to deepen our understanding of the importance of mineral elements in crop growth and production. We extend our heartfelt gratitude to the authors, whose contributions are instrumental in defining the quality of this research topic. We also express our profound appreciation to the reviewers, whose invaluable insights have helped enhance the quality of these papers.

Gang Li, Dong-Xing Guan, and Daniel Menezes-Blackburn
Editors

agronomy

Editorial

The Importance of Mineral Elements for Sustainable Crop Production

Dong-Xing Guan [1], Daniel Menezes-Blackburn [2] and Gang Li [3,4,5,*]

1 College of Environmental and Resource Sciences, Zhejiang University, Hangzhou 310058, China
2 Department of Soils, Water and Agricultural Engineering SQU, Sultan Qaboos University, Al-Khoud, Muscat 123, Oman
3 CAS Key Lab of Urban Environment and Health, Institute of Urban Environment, Chinese Academy of Sciences, Xiamen 361021, China
4 Zhejiang Key Laboratory of Urban Environmental Processes and Pollution Control, CAS Haixi Industrial Technology Innovation Center in Beilun, Ningbo 315830, China
5 Jiaxing Key Lab of Soil Health, Yangtze Delta Region Healthy Agriculture Institute, Jiaxing 314503, China
* Correspondence: gli@iue.ac.cn

1. Introduction

By 2050, the global population is projected to reach 9.7 billion people, necessitating a substantial increase in food production [1]. Mineral elements, including macro- and micronutrients, are essential for crops to complete their growth cycles and produce the yields necessary to meet demand. Deficiencies in minerals such as nitrogen (N), phosphorus (P), and potassium (K) severely constrain plant growth and limit agricultural productivity worldwide [2]. While mineral fertilizers supported Green Revolution yield improvements, their excessive use degrades soil and water quality [3]. Sustainably meeting future nutritional needs requires optimizing plant mineral nutrition to improve productivity, nutrient use efficiency, and stress resilience, while stabilizing soil resources.

This editorial synthesizes insights from a recent special issue of Agronomy themed by the role of mineral elements in crop growth and production. The collection highlights emerging strategies for tailoring mineral element applications, enhancing acquisition, and integrating plant-soil dynamics. Balancing these perspectives is key to leveraging plant nutrition for the sustainable intensification of agriculture.

Citation: Guan, D.-X.; Menezes-Blackburn, D.; Li, G. The Importance of Mineral Elements for Sustainable Crop Production. *Agronomy* **2024**, *14*, 209. https://doi.org/10.3390/agronomy14010209

Received: 11 January 2024
Accepted: 16 January 2024
Published: 17 January 2024

2. Optimizing the Application of Essential Macro- and Micro-Nutrients

The balanced application of essential macro-nutrients and micro-nutrients is crucial for crops to fully express their genetic potential and achieve optimal growth and yield. Research has shown that the synergistic effects of micro-nutrients can significantly enhance crop performance. For instance, Safdar et al. [4] presented a two-year field study examining the effects of sole and combined soil application of boron (B) and zinc (Zn) on oilseed rape under semiarid conditions. They demonstrated that the combined application of B and Zn in oilseed rape led to marked improvements in the yield, oil content, and quality parameters. This synergy likely stems from the complementary roles that these nutrients play during critical reproductive and developmental phases, such as pollen viability and seed formation. As the study's geographical and temporal scope was limited to a semi-arid climate over two growing seasons, it is crucial to extend this research across diverse environments and crop genotypes. This would help to develop more comprehensive and universally applicable guidelines for optimal B and Zn application rates and ratios.

Głowacka et al. [5] explored the interactive effects of N and sulfur (S) fertilization in soybean production. They found that a balanced application of these nutrients significantly increased productivity, seed protein content, and S-containing amino acids compared with N application alone. This research suggests that S plays a critical role in influencing N utilization efficiency. The optimal N:S ratio appeared to differ across the different soybean growth stages, indicating a need for more nuanced application strategies. Further research is necessary to understand the interactive effects of N and S across a broader spectrum of fertilizer application rates, timings, and environmental conditions.

3. Enhancing Nutrient Acquisition and Utilization Efficiency

Efficient acquisition and utilization of nutrients are paramount for reducing reliance on external inputs and promoting sustainable agricultural practices. Innovative genetic and microbial approaches have shown promise for enhancing crop nutrient capture and use efficiency. Hu et al. [6] investigated the overexpression of a phosphate transporter gene, *OsPHT1;4*, in rice. They found that this genetic modification significantly boosted P uptake, utilization efficiency, grain yields, and overall biomass under conditions where P was deficient. This finding suggests that the targeted manipulation of nutrient transporters provides a promising route to enhance crop P efficiency.

Lin et al. [7] demonstrated the potential of microbiome engineering by showing that inoculating rice with a selenium (Se)-tolerant bacterial strain increased plant Se accumulation. This finding opens exciting possibilities for enhancing the nutritional value of crops through the manipulation of the soil microbiome. Translating this increase in Se content into improved human nutrition requires further research and careful dietary integration.

Li et al. [8] conducted a detailed investigation into the effects of varying P supply on soybeans using a split-root system. In a related study, Li et al. [9] further researched the optimal P levels necessary for effective nodulation, N fixation activity, and plant N accumulation in soybeans. Both studies highlighted the complex interplay between P supply and soybean growth. They underscored the necessity of a more sophisticated approach to nutrient management, one that considers the temporal variations in crop needs and the intricate mechanisms by which plants acquire and utilize nutrients. These findings represent a significant step forward, and further research is needed to understand how these insights might translate across different legume species under various environmental conditions.

4. Matching Nutrition to Plant-Soil Environments

Understanding and matching nutrition to specific plant-soil environments is essential for optimizing crop health and yields while maintaining environmental sustainability. Kumar et al. [10] investigated the effects of tillage-based crop establishment methods and irrigation approaches on wheat production. Their research revealed that these management practices significantly influence soil biological properties, including the microbial community structure and the abundance of fungi and bacteria. These biological shifts, in turn, affect soil fertility, moisture availability, and ultimately, crop yields. This study underscores the importance of taking a broader agro-ecological perspective that integrates soil, water, and crop management factors. Such an approach is vital for balancing productivity gains with soil and environmental health. Further studies are required to identify combinations of practices that maximize synergies across diverse wheat cropping contexts.

Brodowska et al. [11] examined how K fertilization, applied in combination with N, affects the levels of trace elements in maize-cultivated soils, a crucial aspect of plant nutrition often overlooked. Their study sheds light on the complex interactions between different nutrients and their collective effects on crop health and productivity. These interactions underscore the need for a comprehensive approach to nutrient management that considers the myriad ways in which different elements interact within the plant-soil system.

Ibrahim et al. [12] provided a comprehensive review of P mobilization strategies in plant-soil environments, offering insights into various biotechnological, chemical, and agronomic options. This review discusses plant root adaptations, ligand synthesis, enzyme applications, and rhizosphere modification, among others. These strategies have the potential to enhance crop P acquisition, while reducing reliance on inorganic P inputs and improving soil health. Moving these strategies from conceptual feasibility to cost-effective field application will require continued research and innovation.

Zhao et al. [13] introduced a novel in situ probe designed to analyze the availability of multiple elements such as S, P, and arsenic at the soil/sediment-water interface and in the rice rhizosphere. Based on the diffusive gradients in thin-films technique, this tool represents a significant advance in soil science, offering the potential to dramatically improve our understanding of nutrient dynamics, guide more precise fertilizer application, and ensure environmental safety [14]. The development and application of such high-resolution measurement technologies are crucial for advancing precision agriculture and sustainable nutrient management practices [12,15].

5. Balancing Mineral Nutrition Status for Sustainable Production

The collective research presented in this special issue underscores that while mineral nutrients are indispensable for crop productivity, their management must be strategically balanced to yield gains with environmental sustainability. Continually integrating perspectives across plant nutrition, genetics, soil science, agronomy, and ecology will be essential going forward. Key priorities include: leveraging synergies between mineral elements while avoiding excesses or imbalances; building adaptive, site-specific application guidelines tailored to crop needs over growth cycles; enhancing mineral uptake efficiency via crop breeding and microbiome manipulation; fostering integrated soil fertility and health through diversified practices; and developing tools to monitor plant status and predict soil fertility responses. A systems-level approach balancing productivity, profitability, and environmental stewardship will be critical to sustainably meeting future food demands. The research compiled in this special issue provides key insights to guide mineral nutrition management in this direction.

6. Conclusions

In conclusion, this special issue highlights the essential roles mineral elements play in crop growth cycles and the need to continually refine nutrition management. Optimizing application, enhancing efficiency, and integrating plant-soil dynamics will be crucial for leveraging mineral nutrition for sustainable intensification. Continued systems-level research to balance productivity gains with environmental protection is the cornerstone for addressing future nutritional demands. As the world moves towards sustainable agricultural practices, the insights and strategies discussed in this special issue will provide a valuable foundation for guiding mineral nutrition management in the right direction.

Author Contributions: D.-X.G.: Writing—original draft, Writing—review & editing; D.M.-B.: Writing—review & editing. G.L.: Writing—review & editing. All authors have read and agreed to the published version of the manuscript.

Funding: This work is supported by the Key Project of Science and Technology Innovation in Ningbo City (No.: 2023Z114; 2022Z169).

Data Availability Statement: The data presented in this study are available on request from the corresponding author.

Conflicts of Interest: The authors declare no conflict of interest.

References

1. United Nations Department of Economic and Social Affairs, Population Division. World Population Prospects 2022: Summary of Results. 2022, UN DESA/POP/2022/TR/NO. 3. Available online: https://www.un.org/development/desa/pd/content/World-Population-Prospects-2022 (accessed on 10 January 2024).
2. Mueller, N.D.; Gerber, J.S.; Johnston, M.; Ray, D.K.; Ramankutty, N.; Foley, J.A. Closing yield gaps through nutrient and water management. *Nature* **2012**, *490*, 254–257. [PubMed]
3. Rashmi, I.; Roy, T.; Kartika, K.S.; Pal, R.; Coumar, V.; Kala, S.; Shinoji, K.C. Organic and Inorganic Fertilizer Contaminants in Agriculture: Impact on Soil and Water Resources. In *Contaminants in Agriculture: Sources, Impacts and Management*; Naeem, M., Ansari, A.A., Gill, S.S., Eds.; Springer Nature Switzerland AG: Cham, Switzerland, 2020; pp. 3–41.

4. Safdar, M.E.; Qamar, R.; Javed, A.; Nadeem, M.A.; Javeed, H.M.R.; Farooq, S.; Głowacka, A.; Michałek, S.; Alwahibi, M.S.; Elshikh, M.S.; et al. Combined application of boron and zinc improves seed and oil yields and oil quality of oilseed rape (*Brassica napus* L.). *Agronomy* **2023**, *13*, 2020. [CrossRef]
5. Głowacka, A.; Jariene, E.; Flis-Olszewska, E.; Kiełtyka-Dadasiewicz, A. The effect of nitrogen and sulphur application on soybean productivity traits in temperate climates conditions. *Agronomy* **2023**, *13*, 780. [CrossRef]
6. Hu, Z.; Huang, X.; Wang, X.; Xia, H.; Liu, X.; Sun, Y.; Sun, S.; Hu, Y.; Cao, Y. Overexpression of *OsPHT1;4* increases phosphorus utilization efficiency and improves the agronomic traits of rice cv. Wuyunjing 7. *Agronomy* **2022**, *12*, 1332. [CrossRef]
7. Lin, X.-R.; Chen, H.-B.; Li, Y.-X.; Zhou, Z.-H.; Li, J.-B.; Wang, Y.-Q.; Zhang, H.; Zhang, Y.; Han, Y.-H.; Wang, S.-S. *Priestia* sp. LWS1 is a selenium-resistant plant growth-promoting bacterium that can enhance plant growth and selenium accumulation in *Oryza sativa* L. *Agronomy* **2022**, *12*, 1301. [CrossRef]
8. Li, H.; Wang, X.; Liang, Q.; Lyu, X.; Li, S.; Gong, Z.; Dong, S.; Yan, C.; Ma, C. Regulation of phosphorus supply on nodulation and nitrogen fixation in soybean plants with dual-root systems. *Agronomy* **2021**, *11*, 2354.
9. Li, H.; Wang, L.; Zhang, Z.; Yang, A.; Liu, D. Effect of phosphorus supply levels on nodule nitrogen fixation and nitrogen accumulation in soybean (*Glycine max* L.). *Agronomy* **2022**, *12*, 2802.
10. Kumar, R.; Naresh, R.K.; Bhatt, R.; Chandra, M.S.; Kumar, D.; Alamri, S.; Siddiqui, M.H.; Alfagham, A.T.; Kalaji, H.M. Tillage crop establishment and irrigation methods improve the productivity of wheat (*Triticum aestivum*): Water use studies, and the biological properties and fertility status of soil. *Agronomy* **2023**, *13*, 1839.
11. Brodowska, M.S.; Wyszkowski, M.; Bujanowicz-Haraś, B. Mineral fertilization and maize cultivation as factors which determine the content of trace elements in soil. *Agronomy* **2022**, *12*, 286. [CrossRef]
12. Ibrahim, M.; Iqbal, M.; Tang, Y.-T.; Khan, S.; Guan, D.-X.; Li, G. Phosphorus mobilization in plant–soil environments and inspired strategies for managing phosphorus: A review. *Agronomy* **2022**, *12*, 2539. [CrossRef]
13. Zhao, M.; Liu, J.; Zhang, C.; Liang, X.; E, Q.; Liu, R.; Zhao, Y.; Liu, X. Development and applications of an in situ orobe for multi-element high-resolution measurement at soil/sediment-water interface and rice rhizosphere. *Agronomy* **2021**, *11*, 2383. [CrossRef]
14. Guan, D.-X.; He, S.-X.; Li, G.; Teng, H.H.; Ma, L.Q. Application of diffusive gradients in thin-films technique for speciation, bioavailability, modeling and mapping of nutrients and contaminants in soils. *Crit. Rev. Environ. Sci. Technol.* **2022**, *52*, 3035–3079. [CrossRef]
15. Li, X.-Y.; Li, S.-Q.; Jiang, Y.-F.; Yang, Q.; Zhang, J.-C.; Kuzyakov, Y.; Teng, H.H.; Guan, D.-X. Multi-imaging platform for rhizosphere studies: Phosphorus and oxygen fluxes. *J. Environ. Manag.* **2024**, *351*, 119763. [CrossRef] [PubMed]

agronomy

MDPI

Article

Combined Application of Boron and Zinc Improves Seed and Oil Yields and Oil Quality of Oilseed Rape (*Brassica napus* L.)

Muhammad Ehsan Safdar [1], Rafi Qamar [1], Amara Javed [1], Muhammad Ather Nadeem [1], Hafiz Muhammad Rashad Javeed [2], Shahid Farooq [3,*], Aleksandra Głowacka [4], Sławomir Michałek [5,*], Mona S. Alwahibi [6], Mohamed S. Elshikh [6] and Mohamed A. A. Ahmed [7,8]

1 Department of Agronomy, College of Agriculture, University of Sargodha, Sargodha 40100, Pakistan; mehsan.safdar@uos.edu.pk (M.E.S.); rafi.qamar@uos.edu.pk (R.Q.); drafi1573@gmail.com (A.J.); ather.nadeem@uos.edu.pk (M.A.N.)
2 Department of Environmental Sciences, COMSATS University Islamabad, Vehari Campus, Vehari 61100, Pakistan; rashadjaveed@cuivehari.edu.pk
3 Department of Plant Protection, Faculty of Agriculture, Harran University, Şanlıurfa 63050, Turkey
4 Department of Plant Cultivation Technology and Commodity Sciences, University of Life Sciences in Lublin, 13 Akademicka Street, 20-950 Lublin, Poland; aleksandra.glowacka@up.lublin.pl
5 Department of Botany and Plant Physiology, University of Life Sciences in Lublin, 13 Akademicka Street, 20-950 Lublin, Poland
6 Department of Botany and Microbiology, College of Science, King Saud University, Riyadh 11451, Saudi Arabia; malwhibi@ksu.edu.sa (M.S.A.); melshikh@ksu.edu.sa (M.S.E.)
7 Plant Production Department (Horticulture—Medicinal and Aromatic Plants), Faculty of Agriculture (Saba Basha), Alexandria University, Alexandria 21531, Egypt; drmohamedmarey19@alexu.edu.eg
8 School of Agriculture, Yunnan University, Chenggong District, Kunming 650091, China
* Correspondence: shahid@harran.edu.tr (S.F.); slawomir.michalek@up.lublin.pl (S.M.)

Citation: Safdar, M.E.; Qamar, R.; Javed, A.; Nadeem, M.A.; Javeed, H.M.R.; Farooq, S.; Głowacka, A.; Michałek, S.; Alwahibi, M.S.; Elshikh, M.S.; et al. Combined Application of Boron and Zinc Improves Seed and Oil Yields and Oil Quality of Oilseed Rape (*Brassica napus* L.). *Agronomy* **2023**, *13*, 2020. https://doi.org/10.3390/agronomy13082020

Academic Editors: Gang Li, Dong-Xing Guan and Daniel Menezes-Blackburn

Received: 11 June 2023
Revised: 25 July 2023
Accepted: 28 July 2023
Published: 29 July 2023

Abstract: Oilseed crops require several micronutrients to support their physiological functions and reproductive phases. A deficiency of these nutrients can significantly reduce the yield and oil quality of oilseed crops. Soil application of micronutrients can reduce their deficiency and improve plant growth, yield, and oil quality. Oilseed rape (*Brassica napus* L.) is an important oilseed crop that produces oil with low levels of saturated fat and high levels of beneficial omega-3 fatty acids, which renders it a widely used cooking oil. However, the yield and oil quality of oilseed rape are significantly affected by the deficiency of boron (B) and zinc (Zn). This two-year field study determined the influence of sole and combined soil application of B and Zn on the physiological attributes of plants, seed and oil yields, and oil quality under semiarid climatic conditions. Nine different B and Z combinations, i.e., B0 + Zn0 (control), B0 + Zn8, B0 + Zn10, B1 + Zn0, B1 + Zn8, B1 + Zn10, B2 + Zn0, B2 + Zn8, and B2 + Zn10 (kg ha^{-1}), were included in the study. Sole and combined application of B and Zn significantly altered physiological attributes, seed and oil yields, and oil quality. The highest values for plant height, number of siliques per plant, number of seeds per silique, 1000-seed weight, seed and oil yields, oil quality (higher stearic acid, palmitic acid, oleic acid, linoleic acid, linolenic acid, and lower erucic acid), and physiological traits (protein concentration, soluble sugar concentration, chlorophyll concentration, photosynthesis and transpiration rates, and stomatal conductance) were recorded with the combined application of 2 + 8 kg ha^{-1} B and Zn, respectively, during both years of this study. The lowest values of yield- and oil-quality-related traits and physiological attributes were recorded for the control treatment. A dose-dependent improvement was recorded in B and Zn contents in leaves, and the highest values were recorded with the combined soil application of 2 + 10 kg ha^{-1} B + Zn, respectively. It can be concluded that 2 + 8 kg ha^{-1} B + Zn should be applied to oilseed rape for higher seed and oil yields and better oil quality under semiarid climatic conditions.

Keywords: antioxidant enzymes; oil profile; seed yield; photosynthesis; semiarid climate

1. Introduction

Oilseeds, alongside sugar and cereal crops, are essential components in ensuring a well-balanced and nutritious diet for humans. Vegetable oil is a significant source of essential fatty acids and vitamin E, which are vital for the proper physiological functioning of the human body [1,2]. The production of edible oil is of utmost importance in fulfilling the domestic demand and industrial requirements of a country. Additionally, it holds promising potential to emerge as a significant contributor to employment opportunities [3]. Pakistan is currently importing 2.917 million tons of edible oil, with a total value of PKR 574.199 billion (USD 3.419 billion) annually. Edible oil production in Pakistan during 2020–2021 was 0.374 million metric tons. The anticipated total yield of edible oil extracted from various crops was 3.291 million metric tons [3]. Due to a significant shortage of edible oil in Pakistan, it is imperative for the government to augment the cultivation of oilseed crops to address the edible oil shortage issue [3].

Oilseed rape (*Brassica napus* L.) is an important oilseed crop on a global scale [4]. It is positioned in third place following soybean and palm oil in terms of area under cultivation and holds fifth position in terms of oil production [5]. Oilseed rape oil exhibits superior nutritional properties due to its lower content of erucic acid and saturated fats, which are present in concentrations of 2% and 6%, respectively [6]. The seeds typically contain approximately 40–45% oil [7], with 6–14% linolenic acid and 50–66% oleic acid [8]. The oil has the required profile of saturated fatty acids (7%), higher unsaturated fatty acid contents, i.e., oleic acid (~61%) and linoleic acids (8%) [9], and lower erucic acid, glucosinolates, and cholesterol. Therefore, this oil is considered safe for human consumption [10,11]. It is cultivated on 2418 thousand hectares in Pakistan, which produce 2256 thousand tons of seeds and 374 thousand tons of oil [3]. It is an inexpensive oil that helps individuals recover from malnutrition and improves their health, making it ideal for populations residing in developing nations [12]. Climate extremes, such as heat and drought stress [13–15], soil salinity [16], limited light availability [17], and waterlogging [18], are the major constraints significantly reducing the yield and oil quality of oilseed rape.

Insufficient availability of nutrients, particularly micronutrients, is another significant constraint in oilseed rape production [19]. Micronutrients play a crucial role during the vegetative and reproductive growth of plants [20]. High-yielding cultivars produce low yields despite NPK application, which can be attributed to the insufficient utilization of micronutrients [21]. The optimal yield potential of high-yielding cultivars can be realized through the application of micronutrients in conjunction with macronutrients [22]. Chaudry et al. [23] reported that the application of micronutrients, specifically zinc (Zn) and boron (B), significantly increased wheat yield compared to the control group, whether applied individually or in combination, while Mandal et al. [24] indicated a significant correlation between the application of fertilizer and the physiological growth process.

Zinc is a vital micronutrient required for the growth and development of crop plants. It is a constituent of carbonic anhydrase and elicits aldolase, both of which are imperative for carbon metabolism [25]. Additionally, it is a constituent of diverse biomolecules, including lipids and proteins, and functions as a cofactor for auxins, thereby exerting a significant influence on nucleic acid degradation [26]. Zinc is an essential constituent of numerous enzymes and is compulsory for their activation. Consequently, Zn deficiency limits carbohydrate digestion, induces injuries to the pollen tube, and reduces yield [27]. The primary cause of Zn deficiency in crops is the reduced solubility of Zn in soils, which is somewhat more significant than the reduction in the overall amount of Zn [28]. The efficacy of Zn application in improving crop yield and quality has been reported in earlier studies [29]. Soil application of Zn to oilseed rape has been found to increase branching, the number of pods, and seed production [30]. A significant proportion (80% of rainfed area) of Pakistani soils exhibits Zn deficiency due to higher soil pH levels [31]. Zinc deficiency in the topsoil significantly hampers agricultural productivity. The decrease in Zn levels in plants induces the generation of reactive oxygen species (ROS) within plant tissues, which disrupts the integrity of cell membranes and impedes the normal functioning of cells [32].

Boron is another essential micronutrient and necessary for the growth of crop plants. However, it can be observed that B levels are rapidly decreasing in Yermosols or Aridisols [33]. Boron deficiency is the second most important micronutrient constraint in crop production after zinc [34]. Boron is an important component for various biological processes in plants, including but not limited to the growth and development of pollen tubes, the maintenance of membrane integrity, seed production, and pollination [35]. The primary roles of B include the breakdown of nucleic acids, carbohydrates, proteins, indole acetic acid, and phenol, which are involved in the synthesis of plant cell walls and the maintenance of membrane integrity [36]. Additionally, B plays a key role in cellular division and the control of carbohydrate and protein metabolism, and these processes influence the reproductive phase and development of seeds [37]. Boron deficiency in soil leads to the erratic growth of seedlings and reduced photosynthesis [38]. Furthermore, B deficiency limits root elongation and deforms flowers and fruits due to inadequate cell division in the meristematic region. Conversely, a sufficient B supply improves root development [39]. Nevertheless, an excess or deficiency of B impedes physiological and morphological functions in plants [40].

Zinc is a crucial micronutrient for human beings and plays a significant role in various biological processes, including protein, lipid, and nucleic acid metabolism and gene transcription. Zinc plays important roles in reproduction, immune response, and wound healing [41]. Therefore, Zn must be included in the diet for these processes. Boron is important for the promotion of bone health, the facilitation of hormonal equilibrium, and the provision of defense against oxidative damage and inflammation in both plant and animal organisms [42]. Although B is not essential for human health, its deficiency could negatively affect these processes. An enhancement in these micronutrient levels in daily dietary intake is particularly important in rural areas of developing nations, where populations are already at risk of micronutrient deficiencies [43]. Hence, the introduction of fortified foods with micronutrients is necessary for these areas [44]. Enhancing the bioavailability of micronutrients in crucial food grains through biofortification could be a promising strategy to cope with micronutrient deficiency. The integration of micronutrient-efficient cultivars and appropriate agronomic techniques presents a promising opportunity to enhance the yield and quality of arable crops [45,46]. Nonetheless, the implementation of agronomic strategies is comparatively more rapid and dependable in contrast to genetic biofortification [21,47]. The implementation of biofortification techniques in oilseed crops, specifically targeting the incorporation of nutritionally significant elements such as B and Zn, has emerged as a promising strategy to address the issue of micronutrient deficiency in developing nations [48]. Agronomic biofortification of micronutrients can be achieved through three primary approaches, i.e., seed priming, foliar, and soil application. These methods are characterized by their rapid nutrient delivery, economic feasibility, and ease of implementation [21,47]. Soil application of micronutrients is a widely employed agronomic technique to enhance both the yield and nutritional quality of food crops [33,49]. Nevertheless, the excessive application of micronutrients can lead to crop toxicity by disrupting the equilibrium between soil solution and adsorption sites [50,51]. Usman and Mohamed [52] observed that the application of Zn, either alone or in combination with B elevated Zn levels in plants. An increased Zn concentration leads to higher flowering and a reduction in fruit drop [53].

High-yielding cultivars of oilseed rape have been introduced in Pakistan to improve the production of edible oil in the country. However, there is a significant gap between the potential yield of these cultivars and the actual yield. Numerous studies have investigated the impact of B and Zn on crop yields, either individually or in combination. However, the potential synergistic or antagonistic effects of soil-applied B and Zn (either alone or in combination) on the growth and yield of oilseed rape have never been studied. Additionally, the underlying physiological mechanisms have not been explored in the semiarid environments characterized by calcareous soils. Hence, the current study investigated the impacts of soil-applied B and Zn (alone or in combination) on the growth, physiological, and biochemical attributes; seed and oil yields; and oil quality of oilseed rape. It was hypothesized that the growth, physiological, and biochemical attributes; seed and oil yields; and oil quality of oilseed rape would be significantly altered by the soil application of B and Zn. It was further hypothesized that the combined application of B and Zn would improve the growth, physiological, and biochemical attributes; seed and oil yields; and oil quality of oilseed rape compared to their individual application. The results of this study would help to improve the seed and oil yields and oil quality of oilseed rape in semiarid environments characterized by calcareous soils.

2. Materials and Methods

2.1. Experimental Site

The experiment was carried out at the Agronomic Farm, University of Sargodha, Pakistan (32.08° N, 72.67° E, 193 m asl), for two consecutive seasons, i.e., 2020–2021 and 2021–2022. The experimental site is characterized by a subtropical semiarid climate, with an average annual precipitation of 400 mm. The average minimum temperature observed during the coldest month is 14.3 °C, whereas the average minimum temperature during the warmest month is 39.2 °C. Physicochemical analysis of the experimental soil was conducted for two consecutive years, and the results are presented in Table 1. The experimental soil had a higher soil pH and was deficient in boron and zinc, both of which limit the growth and productivity of oilseed rape.

Table 1. Physiochemical properties of experimental soil during 2020–2021 and 2021–2022.

Soil Properties	Values		Analytical Method and Reference
	2020–2021	2021–2022	
Physical composition [54]			
Sand (g kg^{-1})	470 ± 3.2	468 ± 3.3	Bouyoucos hydrometer method [54]
Silt (g kg^{-1})	239 ± 2.4	239 ± 2.2	
Clay (g kg^{-1})	289 ± 1.5	289 ± 1.4	
Textural class	Loam–clay loam		
Chemical composition			
Saturation %	40.22 ± 1.18	40.72 ± 1.15	[55]
pH	7.6 ± 0.04	7.7 ± 0.02	[56]
ECe (µS cm^{-1})	15.42 ± 22.2	16.82 ± 28.76	[56]
Soil organic matter (g kg^{-1})	7.40 ± 0.61	7.40 ± 0.32	Walkley and Black method [57]
Total soil N (mg kg^{-1})	4.09 ± 8.14	4.14 ± 7.32	Modified Kjeldahl method [58]
Extractable P (mg kg^{-1} soil)	7.39 ± 0.11	7.73 ± 0.31	Olsen's method [59]
Available potassium (mg kg^{-1})	271 ± 12.12	273 ± 11.14	Flame photometric method [60]
Hot-water-soluble boron (mg kg^{-1})	0.33 ± 0.08	0.37 ± 0.12	Hot water extraction [61]
DTPA-extractable zinc (mg kg^{-1})	0.46 ± 0.11	0.51 ± 0.14	DTPA soil test [62]

The values presented are means (n = 4) ± standard errors of the means.

2.2. Experimental Details and Crop Husbandry

Experimental field was irrigated prior to seedbed preparation, and fine seedbed was prepared once soil reached an appropriate moisture level for tillage practices. Two with moldboard plough, followed by two cultivations utilizing a narrow tine cultivator were performed. Finally, two plankings were conducted to prepare the fine seedbed. The seeds of oilseed rape cultivar 'G 97' were procured from a local market, treated with fungicide ((thiophenate methyl) 2.5 g per kg seed), and sown using a manual drill at a depth of 2 cm by keeping seed rate at 2.5 kg ha^{-1}. The rows and plants were maintained at 0.45 m and 0.1 m distances, respectively. Sowing was conducted during the final week of October each year. Thinning was carried out at the 2 to 4 true leaf stage, approximately 3 to 4 weeks after sowing for maintaining plant-to-plant distance. Nine different combinations of B and Zn, i.e., B0 + Zn0 (control), B0 + Zn8, B0 + Zn10, B1 + Zn0, B1 + Zn8, B1 + Zn10, B2 + Zn0, B2 + Zn8, and B2 + Zn10 (kg ha^{-1}), were included in this study.

The experiment was conducted according to randomized complete block design with four replications, and a net plot size of 4 m × 2.25 m. The first irrigation was applied 30 days after sowing. Subsequent irrigations were applied at flowering, silique formation, and seed formation. Tensiometer (Model RM 627) was utilized to maintain plant available soil moisture at 70% during the study.

Granular zinc sulfate (33%) and boric acid (17.5%) were used as sources of Zn and B, respectively. The calculated amounts of B and Zn according to the treatments were applied at the time of sowing. Nitrogen (N), phosphorus (P), and potassium (K) were applied at a rate of 23, 23, and 12 kg ha^{-1}, respectively, utilizing urea, diammonium phosphate, and sulfate of potash as the sources. The entire amount of K and P and one-third N were applied during crop sowing. The remaining N was applied in three splits at flowering, silique development, and seed formation.

Weeding was carried out twice to manage the weed species. Manual weeding was conducted twice within a period of 2 to 5 weeks after sowing, prior to the closure of the crop canopy. All cultural and management practices were consistently implemented in accordance with the crop requirements across all plots. The siliques were manually detached from the plants at maturity and subsequently threshed in mid-April.

2.3. Data Collection

Plant height, yield, and yield-related traits; B and Zn contents in leaves; seed and oil yields; oil quality traits; and physiological and biochemical traits were recorded according to standard procedures described below.

2.4. Plant Height and Yield-Related Traits

The heights of ten mature randomly selected plants in each treatment were measured using a meter rod. Yield components, i.e., number of siliques per plant, number of seeds per silique, and 1000-seed weight, were recorded at maturity. The siliques were manually separated from the plants, counted, and subsequently subjected to manual threshing to record number of seeds per silique. The weight of 1000 seeds was measured using an analytical balance (Model Number HC2204). The seed yield was measured by harvesting and threshing all plants in each experimental yield. The seeds obtained from each plot after threshing were weighed on electronic balance (PL 3200+ L Japan), and the weight per plot was subsequently converted into tons per hectare using unitary method.

2.5. B and Zn Contents in Leaves

The leaves were collected, rinsed with distilled water, and dried in an oven at 65 °C for 48 h to determine B and Zn concentrations. The leaves were ground using an electric grinder. Boron was determined by converting the leaves into ash, utilizing a muffle furnace operating at 550 °C for 6 h. The resulting ash was treated with 0.36 N H_2SO_4. Azomethine-H method facilitated the determination of B concentration on a spectrophotometer at 420 nm [63]. Leaf powder (0.5 g) was digested in a mixture of HNO_3 and $HClO_4$ with a ratio of 2:1 for 6 h to determine Zn contents [64]. The distilled water was used to bring the volume of the mixture to 25 mL. The concentration of Zn was recorded by utilizing atomic absorption spectroscopy (AA-6300, Shimadzu, Japan).

2.6. Estimation of Oil Yield and Oil Quality Traits

The seeds were oven-dried at 45 °C for 24 h and ground to powder. Oil extraction was conducted using 3.5 g seed powder. The excess fat was removed from the samples by using Soxhlet apparatus maintained at 60 °C for 8 h. Each specimen required 180 g petroleum ether solution. Afterward, seed specimens were subjected to a thermal treatment of 50 °C for 24 h to extract the oil. Oil yield was computed through the multiplication of the seed yield by the oil content. The fatty acid composition, encompassing palmitic acid (C16:0), stearic acid (C18:0), oleic acid (C18:1), linolenic acid (C18:3), linoleic acid (C18:2), and erucic acid (C22:1), was determined by using gas chromatography of methyl esters [65].

2.7. Determination of Protein and Soluble Sugar Contents

Leaf soluble protein and total soluble sugars (TSSs) were quantified by using fresh leaf extract (0.1 g) in potassium phosphate buffer (50 mM and a pH value of 7.5). The extract was centrifuged at 15,500 rpm (25,155 relative centrifugal force (RCF) or g force) for 15 min at 4 °C after filtering through four layers of cheese cloth. Supernatant was collected and kept at 4 °C. Bradford [66] protein dye-binding technique was used to determine the amount of leaf soluble protein using bovine serum albumin as the reference protein. TSSs were analyzed in a Cecil CE 2021 spectrophotometer using an anthrone reagent [67].

2.8. Estimation of Chlorophyll Contents, Photosynthesis, Transpiration Rate, and Stomatal Conductance

Chlorophyll contents were determined from the fully flourishing third younger leaf by using chlorophyll meter (Model, SPAD-502: Konica Minolta Sensing: Inc., Osaka, Japan) [68]. Four young leaves were randomly chosen 30 days after sowing and sequentially placed in an infrared gas analyzer (IRGA). Stomatal conductance, photosynthesis, and transpiration rates were measured between 11:00 and 12:00 a.m. The IRGA chamber was programmed to take readings under the conditions specified by Zekri [69] and Moya et al. [70]. The conditions were 403.3 mmol m^{-2} S^{-1} molar flow rate, 99.90 KPa atmospheric pressure, 6.0 to 8.9 millibar vapor pressure, 1711 mol m^{-2} S^{-1} photosynthetically active radiation, leaf temperature of 28.40 to 32.40 °C, ambient temperature of 22.40 to 27.90 °C, and ambient CO_2 concentration of 352 mol mol^{-1} [71].

2.9. Antioxidant Enzymes

2.9.1. Extraction

The enzyme extraction procedure was conducted at 4 °C using 200 mg leaf samples. The samples were homogenized in a pre-chilled mortar and pestle using 3 mL of ice-cold 50 mM sodium phosphate buffer (pH 7.0), 0.1 mM EDTA, and 1% *w*/*v* polyvinyl pyrrolidone (PVP). The mixture was centrifuged at 15,000 rpm (25,155 RCF or g force) for 20 min at 4 °C. The supernatant was utilized as a raw enzyme extract. The enzyme assays were conducted under ambient conditions, and the enzymatic activity was quantified using a spectrophotometer.

2.9.2. Peroxidase (POD) Activity

A procedure reported by Ullah et al. [72] was slightly modified to determine peroxidase (POD) activity. The 1 mL reaction mixture contained 40 mM phosphate buffer, 15 mM guaiacol, and 5 mM H_2O_2 (pH 6.8). After the reaction mixture settled, reaction was initiated by adding H_2O_2, and the increase in absorbance at 470 nm was measured for 1 min. The POD activity was measured in accordance with its 25 mM^{-1} cm^{-1} extinction coefficient.

2.9.3. Superoxide Dismutase (SOD) Activity

The SOD activity was determined by inhibiting the photochemical reduction of nitroblue tetrazolium (NBT) [73]. A superoxide-generating system of 14.3 mM methionine, 82.5 mM NBT, and 2.2 mM riboflavin was added to the reaction mixture (3 mL) of 50 mM phosphate buffer (pH 7.8) and 0.1 mM EDTA. The reaction was initiated by adding 100 cc of unprocessed enzyme. The free-radical-induced NBT reduction was tested in a reaction medium containing all the components except the enzyme. Six 15W fluorescent bulbs were used as the light source, and the tubes were kept there for 30 min. Turning off the light halted the reaction. The entire mixture of reactants was incubated in the dark as a dark blank, along with 100 mL of enzyme extract. The reduction in NBT was evaluated by monitoring the shift in absorbance at 560 nm. To calculate enzyme units, measurements from the dark blank were used. One unit of SOD was defined as the quantity of the enzyme that resulted in a 50% inhibition of NBT reduction under the test conditions. The unit of measurement for enzyme activity was mg^{-1} protein.

2.10. Data Analysis

The collected data of all the recorded traits were analyzed by using one-way analysis of variance (ANOVA). The normality was tested by Shapiro–Wilk normality test, which indicated a normal distribution. Therefore, statistical analysis was performed on original data. The differences among years were analyzed by *t*-test, which were significant. Therefore, data from each year were analyzed, presented, and interpreted separately. SAS software (Version 9.1; SAS Institute, Cary, NC, USA) was used for statistical analysis. Least significant difference post hoc test at 95% probability was used to separate treatment means where ANOVA denoted significant differences [74]. SigmaPlot was used for the graphical presentation of the data (SigmaPlot 2008).

3. Results

3.1. Growth, Yield, and Boron and Zinc Contents

Growth, yield, yield-related attributes, and B and Zn contents in leaves were significantly ($p < 0.05$) affected by sole and combined application of B and Zn during 2020–2021 and 2021–2022 (Tables 2 and 3). Taller plants (10 and 13% during 2020–2021 and 2021–2022, respectively) were recorded with B2 + Zn8 treatment compared to control treatments (Table 2). Nevertheless, B1 + Zn8, B1 + Zn10, and B2 + Zn0 treatments were on par with B2 + Zn8 during first year. The lowest plant height was noted for the control treatment during both years of this study (Table 2).

Table 2. The impact of soil-applied boron and zinc on plant height, number of siliques per plant, number of seeds per silique, and 1000-seed weight of oilseed rape grown under field conditions during 2020–2021 and 2021–2022.

Treatments	Plant Height (cm)		Number of Siliques $Plant^{-1}$		Number of Seeds $Silique^{-1}$		1000-Seed Weight (g)	
	2020–2021	2021–2022	2020–2021	2021–2022	2020–2021	2021–2022	2020–2021	2021–2022
Control	122.20 ± 2.46 E	117.97 ± 2.21 F	200.21 ± 2.60 F	194.63 ± 4.21 G	13.02 ± 1.04 E	10.94 ± 0.92 E	2.63 ± 0.09 C	2.44 ± 0.11 E
B0 + Zn8	125.67 ± 0.87 DE	126.17 ± 1.40 E	260.74 ± 2.89 AB	253.41 ± 9.42 B	15.14 ± 0.48 CD	15.23 ± 0.58 C	2.86 ± 0.15 BC	2.64 ± 0.09 D
B0 + Zn10	127.43 ± 1.59 CD	128.37 ± 1.78 CD	204.94 ± 2.08 E	198.45 ± 3.02 F	15.21 ± 0.50 CD	15.07 ± 0.66 C	3.26 ± 0.41 AB	2.84 ± 0.11 BC
B1 + Zn0	129.13 ± 1.30 BCD	124.77 ± 1.45 E	237.45 ± 4.16 CD	243.78 ± 5.23 C	15.33 ± 0.69 CD	15.04 ± 0.71 C	3.23 ± 0.38 AB	2.94 ± 0.08 B
B1 + Zn8	132.20 ± 2.04 AB	126.57 ± 1.15 DE	249.14 ± 7.37 BC	244.15 ± 4.41 C	16.22 ± 0.71 BC	16.15 ± 1.02 BC	3.03 ± 0.31 ABC	2.64 ± 0.09 D
B1 + Zn10	131.15 ± 1.01 ABC	124.97 ± 1.49 E	230.47 ± 1.15 D	230.32 ± 8.78 D	18.21 ± 0.50 B	17.33 ± 0.48 B	2.93 ± 0.45 ABC	2.74 ± 0.15 CD
B2 + Zn0	132.40 ± 1.16 AB	132.57 ± 2.35 B	232.97 ± 6.36 CD	218.95 ± 7.86 E	13.44 ± 0.64 E	11.47 ± 0.71 DE	3.00 ± 0.32 ABC	2.84 ± 0.12 BC
B2 + Zn8	135.60 ± 1.33 A	134.97 ± 2.12 A	268.73 ± 11.18 A	274.32 ± 10.31 A	21.45 ± 0.66 A	21.63 ± 1.95 A	3.30 ± 0.12 A	3.34 ± 0.11 A
B2 + Zn10	129.60 ± 1.89 BCD	129.97 ± 1.90 C	229.44 ± 6.69 D	241.96 ± 4.54 C	14.94 ± 0.82 DE	12.86 ± 0.55 D	3.10 ± 0.25 AB	2.74 ± 0.13 CD
$LSD_{0.05}$	4.72	2.02	17.20	3.43	2.18	1.71	0.43	0.10

The values following B and Zn indicate the amount of applied B and Zn kg ha^{-1}. The values are means ± standard errors. Means followed by different letters within a column are significantly different ($p \leq 0.05$) from each other.

Table 3. The impact of soil-applied boron and zinc on seed yield and boron and zinc contents in leaves of oilseed rape grown under field conditions during 2020–2021 and 2021–2022.

Treatments	Seed Yield (t ha^{-1})		B Content in Leaves (mg kg^{-1})		Zn Content in Leaves (mg kg^{-1})	
	2020–2021	2021–2022	2020–2021	2021–2022	2020–2021	2021–2022
Control	1.82 ± 0.05 D	1.79 ± 0.11 D	10.25 ± 0.09 G	10.23 ± 0.21 H	27.46 ± 0.87 I	27.42 ± 1.14 I
B0 + Zn8	2.01 ± 0.06 CD	1.79 ± 0.10 D	11.13 ± 0.14 F	11.09 ± 0.22 G	32.13 ± 1.02 F	32.09 ± 0.98 F
B0 + Zn10	2.13 ± 0.13 BC	1.99 ± 0.09 BC	11.37 ± 0.17 E	11.33 ± 0.15 F	34.18 ± 0.55 D	34.16 ± 0.55 D
B1 + Zn0	1.85 ± 0.09 D	1.73 ± 0.15 D	14.12 ± 0.15 D	14.08 ± 0.15 E	29.66 ± 0.45 H	29.63 ± 0.34 H
B1 + Zn8	2.02 ± 0.17 CD	2.01 ± 0.08 BC	14.24 ± 0.17 D	14.21 ± 0.18 E	33.53 ± 0.25 E	33.51 ± 0.52 E
B1 + Zn10	2.07 ± 0.19 BC	1.92 ± 0.19 CD	14.57 ± 0.22 C	14.53 ± 0.25 D	35.58 ± 1.22 C	35.56 ± 0.44 C
B2 + Zn0	2.26 ± 0.23 B	2.17 ± 0.12 B	17.09 ± 0.14 B	17.04 ± 0.10 C	30.13 ± 0.48 G	30.08 ± 0.38 G
B2 + Zn8	2.88 ± 0.24 A	2.44 ± 0.14 A	17.22 ± 0.13 B	17.19 ± 0.23 B	37.46 ± 1.21 B	37.43 ± 1.14 B
B2 + Zn10	2.10 ± 0.17 BC	2.03 ± 0.15 C	17.46 ± 0.11 A	17.42 ± 0.11 A	39.42 ± 1.15 A	39.39 ± 1.11 A
$LSD_{0.05}$	0.20	0.19	0.13	0.10	0.11	0.12

The values following B and Zn indicate the amount of applied B and Zn kg ha^{-1}. The values are means ± standard errors. Means followed by different letters within a column are significantly different ($p \leq 0.05$) from each other.

Soil application of B2 + Zn8 improved the number of siliques per plant by 25 and 29% during 2020–2021 and 2021–2022, respectively, compared to the control treatment. The B0 + Zn8 treatment followed B2 + Zn8 for the number of siliques per plant, while the control treatment resulted in the lowest number of siliques per plant during both years. The highest and lowest number of seeds per silique were recorded for the B2 + Zn8 and control treatments, respectively. The B2 + Zn8 treatment improved the number of seeds per silique by 38% and 52% compared to the control during 2020–2021 and 2021–2022, respectively (Table 2). Similarly, B1 + Zn10 followed B2 + Zn8 for the number of seeds per silique during both years. The highest and lowest 1000-seed weight and grain yields were observed for the B2 + Zn8 and control treatments, respectively, during both years of this study. Soil application of B2 + Zn8 improved 1000-seed weight by 20% and 27% during 2020–2021 and 2021–2022, respectively, compared to the control treatment (Table 2). The seed yield of plants fertilized with B2 + Zn8 was improved by 37% and 27% during 2020–2021 and 2021–2022, respectively, compared to the control (Table 3).

The B and Zn contents in the leaves were linearly increased with their application doses. The highest and lowest B and Zn contents in the leaves were noted for the B2 + Zn10 and control treatments, respectively, during both years (Table 3). Soil application of B2 + Zn8 resulted in the second-highest values for B and Zn contents in leaves during both years of this study (Table 3).

3.2. *Oil Yield and Quality*

Combined and sole application of B and Zn significantly improved oil yield and quality during both years of study (Tables 4 and 5). The highest and lowest values for oil yield, stearic acid, palmitic acid, and oleic acid were recorded for the B2 + Zn8 and control treatments, respectively, during both years of the study (Table 4).

Soil application of B2 + Zn8 resulted in the highest (54% and 53% higher than control treatment during 2020–2021 and 2021–2022, respectively) oil yield during both years (Table 4). Similarly, stearic acid was improved by 31% and 30%, palmitic acid by 22% and 21%, oleic acid by 33% and 34%, linoleic acid by 26% and 24%, and linolenic acid by 38% and 40% with soil-applied B2 + Zn8 during 2020–2021 and 2021–2022, respectively, compared to the control treatment (Table 5). The B2 + Zn10 and B2 + Zn0 treatments followed B2 + Zn8 for linoleic acid and linolenic acid during both years of this study. However, erucic acid was significantly ($p \leq 0.05$) reduced by 62% and 63% with the application of B2 + Zn8 during 2020–2021 and 2021–2022, respectively (Table 5).

Table 4. The impact of soil-applied boron and zinc on oil yield, stearic acid, palmitic acid, and oleic acid in the oil of oilseed rape grown under field conditions during 2020–2021 and 2021–2022.

Treatments	Oil Yield (t ha^{-1})		Stearic Acid (mg g^{-1})		Palmitic Acid (mg g^{-1})		Oleic Acid (mg g^{-1})	
	2020–2021	2021–2022	2020–2021	2021–2022	2020–2021	2021–2022	2020–2021	2021–2022
Control	0.54 ± 0.07 D	0.55 ± 0.02 H	4.53 ± 0.45 G	4.51 ± 0.32 G	12.24 ± 0.67 H	12.23 ± 0.91 H	152.93 ± 5.78 E	150.93 ± 15.65 E
B0 + Zn8	0.67 ± 0.06 C	0.58 ± 0.02 G	5.11 ± 0.42 F	5.09 ± 0.39 F	13.46 ± 0.30 G	13.43 ± 0.32 G	181.88 ± 4.58 D	179.88 ± 11.45 D
B0 + Zn10	0.76 ± 0.09 B	0.69 ± 0.07 E	5.24 ± 0.38 F	5.22 ± 0.44 F	13.83 ± 0.19 F	13.81 ± 0.14 F	184.08 ± 5.66 D	181.86 ± 12.60 D
B1 + Zn0	0.72 ± 0.07 BC	0.78 ± 0.11 B	5.76 ± 0.31 E	5.74 ± 0.22 E	14.43 ± 0.21 D	14.41 ± 0.10 D	195.09 ± 11.23 CD	193.09 ± 10.91 CD
B1 + Zn8	0.77 ± 0.09 B	0.79 ± 0.12 B	5.92 ± 0.23 DE	5.89 ± 0.24 DE	14.66 ± 0.29 C	14.63 ± 0.09 C	196.75 ± 13.61 CD	195.75 ± 9.87 CD
B1 + Zn10	0.66 ± 0.05 C	0.64 ± 0.04 F	6.13 ± 0.15 C	6.11 ± 0.14 C	14.13 ± 0.14 E	14.11 ± 0.15 E	191.86 ± 9.90 CD	190.86 ± 11.56 CD
B2 + Zn0	0.77 ± 0.04 B	0.76 ± 0.02 C	6.33 ± 0.19 B	6.31 ± 0.18 B	15.48 ± 0.16 A	15.46 ± 0.22 A	215.15 ± 12.23 AB	213.15 ± 14.56AB
B2 + Zn8	1.18 ± 0.08 A	1.19 ± 0.07 A	6.54 ± 0.17 A	6.52 ± 0.20 A	15.63 ± 0.17 A	15.61 ± 0.32 A	229.24 ± 10.65 A	227.24 ± 10.71 A
B2 + Zn10	0.73 ± 0.07 BC	0.72 ± 0.02 D	6.03 ± 0.14 CD	6.02 ± 0.12 CD	15.15 ± 0.23 B	15.13 ± 0.18 B	203.04 ± 13.67 BC	201.04 ± 12.44 BC
$LSD_{0.05}$	0.07	0.01	0.19	0.20	0.17	0.19	16.45	17.40

The values following B and Zn indicate the amount of applied B and Zn kg ha^{-1}. The values are means ± standard errors. Means followed by different letters within a column are significantly different ($p \leq 0.05$) from each other.

Table 5. The impact of soil-applied boron and zinc on linoleic acid, linolenic acid, and erucic acid contents in the oil of oilseed rape crop grown under field conditions during 2020–2021 and 2021–2022.

Treatments	Linoleic Acid (mg g^{-1})		Linolenic Acid (mg g^{-1})		Erucic Acid (mg g^{-1})	
	2020–2021	2021–2022	2020–2021	2021–2022	2020–2021	2021–2022
Control	40.64 ± 2.24 G	39.62 ± 3.12 F	16.33 ± 3.35 D	15.33 ± 2.12 D	0.87 ± 0.03 A	0.85 ± 0.08 A
B0 + Zn8	45.39 ± 1.12 F	44.37 ± 1.01 E	22.36 ± 1.12 C	21.36 ± 1.06 C	0.74 ± 0.14 ABC	0.72 ± 0.10 AB
B0 + Zn10	46.50 ± 1.15 EF	45.50 ± 1.90 DE	22.77 ± 1.07 C	22.71 ± 2.21 BC	0.82 ± 0.11 AB	0.80 ± 0.06 A
B1 + Zn0	49.51 ± 1.05 CD	47.51 ± 2.12 CD	23.38 ± 0.45 C	22.38 ± 2.09 BC	0.64 ± 0.15 BCD	0.63 ± 0.13 BC
B1 + Zn8	50.39 ± 2.12 BCD	49.39 ± 2.02 BC	24.63 ± 2.45 ABC	23.63 ± 2.11 ABC	0.53 ± 0.13 CD	0.51 ± 0.11 C
B1 + Zn10	48.29 ± 1.98 DE	46.29 ± 2.33 DE	23.06 ± 0.34C	22.06 ± 0.91 C	0.69 ± 0.11 BCD	0.67 ± 0.07 BC
B2 + Zn0	52.39 ± 2.45 AB	51.39 ± 2.11 AB	26.33 ± 1.50 A	25.33 ± 1.05 A	0.49 ± 0.02 E	0.47 ± 0.03 D
B2 + Zn8	54.75 ± 1.65 A	52.75 ± 1.01 A	26.69 ± 1.91 A	25.69 ± 1.11 A	0.33 ± 0.10 E	0.31 ± 0.07 D
B2 + Zn10	51.83 ± 1.16 BC	50.83 ± 0.94 AB	25.63 ± 1.08 AB	24.63 ± 1.19 AB	0.40 ± 0.09 E	0.38 ± 0.08 D
$LSD_{0.05}$	2.49	2.52	2.53	2.51	0.22	0.24

The values following B and Zn indicate the amount of applied B and Zn kg ha^{-1}. The values are means ± standard errors. Means followed by different letters within a column are significantly different ($p \leq 0.05$) from each other.

3.3. Physiological Parameters

Sole and combined application of B and Zn significantly altered protein, soluble sugar, and chlorophyll concentrations. Protein, soluble sugar, and chlorophyll concentrations in leaves were significantly increased under the combined application of B and Zn. The highest values of these traits were recorded with the application of B2 + Zn8 during both study years, while the control treatment resulted in the lowest values (Table 6).

Table 6. The impact of soil-applied boron and zinc on protein concentration, soluble sugar concentration, chlorophyll concentration, and photosynthesis rate of oilseed rape grown under field conditions during 2020–2021 and 2021–2022.

Treatments	Protein Concentration ($mg\ g^{-1}$ FW)		Soluble Sugar Concentration ($mg\ g^{-1}$ FW)		Chlorophyll Concentration (SPAD Value)		Photosynthesis Rate ($\mu mol\ CO_2\ m^{-2}\ s^{-1}$)	
	2020–2021	2021–2022	2020–2021	2021–2022	2020–2021	2021–2022	2020–2021	2021–2022
Control	20.19 ± 0.53 F	20.17 ± 0.90 F	20.19 ± 1.03 F	20.17 ± 0.88 F	34.65 ± 1.41 G	34.61 ± 1.12 G	8.51 ± 0.31 C	8.49 ± 0.78 C
B0 + Zn8	21.62 ± 0.88 E	21.59 ± 0.67 E	21.62 ± 0.43 E	21.59 ± 0.41 E	36.58 ± 1.14 F	36.55 ± 1.64 F	9.48 ± 1.03 BC	9.46 ± 0.98 BC
B0 + Zn10	21.86 ± 0.73 E	21.83 ± 0.55 E	21.86 ± 0.33 E	21.83 ± 0.52 E	38.92 ± 1.01 DE	38.88 ± 1.21 DE	9.88 ± 0.90 BC	9.84 ± 0.78 BC
B1 + Zn0	22.36 ± 0.21 CD	22.33 ± 0.15 CD	22.36 ± 0.19 CD	22.33 ± 0.19 CD	39.60 ± 1.71 DE	39.58 ± 1.49 DE	10.18 ± 0.23 B	10.15 ± 0.31 B
B1 + Zn8	22.51 ± 0.18 C	22.49 ± 0.11 C	22.51 ± 0.17 C	22.49 ± 0.24 C	40.37 ± 1.14 D	40.34 ± 1.12 D	10.62 ± 0.45 B	10.60 ± 0.41 B
B1 + Zn10	22.20 ± 0.61 D	22.18 ± 0.13 D	22.20 ± 0.24 D	22.18 ± 0.19 D	38.25 ± 2.31 EF	38.21 ± 1.01 DE	9.64 ± 0.97 BC	9.62 ± 0.89 BC
B2 + Zn0	24.10 ± 0.22 AB	24.07 ± 0.22 AB	24.10 ± 0.25 AB	24.07 ± 0.22 AB	45.36 ± 1.43 B	45.33 ± 1.93 B	13.62 ± 1.12 A	13.58 ± 1.45 A
B2 + Zn8	24.16 ± 0.11 A	24.13 ± 0.09 A	24.16 ± 0.17 A	24.13 ± 0.11 A	48.25 ± 2.12 A	48.20 ± 1.10 A	14.18 ± 0.94 A	14.16 ± 1.33 A
B2 + Zn10	23.89 ± 0.20 B	23.86 ± 0.11 B	23.89 ± 0.20 B	23.86 ± 0.08 B	43.27 ± 1.78 C	43.25 ± 1.22 C	12.84 ± 1.78 A	12.81 ± 1.83 A
$LSD_{0.05}$	0.26	0.27	0.26	0.27	1.81	1.80	1.75	1.73

The values following B and Zn indicate the amount of applied B and Zn kg ha^{-1}. The values are means ± standard errors. Means followed by different letters within a column are significantly different ($p \leq 0.05$) from each other.

Soil applied B at 2 kg ha^{-1} alone or in combination with Zn significantly improved the photosynthesis rate. The highest increase in the photosynthesis rate was noted with B2 + Zn8 during both years compared to the control treatment (Table 6). Stomatal conductance and transpiration rate were significantly improved by 91% and 93% and 48% and 49% with B2 + Zn8 compared to the control during 2020–2021 and 2021–2022, respectively (Figures 1 and 2).

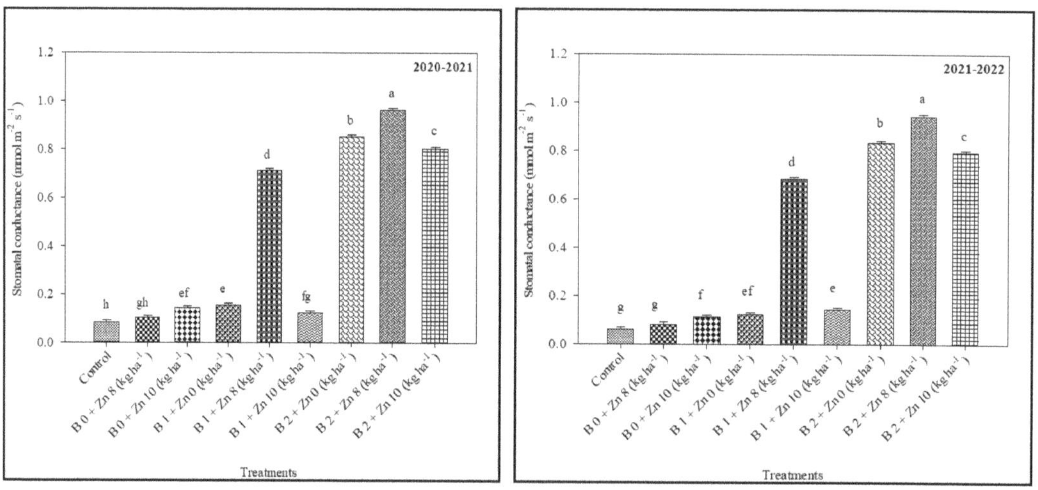

Figure 1. The influence of sole and combined application of boron and zinc on stomatal conductance of oilseed rape plants grown under field conditions during 2020–2021 and 2021–2022. The error bars are standard errors of the means (n = 4). Different letters on the bars indicate significant differences among treatment means ($p \leq 0.05$).

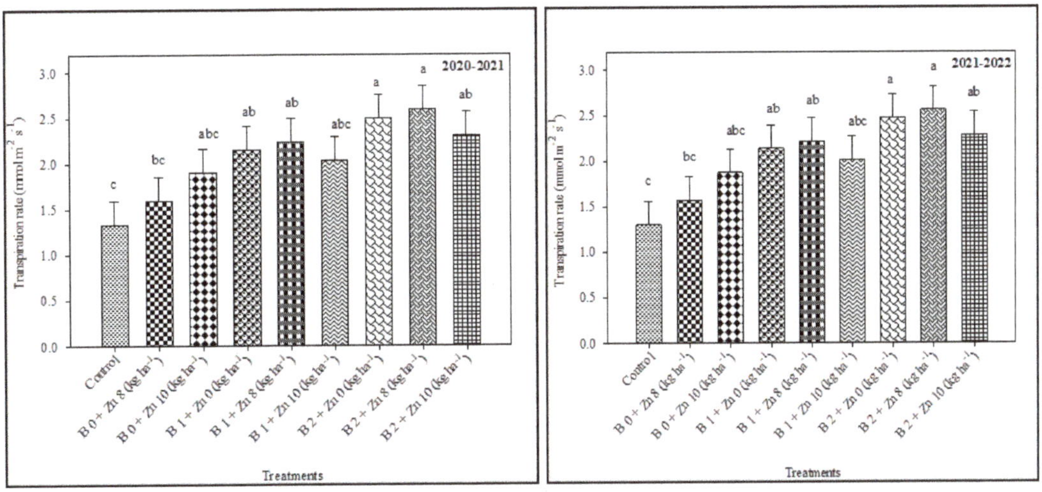

Figure 2. The influence of sole and combined application of boron and zinc on transpiration rate of oilseed rape plants grown under field conditions during 2020–2021 and 2021–2022. The error bars are standard errors of the means (n = 4). Different letters on the bars indicate significant differences among treatment means ($p \leq 0.05$).

3.4. Antioxidant Enzyme Activities

Boron and zinc deficiency caused significant oxidative stress in plants, leading to higher activities of POD and SOD enzymes (Figures 3 and 4). The highest and lowest activities of POD and SOD were noted with the control and B2 + Zn8 treatments, respectively. The activities of both enzymes were decreased with an increasing concentration of B and Zn, indicating that plants did not experience oxidative stress when both nutrients were available in sufficient amounts.

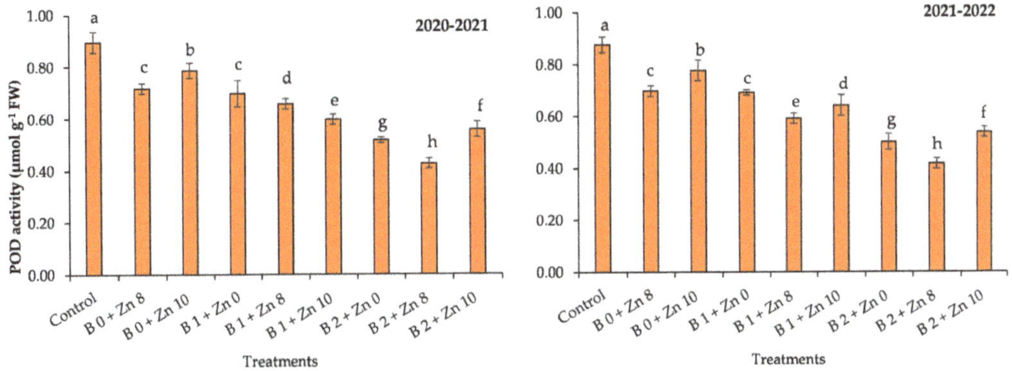

Figure 3. The influence of sole and combined application of boron and zinc on the activity of POD enzyme in oilseed rape plants grown under field conditions during 2020–2021 and 2021–2022. The error bars are standard errors of the means (n = 4). Different letters on the bars indicate significant differences among treatment means ($p \leq 0.05$).

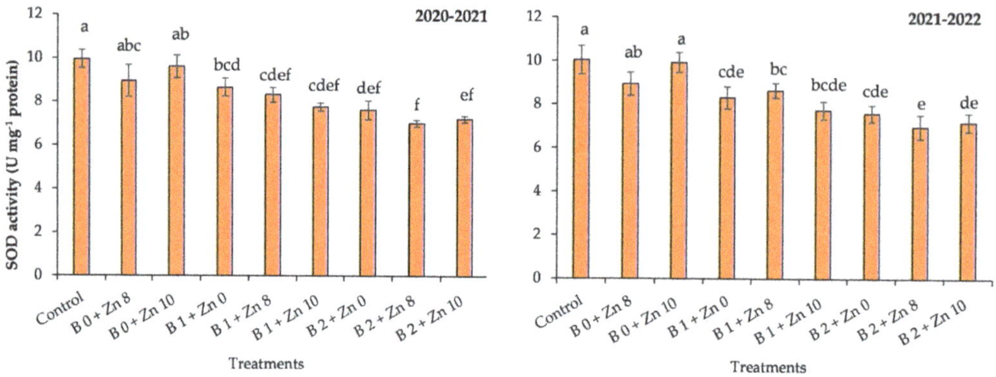

Figure 4. The influence of sole and combined application of boron and zinc on the activity of SOD enzyme in oilseed rape plants grown under field conditions during 2020–2021 and 2021–2022. The error bars are standard errors of the means (n = 4). Different letters on the bars indicate significant differences among treatment means ($p \leq 0.05$).

4. Discussion

Micronutrients are required in low amounts by plants, and their deficiencies exert significant negative impacts on the growth, yield, and quality of the produce. Sole and combined application of B and Zn significantly affected growth, seed yield, yield attributes, B and Zn contents in leaves, and oil yield and quality, as well as the physiological and antioxidant characteristics of oilseed rape, as hypothesized. Similarly, the combined application of B and Zn resulted in higher improvements in these traits compared to their individual application. The highest improvements in all traits were recorded with the B2 + Zn8 treatment during both years of this study. Overall, the combined application of B and Zn significantly increased the yield and yield-related attributes of oilseed rape in the current study. This could be attributed to the higher synthesis of assimilates, their improved mobility to potential sinks, and enhanced efficiency in pollination and seed development. Kanwal et al. [75] reported that optimum B and Zn application to the root zone caused a significant and consistent increase in the yield and yield-related traits of oilseed rape. Our results are in agreement with Aref [76] and Rehman et al. [33] who reported that the synergistic impact of these two micronutrients improved plant height, seed yield, and yield characteristics under low B and Zn availability in the soil. Plants uptake the required amount of B if B and Zn are supplemented in B-deficient soils, which improves plant development [77]. Additionally, the soil application of B and Zn resulted in a higher photosynthetic rate and chlorophyll content, ultimately leading to an increase in dry matter and subsequent growth and production [33,78]. Boron plays a crucial role in the synthesis of chlorophyll molecules required for photosynthesis. Sufficient B concentration in the current study probably facilitated the synthesis of chlorophyll, thereby resulting in enhanced leaf pigmentation and vitality in the current study. Zinc is involved in the activation of enzymes that are essential for the process of chlorophyll synthesis. The higher chlorophyll and photosynthesis under the B2 + Zn8 treatment are thought to be the result of enzyme activation, which resulted in better chlorophyll synthesis and subsequently improved photosynthesis.

The application of optimum B and Zn doses significantly improves the growth and yield of crop plants [79]. B and Zn application has been found to have a significant impact on the translocation of photoassimilates from the source (i.e., leaves) to other plant parts and the promotion of pollen tube elongation, ultimately resulting in an increased number of siliques [80,81]. Rehman et al. [49] and Potarzycki and Grzebisz [82] reported that B + Zn application improved the number of siliques per plant due to the higher number of flowers, the appropriate development of pollen and the pollen tube, pollination, and seed formation. It is well described that the application of B and Zn increases the number of seeds per silique and 1000-seed weight [33,83]. Several research studies have reported that the combined application of B and Zn can enhance seed production by increasing yield characteristics [33,84]. The deficiency of these nutrients impedes the development of the petiole and peduncle cells, resulting in diminished growth, seed yield, and yield characteristics [85]. Furthermore, a lower plant height at varying B + Zn concentrations may indicate a sensitive differentiation between deficiency and toxicity, thereby impeding the growth and yield characteristics of the plant without any apparent symptoms [86]. The reduction in plant growth, seed yield, and yield traits observed in the control group or low/higher doses of B + Zn can be attributed to a decrease in enzymatic reactions that regulate cell division and elongation. Conversely, excessive levels of B + Zn can lead to an imbalance of various enzymes, ultimately resulting in a reduction in plant height, seed yield, and yield traits [80]. The application of B2 + Zn10 resulted in enhanced uptake of both B and Zn by the leaves, indicating a synergistic relationship between these nutrients. However, the higher/excessive consumption of B and Zn reduced the growth and yield characteristics.

The observed enhancement in the yield and quality of oil under B2 + Zn8 may be attributed to the participation of micronutrients in the synthesis of elevated levels of fatty acid compounds [87,88]. The combined application of B and Zn may have contributed to the increase in oil content in seeds through the potential impact on the protein content of leaves. Therefore, the optimal utilization of B and Zn has the potential to enhance the seed oil content [29,89]. The combined application of B and Zn resulted in a significant improvement in the yield and quality of oil in the current study. These findings suggest that a synergistic relationship exists between B and Zn [90]. Although B and Zn are crucial for the growth and development of oilseed rape, their application does not directly affect the fatty acid composition of oilseed crops. These micronutrients play a crucial role in facilitating diverse enzymatic processes that are integral to photosynthesis. The increased photosynthetic activity has the potential to result in the elevated production of carbon precursors that can be utilized for the synthesis of fatty acids. The improved fatty acid profile of the oilseed rape oil in the current study can be linked to improved photosynthesis and, subsequently, the production of carbon precursors, which are utilized by plants for the synthesis of fatty acids.

The highest protein concentration, soluble sugar concentration, chlorophyll concentration, photosynthesis rate, stomatal conductance, and transpiration rate were recorded under the B2 + Zn8 treatment during both years. Zinc plays a crucial role in the structural composition and catalytic mechanisms of proteins and enzymes, thereby significantly contributing to plant development and exhibiting a favorable impact on growth. Consequently, Zn deficiency in plants may result in a reduction in protein and soluble sugar concentration in grains and lower amino acids in plant parts [91]. Higher protein concentration and soluble sugar concentration observed under B2 + Zn8 could impede the mobility of antioxidant enzymes, which causes oxidative destruction to protein and soluble sugar [32]. Photosynthesis in plants involves the absorption of light to produce organic compounds. The application of B2 alone increased leaf area, potentially leading to an improvement in the production of indole acetic acid. This hormone promotes chlorophyll concentration and the photosynthetic rate, which might be attributed to the higher photosynthesis rate in the current study [92]. Boron plays a crucial role in the maintenance of structural integrity, particularly in relation to the vascular tissues responsible for the transportation of water. The regulation of B levels has the potential to impact stomatal conductance and the intra-plant movement of water. Sufficient zinc concentrations have the capacity to regulate stomatal conductance and effectively manage transpiration. The optimum Zn and B supply in the current study improved the protein concentration, soluble sugar concentration, chlorophyll concentration, photosynthesis rate, stomatal conductance, and transpiration rate in the current study by improvements in these processes.

Zinc efficiently improves chlorophyll concentration through enzyme activation, photosynthesis rate, and the movement of photosynthate to seeds [93]. Our results corroborate the findings of Aref [76], who found that the synergistic effect of B and Zn increases chlorophyll contents and the photosynthesis rate. In the current experiment, B and Zn application improved chlorophyll contents and the photosynthesis rate, while higher rates of B + Zn reduced chlorophyll concentration, which might be attributed to the toxicity of these nutrients. Similar outcomes were obtained by Akta et al. [94] who concluded that an increase in B and Zn contents decreased the chlorophyll contents in leaves. Lower stomatal conductance and transpiration rates were recorded in Zn- and B-deficient plants during both years in the current study, which might be attributed to the destruction of the vascular bundle [95]. An impairment of the xylem vessels decreases the movement of water from the roots to the leaves, thereby impacting the stomatal conductance and transpiration rate. On the other hand, phloem impairment can impede the transportation of essential nutrients and sugars, thereby exerting an indirect influence on stomatal conductance and photosynthetic processes. Reduced transpiration rate and stomatal conductance under B and Zn deficiency in the current study are due to these processes. Pinho et al. [96] also concluded that a linear relationship persists among stomatal conductance and transpiration rates and micronutrient availability. Han et al. [97] stated that B application (2 kg ha^{-1}) increased stomatal conductance and reduced intercellular CO_2 absorption. Enhancements in physiological characteristics may lead to better crop growth due to the initiation of various physiological processes that result in increased seed production and improved oil quality [28,98,99]. However, B and Zn deficiencies cause yellowing of the foliage and necrosis [100].

In present study, peroxidase (POD) and superoxide dismutase (SOD) activity was lower under B2 + Zn8, indicating that there was low oxidative stress in this treatment. Antioxidant enzymes play an active role in reducing the detrimental effects of ROS species on photosynthesis and photorespiration. The highest POD and SOD activities were recorded for the control treatments, indicating that deficiency of B and Zn caused oxidative stress in plants. Nevertheless, B and Zn application lowered oxidative stress, as evidenced by the reduced activities of these enzymes.

5. Conclusions

It can be concluded that the combined application of B and Zn (2 and 8 kg ha^{-1}, respectively) in soil could be successfully used to improve the yield, yield-related attributes, and oil quality of oilseed rape. Significantly higher oil yield and quality were recorded under the B2 + Zn8 treatment. The combined application of B and Zn (2 and 8 kg ha^{-1}, respectively) significantly improved the physiological characteristics and activities of antioxidant enzymes. Therefore, successful enhancements in the yield, yield-related characteristics, and oil quality of oilseed rape can be achieved through the combined application of B and Zn at rates of 2 and 8 kg ha^{-1}, respectively, under semiarid climatic conditions.

Author Contributions: Conceptualization, M.E.S., M.A.N., S.F., A.G., S.M., M.S.A., M.S.E. and M.A.A.A.; data curation, A.J.; formal analysis, R.Q., A.G., S.M., M.S.E. and M.A.A.A.; funding acquisition, A.G., S.M., M.S.A. and M.S.E.; investigation, A.J.; methodology, M.E.S., R.Q. and H.M.R.J.; project administration, M.A.N.; resources, M.S.A.; software, R.Q., A.J. and H.M.R.J.; supervision, M.E.S.; validation, H.M.R.J.; visualization, R.Q., A.J., S.F. and M.S.E.; writing—original draft, M.E.S.; writing—review and editing, R.Q., M.A.N., S.F., A.G., S.M., M.S.A., M.S.E. and M.A.A.A. All authors have read and agreed to the published version of the manuscript.

Funding: The APC was funded by the University of Life Sciences in Lublin, from a subsidy granted by the Minister of Education and Science of Poland. The authors extend their appreciation to the Researchers Supporting Project (number: RSP2023R173), King Saud University, Riyadh, Saudi Arabia.

Data Availability Statement: All data are provided within the manuscript.

Acknowledgments: The authors would like to acknowledge the Department of Agronomy, the College of Agriculture, the University of Sargodha, for funding to conduct this study. The authors extend their appreciation to the Researchers Supporting Project (number: RSP2023R173), King Saud University, Riyadh, Saudi Arabia.

Conflicts of Interest: The authors declare no conflict of interest.

References

1. de Oliveira Neto, S.S.; Zeffa, D.M.; Freiria, G.H.; Zoz, T.; da Silva, C.J.; Zanotto, M.D.; Sobrinho, R.L.; Alamri, S.A.; Okla, M.K.; AbdElgawad, H. Adaptability and Stability of Safflower Genotypes for Oil Production. *Plants* **2022**, *11*, 708. [CrossRef] [PubMed]
2. Hussain, M.; Ul-Allah, S.; Farooq, S. Sesame (*Sesamum indicum* L.). In *Neglected and Underutilized Crops*; Elsevier: Amsterdam, The Netherlands, 2023; pp. 733–755.
3. GOP. *Economic Survey of Pakistan*; Economic Advisory Wing: Islamabad, Pakistan, 2021.
4. Singh, V.; Pandey, S.; Synrem, G. Line x Tester Analysis for Seed Yield and Its Components in Toria (*Brassica campestris* L. Var. Toria). *J. Pharmacogn. Phytochem* **2022**, *11*, 393–398.
5. FAO. FAOSTAT 2023. Available online: www.faostat.fao.org (accessed on 11 January 2023).
6. Gunstone, F.D. *Vegetable Oils in Food Technology*; Gunstone, F.D., Ed.; Wiley: Hoboken, NJ, USA, 2011; ISBN 9781444332681.
7. Barth, C.A. Rapeseed for Human Nutrition–Present Knowledge and Future Options. In Proceedings of the 12th International Rapeseed Congress, Wuhan, China, 26–30 March 2007; pp. 3–5.
8. Ghazani, S.M.; Marangoni, A.G. Minor Components in Canola Oil and Effects of Refining on These Constituents: A Review. *J. Am. Oil Chem. Soc.* **2013**, *90*, 923–932. [CrossRef]
9. Li, X.; Yu, X.; Yang, H.; Wang, J.; Li, Z.; Bai, C.; Wang, J.; Wang, B.; Zhou, G.; Kuai, J. Physiological Response Mechanism of Oilseed Rape to Abiotic Stress and the Stress-Resistant Cultivation Regulation. In *Sustainable Crop Productivity and Quality Under Climate Change*; Elsevier: Amsterdam, The Netherlands, 2022; pp. 207–234.
10. Molazem, D.; Azimi, J.; Dideban, T. Measuring the Yield and Its Components, in the Canola in Different Planting Date and Plant Density of the West Guilan. *Int. J. Agric. Crop Sci.* **2013**, *6*, 869–872.
11. Wynne, K.; Neely, C.B.; Adams, C.; Kimura, E.; DeLaune, P.B.; Hathcoat, D.; Gerrish, B. Testing Row Spacing and Planting Rate for Fall-planted Spring Canola in the Southern United States. *Agron. J.* **2020**, *112*, 1952–1962. [CrossRef]
12. Dirwai, T.L.; Senzanje, A.; Mabhaudhi, T. Calibration and Evaluation of the FAO AquaCrop Model for Canola (*Brassica napus*) under Varied Moistube Irrigation Regimes. *Agriculture* **2021**, *11*, 410. [CrossRef]
13. Weymann, W.; Böttcher, U.; Sieling, K.; Kage, H. Effects of Weather Conditions during Different Growth Phases on Yield Formation of Winter Oilseed Rape. *Field Crop. Res.* **2015**, *173*, 41–48. [CrossRef]
14. Moradi Aghdam, A.; Sayfzadeh, S.; Shirani Rad, A.H.; Valadabadi, S.A.; Zakerin, H.R. The Assessment of Water Stress and Delay Cropping on Quantitative and Qualitative Traits of Rapeseed Genotypes. *Ind. Crops Prod.* **2019**, *131*, 160–165. [CrossRef]
15. Gan, Y.; Angadi, S.V.; Cutforth, H.; Potts, D.; Angadi, V.V.; McDonald, C.L. Canola and Mustard Response to Short Periods of Temperature and Water Stress at Different Developmental Stages. *Can. J. Plant Sci.* **2004**, *84*, 697–704. [CrossRef]
16. Ebrahimian, E.; Bybordi, A.; Seyyedi, S.M. How Nitrogen and Zinc Levels Affect Seed Yield, Quality, and Nutrient Uptake of Canola Irrigated with Saline and Ultra-Saline Water. *Commun. Soil Sci. Plant Anal.* **2017**, *48*, 345–355. [CrossRef]
17. Procko, C.; Crenshaw, C.M.; Ljung, K.; Noel, J.P.; Chory, J. Cotyledon-Generated Auxin Is Required for Shade-Induced Hypocotyl Growth in *Brassica rapa*. *Plant Physiol.* **2014**, *165*, 1285–1301. [CrossRef]
18. Kadioglu, H.; Hatterman-Valenti, H.; Jia, X.; Chu, X.; Aslan, H.; Simsek, H. Groundwater Table Effects on the Yield, Growth, and Water Use of Canola (*Brassica napus* L.) Plant. *Water* **2019**, *11*, 1730. [CrossRef]
19. Borie, F.; Rubio, R.; Rouanet, J.L.; Morales, A.; Borie, G.; Rojas, C. Effects of Tillage Systems on Soil Characteristics, Glomalin and Mycorrhizal Propagules in a Chilean Ultisol. *Soil Tillage Res.* **2006**, *88*, 253–261. [CrossRef]
20. Stevenson, F.J. Organic Matter-Micronutrient Reactions in Soil. In *Micronutrients in Agriculture*; Soil Science Society of America, Inc.: Madison, WI, USA, 2018; pp. 145–186. ISBN 9780891188780.
21. Rehman, A.; Farooq, M.; Ozturk, L.; Asif, M.; Siddique, K.H.M. Zinc Nutrition in Wheat-Based Cropping Systems. *Plant Soil* **2018**, *422*, 283–315. [CrossRef]
22. Nataraja, T.H.; Halepyati, A.S.; Pujari, B.T.; Desai, B.K. Influence of Phosphorus Levels and Micronutrients on Physiological Parameters of Wheat (*Triticum durum* Dcsf.). *Karnataka J. Agric. Sci.* **2006**, *19*, 685–687.
23. Chaudry, E.H.; Timmer, V.; Javed, A.S.; Siddique, M.T. Wheat Response to Micronutrients in Rainfed Areas of Punjab. *Soil Environ.* **2007**, *26*, 97–101.
24. Mandal, A.; Patra, A.; Singh, D.; Swarup, A.; Ebhinmasto, R. Effect of Long-Term Application of Manure and Fertilizer on Biological and Biochemical Activities in Soil during Crop Development Stages. *Bioresour. Technol.* **2007**, *98*, 3585–3592. [CrossRef]
25. Tsonev, T.; Cebola Lidon, F.J. Zinc in Plants-an Overview. *Emir. J. Food Agric.* **2012**, *24*, 322–333.
26. Umair Hassan, M.; Aamer, M.; Umer Chattha, M.; Haiying, T.; Shahzad, B.; Barbanti, L.; Nawaz, M.; Rasheed, A.; Afzal, A.; Liu, Y.; et al. The Critical Role of Zinc in Plants Facing the Drought Stress. *Agriculture* **2020**, *10*, 396. [CrossRef]

27. Tang, L.; Hamid, Y.; Liu, D.; Shohag, M.J.I.; Zehra, A.; He, Z.; Feng, Y.; Yang, X. Foliar Application of Zinc and Selenium Alleviates Cadmium and Lead Toxicity of Water Spinach—Bioavailability/Cytotoxicity Study with Human Cell Lines. *Environ. Int.* **2020**, *145*, 106122. [CrossRef]
28. Cakmak, I. Enrichment of Cereal Grains with Zinc: Agronomic or Genetic Biofortification? *Plant Soil* **2008**, *302*, 1–17. [CrossRef]
29. Shoja, T.; Majidian, M.; Rabiee, M. Effects of Zinc, Boron and Sulfur on Grain Yield, Activity of Some Antioxidant Enzymes and Fatty Acid Composition of Rapeseed (*Brassica napus* L.). *Acta Agric. Slov.* **2018**, *111*, 73–84. [CrossRef]
30. Afsahi, K.; Nazari, M.; Omidi, H.; Shekari, F.; Bostani, A.A. The Effects of Different Methods of Zinc Application on Canola Seed Yield and Oil Content. *J. Plant Nutr.* **2020**, *43*, 1070–1079. [CrossRef]
31. Rafique, E.; Rashid, A.; Ryan, J.; Bhatti, A.U. Zinc Deficiency in Rainfed Wheat in Pakistan: Magnitude, Spatial Variability, Management, and Plant Analysis Diagnostic Norms. *Commun. Soil Sci. Plant Anal.* **2006**, *37*, 181–197. [CrossRef]
32. Cakmak, I. Tansley Review No. 111 Possible Roles of Zinc in Protecting Plant Cells from Damage by Reactive Oxygen Species. *New Phytol.* **2000**, *146*, 185–205. [CrossRef]
33. Rehman, A.u.; Wang, X.; Hussain, A.; Qamar, R.; Khalofah, A.; Hussain, M. Boron Application in Yermosols Improves Grain Yield and Quality of Chickpea (*Cicer arietinum* L.). *J. King Saud Univ. Sci.* **2022**, *34*, 101768. [CrossRef]
34. Ahmad, W.; Zia, M.H.; Malhi, S.S.; Niaz, A.; Ullah, S. Boron Deficiency in Soils and Crops: A Review. *Crop Plant* **2012**, *2012*, 65–97.
35. Oosterhuis, D.M.; Zhao, D. Effect of Boron Deficiency on the Growth and Carbohydrate Metabolism of Cotton. In *Plant Nutrition*; Springer: Berlin/Heidelberg, Germany, 2001; pp. 166–167.
36. Goldbach, H.E.; Yu, Q.; Wingender, R.; Schulz, M.; Wimmer, M.; Findeklee, P.; Baluška, F. Rapid Response Reactions of Roots to Boron Deprivation. *J. Plant Nutr. Soil Sci.* **2001**, *164*, 173–181. [CrossRef]
37. Miwa, K.; Fujiwara, T. Boron Transport in Plants: Co-Ordinated Regulation of Transporters. *Ann. Bot.* **2010**, *105*, 1103–1108. [CrossRef]
38. Rashid, A.; Ryan, J. Micronutrient Constraints to Crop Production in Soils with Mediterranean-Type Characteristics: A Review. *J. Plant Nutr.* **2004**, *27*, 959–975. [CrossRef]
39. Gupta, U.; Solanki, H. Impact of Boron Deficiency on Plant Growth. *Int. J. Bioassays* **2013**, *2*, 1048–1050.
40. Kastori, R.; Maksimovic, I.; Kraljevic-Balalic, M.; Kobiljski, B. Physiological and Genetic Basis of Plant Tolerance to Excess Boron. *Zb. Matice Prir. Nauk.* **2008**, *114*, 41–51. [CrossRef]
41. McClung, J.P. Iron, Zinc, and Physical Performance. *Biol. Trace Element Res.* **2019**, *188*, 135–139. [CrossRef]
42. Pizzorno, L. Nothing Boring about Boron. *Integr. Med.* **2015**, *14*, 35.
43. Welch, R.M.; Graham, R.D. Breeding for Micronutrients in Staple Food Crops from a Human Nutrition Perspective. *J. Exp. Bot.* **2004**, *55*, 353–364. [CrossRef]
44. Zoveda, F.; Garcia, S.; Pandey, S.; Thomas, G.; Soto, D.; Bianchi, G.; Steinfeld, H.; Faures, J.M.; Griffin, J.; Lipper, L.; et al. *Building a Common Vision for Sustainable Food and Agriculture: Principles and Approaches*; Food and Agriculture Organization of the United Nations: Rome, Italy, 2014; Available online: https://www.fao.org/3/i3940e/i3940e.pdf (accessed on 10 June 2023).
45. Alloway, B.J. Soil Factors Associated with Zinc Deficiency in Crops and Humans. *Environ. Geochem. Health* **2009**, *31*, 537–548. [CrossRef]
46. Khoshgoftarmanesh, A.H.; Schulin, R.; Chaney, R.L.; Daneshbakhsh, B.; Afyuni, M. Micronutrient-Efficient Genotypes for Crop Yield and Nutritional Quality in Sustainable Agriculture. A Review. *Agron. Sustain. Dev.* **2010**, *30*, 83–107. [CrossRef]
47. Rehman, A.; Farooq, M.; Naveed, M.; Nawaz, A.; Shahzad, B. Seed Priming of Zn with Endophytic Bacteria Improves the Productivity and Grain Biofortification of Bread Wheat. *Eur. J. Agron.* **2018**, *94*, 98–107. [CrossRef]
48. Ngozi, U.F. The Role of Biofortification in the Reduction of Micronutrient Food Insecurity in Developing Countries. *Afr. J. Biotechnol.* **2013**, *12*, 5559–5566.
49. Atique-ur-Rehman; Ali, U.I.; Qamar, R.; Rehman, A.; Hussain, M.; Javeed, H.M.R.; Ahmad, S. Boron Foliage Application Mediates Growth, Oil Yield and Quality of Sunflower in Yermosols of Southern Punjab. *Int. J. Agric. Biol.* **2019**, *21*, 209–214. [CrossRef]
50. Alloway, B.J. *Micronutrient Deficiencies in Global Crop Production*; Springer Science & Business Media: Berlin/Heidelberg, Germany, 2008; ISBN 1402068603.
51. He, D.-H.; Lin, Z.-X.; Zhang, X.-L.; Nie, Y.-C.; Guo, X.-P.; Feng, C.-D.; Stewart, J.M. Mapping QTLs of Traits Contributing to Yield and Analysis of Genetic Effects in Tetraploid Cotton. *Euphytica* **2005**, *144*, 141–149. [CrossRef]
52. Usman, A.R.A.; Mohamed, H.M. Effect of Microbial Inoculation and EDTA on the Uptake and Translocation of Heavy Metal by Corn and Sunflower. *Chemosphere* **2009**, *76*, 893–899. [CrossRef]
53. Mousavi, S.R.; Galavi, M.; Ahmadvand, G. Effect of Zinc and Manganese Foliar Application on Yield, Quality and Enrichment on Potato (*Solanum tuberosum* L.). *Asian J. Plant Sci.* **2007**, *6*, 1256–1260. [CrossRef]
54. Bouyoucos, G.J. A Recalibration of the Hydrometer Method for Making Mechanical Analysis of Soils 1. *Agron. J.* **1951**, *43*, 434–438. [CrossRef]
55. U.S. Government Printing Office. *Salinity Laboratory Staff Diagnosis and Improvement of Saline and Alkali Soils, USDA Handbook 60*; U.S. Government Printing Office: Washington, DC, USA, 1954.
56. Dellavalle, N.B. Determination of Soil-Paste PH and Conductivity of Saturation Extract. In *Handbook on Reference Methods for Soil Analysis*; Soil Plant Analysis Council Inc.: Athens, Greece, 1992; pp. 40–43.
57. Walkley, A.; Black, I.A. An Examination of the Degtjareff Method for Determining Soil Organic Matter, and a Proposed Modification of the Chromic Acid Titration Method. *Soil Sci.* **1934**, *37*, 29–38. [CrossRef]

58. Campbell, W.R.; Hanna, M.I. The Determination of Nitrogen by Modified Kjeldahl Methods. *J. Biol. Chem.* **1937**, *119*, 1–7. [CrossRef]
59. Olsen, S.R. *Estimation of Available Phosphorus in Soils by Extraction with Sodium Bicarbonate*; US Department of Agriculture: Washington, DC, USA, 1954.
60. Knudsen, D.; Peterson, G.A.; Pratt, P.F. Lithium, Sodium, and Potassium. *Methods Soil Anal. Part 2 Chem. Microbiol. Prop.* **1983**, *9*, 225–246.
61. Johnson, G.V.; Fixen, P.E. Testing Soils for Sulfur, Boron, Molybdenum, and Chlorine. In *Soil Testing and Plant Analysis*, 3rd ed.; Soil Science Society of America: Madison, WI, USA, 1990; pp. 265–273.
62. Lindsay, W.L.; Martens, D.C. Testing Soils for Copper, Iron, Manganese, and Zinc. *Soil Test. Plant Anal.* **1990**, *3*, 229–264.
63. Bingham, F.T. Boron. *Methods Soil Anal. Part 2 Chem. Microbiol. Prop.* **1983**, *9*, 431–447.
64. Allen, S.E.; Grimshaw, H.M.; Rowland, A.P. *Chemical Analysis. Methods in Plant Ecology*; Moore, P.D., Chapman, S.B., Eds.; Blackwell: London, UK, 1986.
65. Metcalfe, L.D.; Schmitz, A.A.; Pelka, J.R. Rapid Preparation of Fatty Acid Esters from Lipids for Gas Chromatographic Analysis. *Anal. Chem.* **1966**, *38*, 514–515. [CrossRef]
66. Bradford, M.M. A Rapid and Sensitive Method for the Quantitation of Microgram Quantities of Protein Utilizing the Principle of Protein-Dye Binding. *Anal. Biochem.* **1976**, *72*, 248–254. [CrossRef] [PubMed]
67. Yemm, E.W.; Willis, A.J. The Estimation of Carbohydrates in Plant Extracts by Anthrone. *Biochem. J.* **1954**, *57*, 508–514. [CrossRef]
68. Khan, W.; Prithiviraj, B.; Smith, D.L. Photosynthetic Responses of Corn and Soybean to Foliar Application of Salicylates. *J. Plant Physiol.* **2003**, *160*, 485–492. [CrossRef]
69. Zekri, M. Effects of NaCl on Growth and Physiology of Sour Orange and Cleopatra Mandarin Seedlings. *Sci. Hortic.* **1991**, *47*, 305–315. [CrossRef]
70. Moya, J.L. Chloride Absorption in Salt-Sensitive Carrizo Citrange and Salt-Tolerant Cleopatra Mandarin Citrus Rootstocks Is Linked to Water Use. *J. Exp. Bot.* **2003**, *54*, 825–833. [CrossRef]
71. Azhar, N.; Hussain, B.; Ashraf, M.Y.; Abbasi, K.Y. Water Stress Mediated Changes in Growth, Physiology and Secondary Metabolites of Desi Ajwain (*Trachyspermum ammi* L.). *Pak. J. Bot.* **2011**, *43*, 15–19.
72. Ullah, N.; Haq, I.U.; Safdar, N.; Mirza, B. Physiological and Biochemical Mechanisms of Allelopathy Mediated by the Allelochemical Extracts of *Phytolacca latbenia* (Moq.) H. Walter. *Toxicol. Ind. Health* **2015**, *31*, 931–937. [CrossRef] [PubMed]
73. Beyer, W.F.; Fridovich, I. Assaying for Superoxide Dismutase Activity: Some Large Consequences of Minor Changes in Conditions. *Anal. Biochem.* **1987**, *161*, 559–566. [CrossRef]
74. Steel, R.G.D.; Torrie, J.H.; Dickey, D. *Principles and Procedure of Statistics. A Biometrical Approach*, 3rd ed.; McGraw HillBookCo. Inc.: New York, NY, USA, 1997; pp. 352–358.
75. Kanwal, S.; Rahmatullah; Maqsood, M.A.; Bakhat, H.F.S.G. Zinc Requirement of Maize Hybrids and Indigenous Varieties on Udic Haplustalf. *J. Plant Nutr.* **2009**, *32*, 470–478. [CrossRef]
76. Aref, F. Influence of Zinc and Boron Interaction on Residual Available Iron and Manganese in the Soil after Corn Harvest. *Am. Eurasian J. Agric. Environ. Sci.* **2010**, *8*, 767–772.
77. Shaaban, M.M.; El-Fouly, M.M.; Abdel, A.-W.A. Zinc-Boron Relationship in Wheat Plants Grown under Low or High Levels of Calcium Carbonate in the Soil. *Pak. J. Biol. Sci.* **2004**, *7*, 633–639. [CrossRef]
78. Tariq, A.; Anjum, S.A.; Randhawa, M.A.; Ullah, E.; Naeem, M.; Qamar, R.; Ashraf, U.; Nadeem, M. Influence of Zinc Nutrition on Growth and Yield Behaviour of Maize (*Zea mays* L.) Hybrids. *Am. J. Plant Sci.* **2014**, *05*, 2646–2654. [CrossRef]
79. Beckie, H.J.; Francis, A.; Hall, L.M. The Biology of Canadian Weeds. 27. *Avena fatua* L. (Updated). *Can. J. Plant Sci.* **2012**, *92*, 1329–1357. [CrossRef]
80. Silva, C.A.T.; Cagol, A.; Silva, T.R.B.; Nobrega, L.H.P. Boron Application before Sowing of Sunflower Hybrid. *J. Food Agric. Environ.* **2011**, *9*, 580–583.
81. Shrestha, A.; Pradhan, S.; Shrestha, J.; Subedi, M. Role of Seed Priming in Improving Seed Germination and Seedling Growth of Maize (*Zea mays* L.) under Rain Fed Condition. *J. Agric. Nat. Resour.* **2019**, *2*, 265–273. [CrossRef]
82. Potarzycki, J.; Grzebisz, W. Effect of Zinc Foliar Application on Grain Yield of Maize and Its Yielding Compone. *Plant Soil Environ.* **2009**, *55*, 519–527. [CrossRef]
83. Morshedi, A.; Naghibi, H. Effects of Foliar Application of CU and ZN on Yield and Quality of Canola Seed (*Brassica napus* L.). *J. Agric.* **2004**, *11*, 15–22.
84. Yang, M.; Shi, L.; Xu, F.-S.; Lu, J.-W.; Wang, Y.-H. Effects of B, Mo, Zn, and Their Interactions on Seed Yield of Rapeseed (*Brassica napus* L.). *Pedosphere* **2009**, *19*, 53–59. [CrossRef]
85. De Oliveira, R.H.; Dias Milanez, C.R.; Moraes-Dallaqua, M.A.; Rosolem, C.A. Boron Deficiency Inhibits Petiole and Peduncle Cell Development and Reduces Growth of Cotton. *J. Plant Nutr.* **2006**, *29*, 2035–2048. [CrossRef]
86. Fontes, R.L.F.; Medeiros, J.F.; Neves, J.C.L.; Carvalho, O.S.; Medeiros, J.C. Growth of Brazilian Cotton Cultivars in Response to Soil Applied Boron. *J. Plant Nutr.* **2008**, *31*, 902–918. [CrossRef]
87. Bellaloui, N.; Hu, Y.; Mengistu, A.; Kassem, M.A.; Abel, C.A. Effects of Foliar Boron Application on Seed Composition, Cell Wall Boron, and Seed Δ15N and Δ13C Isotopes in Water-Stressed Soybean Plants. *Front. Plant Sci.* **2013**, *4*, 270. [CrossRef]
88. Sher, A.; Sattar, A.; Qayyum, A.; Ijaz, M.; Nawaz, A.; Manaf, A.; Hussain, M. Optimizing the NPK Application in White Mustard (*Sinapis alba* L.) under an Arid Climate in Punjab, Pakistan. *J. Plant Nutr.* **2019**, *42*, 1556–1565. [CrossRef]

89. Kalantar Ahmadi, A.; Shoushi Dezfouli, A.A. Effects of Foliar Application of Micronutrients on Seed Yield and Oil Quality of Canola (*Brassica napus* L. Cv. Hyola401) under Drought Stress Conditions. *Iran. Soc. Crops Plant Breed. Sci.* **2019**, *21*, 237–253. [CrossRef]
90. Bellaloui, N.; Hanks, J.E.; Fisher, D.K.; Mengistu, A. Soybean Seed Composition Is Influenced by Within-Field Variability in Soil Nutrients. *Crop Manag.* **2009**, *8*, 1–12. [CrossRef]
91. McGrath, S.P.; Zhao, F.J. Sulphur Uptake, Yield Responses and the Interactions between Nitrogen and Sulphur in Winter Oilseed Rape (*Brassica napus*). *J. Agric. Sci.* **1996**, *126*, 53–62. [CrossRef]
92. Zhou, T.; Hua, Y.; Huang, Y.; Ding, G.; Shi, L.; Xu, F. Physiological and Transcriptional Analyses Reveal Differential Phytohormone Responses to Boron Deficiency in *Brassica napus* Genotypes. *Front. Plant Sci.* **2016**, *7*, 221. [CrossRef]
93. Wang, N.; Daun, J. Effects of Variety and Crude Protein Content on Nutrients and Anti-Nutrients in Lentils (*Lens culinaris*). *Food Chem.* **2006**, *95*, 493–502. [CrossRef]
94. Akta, L.Y.; Gemici, M.; Türkyalmaz, B.; Güven, A. The Effects of High Boron Concentration on Growing and Protein Metabolism in Wheat. *JA Natoila Eagean Agric. Res.* **2004**, *14*, 88–99.
95. Li, Y.; Hou, L.; Song, B.; Yang, L.; Li, L. Effects of Increased Nitrogen and Phosphorus Deposition on Offspring Performance of Two Dominant Species in a Temperate Steppe Ecosystem. *Sci. Rep.* **2017**, *7*, 40951. [CrossRef]
96. Pinho, L.G.R.; Campostrini, E.; Monnerat, P.H.; Netto, A.T.; Pires, A.A.; Marciano, C.R.; Soares, Y.J.B. Boron Deficiency Affects Gas Exchange and Photochemical Efficiency (JPI Test Parameters) in Green Dwarf Coconut. *J. Plant Nutr.* **2010**, *33*, 439–451. [CrossRef]
97. Han, S.; Chen, L.-S.; Jiang, H.-X.; Smith, B.R.; Yang, L.-T.; Xie, C.-Y. Boron Deficiency Decreases Growth and Photosynthesis, and Increases Starch and Hexoses in Leaves of Citrus Seedlings. *J. Plant Physiol.* **2008**, *165*, 1331–1341. [CrossRef]
98. Khan, M.B.; Farooq, M.; Hussain, M.; Shahnawaz; Shabir, G. Foliar Application of Micronutrients Improves the Wheat Yield and Net Economic Return. *Int. J. Agric. Biol.* **2010**, *12*, 953–956.
99. Marschner, P. *Marschner's Mineral Nutrition of Higher Plants*, 3rd ed.; Academic Press: Cambridge, MA, USA, 2011; ISBN 9780123849052.
100. Monreal, C.M.; DeRosa, M.; Mallubhotla, S.C.; Bindraban, P.S.; Dimkpa, C. The Application of Nanotechnology for Micronutrients in Soil-Plant Systems (VFRC Report 2015/3). *VFRC Rep.* **2015**, *3*, 44.

agronomy

MDPI

Article

The Effect of Nitrogen and Sulphur Application on Soybean Productivity Traits in Temperate Climates Conditions

Aleksandra Głowacka [1,*], Elvyra Jariene [2], Ewelina Flis-Olszewska [1] and Anna Kiełtyka-Dadasiewicz [1]

[1] Department of Plant Production Technology and Commodities Science, University of Life Sciences in Lublin, 20-950 Lublin, Poland
[2] Department of Plant Biology and Food Sciences, Agriculture Academy, Vytautas Magnus University, 44248 Kaunas, Lithuania
* Correspondence: aleksandra.glowacka@up.lublin.pl

Abstract: Both nitrogen and sulphur are important macronutrients necessary for the proper development and yield of soybean. Moreover, sulphur plays a special role in nitrogen metabolism in the plant, and sulphur deficiency leads to a reduction in the utilization of nitrogen from fertilizer. The objective of this study was to assess the effect of nitrogen and sulphur application on the yield and quality traits of soybean seeds. The following factors were analyzed in the experiment: I. Nitrogen application rate: 0, 30 and 60 kg ha^{-1} applied at different times (before sowing and/or at the start of the seed filling stage); II. Sulphur application rate: 0 and 40 kg ha^{-1} applied in two portions: half during the development of lateral shoots and half at the start of flowering. Thus the 14 fertilizer combinations were obtained. Result show that the highest seeds yield was obtained in the combinations with 60 kg N applied 1/2 before sowing + 1/2 after emergence (BBCH 73-75) and 3/4 before sowing + 1/4 after emergence. In these combinations, sulphur did not significantly affect seed yield. In the remaining nitrogen application, sulphur application significantly increased the seed yield. Taking into account the yield and the chemical composition of the soybean seeds, fertilization with 60 kg N ha^{-1} in two portions can be recommended—1/2 or 3/4 before sowing and the remainder during the development of pods and seeds—in combination with sulphur application.

Keywords: nitrogen fertilization; sulphur; crude fat; crud protein; seeds yield; potassium; magnesium; phosphorus

Citation: Głowacka, A.; Jariene, E.; Flis-Olszewska, E.; Kiełtyka-Dadasiewicz, A. The Effect of Nitrogen and Sulphur Application on Soybean Productivity Traits in Temperate Climates Conditions. *Agronomy* **2023**, *13*, 780. https://doi.org/10.3390/agronomy13030780

Academic Editors: Gang Li, Dong-Xing Guan and Daniel Menezes-Blackburn

Received: 9 February 2023
Revised: 2 March 2023
Accepted: 6 March 2023
Published: 8 March 2023

1. Introduction

Soybean is one of the most important crop plants in the world. As it takes up fairly large amounts of essential nutrients, it should be grown on a site with adequate nutrient content [1–3]. One essential element for the growth and development of soybean is nitrogen. Soybean requires a large amount of nitrogen due to the high protein content in the seeds—about 35–40%. Like other legumes, soybean takes up nitrogen from two sources: the atmosphere and mineral fertilizers [4]. Atmospheric nitrogen is fixed by rhizobia. Nitrogen fertilization for legume plants is usually limited to application of starter fertilizer. The level of application varies depending on soil conditions but is usually much lower than standard levels of this nutrient used to fertilize other plants. However, according to many authors, biological nitrogen fixation meets only about 50–60% of soybean's demand for nitrogen, resulting in 80–90% of the maximum yield obtained in the case of adequate nitrogen fertilization [5,6]. Nitrogen application mainly affects crop yield. According to Lorenc-Kozik and Pisulewska [7], a nitrogen application rate that beneficially influences soybean yield varies between 30 kg N ha^{-1} and 60 kg N ha^{-1}. Fertilization of soybean with nitrogen has a beneficial effect on elements of the yield structure such as plant height, pod number per plant, and seed weight per plant. Nitrogen application can also modify the chemical composition of the seeds. Lower levels increase their fat content, while higher levels increase the content of protein [8–10]. In the case of soybean fertilization with nitrogen, not only the level of application is important, but the time of application as well [8,11]. Starter nitrogen fertilizer applied before sowing is aimed at supplying easily available nitrogen from the soil during seedling development and has been shown by numerous studies to increase soybean seed yield [7–9,12] According to [13] soybean has a relatively high demand for N during the seed-filling stage, and biological fixation of N and a low level of starter fertilizer may not supply an adequate amount of N for full exploitation of the plant's yield potential. Therefore, N application during generative growth can increase crop yields.

Another essential nutrient for the normal functioning of plants is sulphur, which is a component of vitamins and amino acids, and thus of proteins. The presence of sulphur is essential for biosynthesis of protein in the seeds. It positively affects not only the quantity but also the quality of crude protein in the harvested crop, especially in the case of legumes [14,15]. Plants of the *Fabaceae* family have a moderate demand for this nutrient [16]. For legumes, depending on the species and climate-soil conditions, the recommended application rate ranges from 20 to 60 kg S ha^{-1} [14,17,18]. It should be remembered, however, that a single portion of this nutrient should not exceed 20 kg S ha^{-1} [14,19]. Many studies [14,16,17,20] indicate a strong interaction between sulphur and nitrogen as essential nutrients for synthesis of amino acids making up proteins. Sulphur plays a special role in nitrogen metabolism in the plant, and sulphur deficiency leads to a reduction in the utilization of nitrogen from fertilizer. Sulphur is an activator of processes regulating C and N metabolism in the plant, and in this way increases the rate of transformation of nitrogen taken up by the plant into protein. Plants that are well supplied with nitrogen and sulphur increase the amount of nitrogen incorporated into organic structures. Sulphur also takes part in fixation of atmospheric nitrogen by rhizobia and in reduction in nitrates to ammonia. Many authors have confirmed the positive effect of sulphur application on fixation of atmospheric nitrogen by the root nodules of legumes and on utilization of mineral nitrogen, and thus on production of plant biomass [16,20,21]. Therefore, adequate supply of these nutrients to plants enables full exploitation of their yield potential. In the absence of sulphur, they produce protein with much lower content of sulphur-containing amino acids, especially methionine, which is one of the most valuable amino acids determining the nutritional value of plants [22–24]. The effectiveness of sulfur fertilization depends on many factors, including the dose and date of application. According to many authors [25,26], oilseeds have a high demand for sulfur, especially from the budding phase to the formation of siliques. The availability of sulfur during this period ensures the proper growth and development of rapeseed. The study conducted by Barczak et al. [19] confirmed that the foliar sulphur application (in the phase of not completely covered interrows until full flowering), as compared with the soil application, showed a better effect on the seed and straw yield size.

Research results confirm the beneficial effect of sulphur on the yield of legume plants, such as narrow-leaved lupin [19], field bean [27], common bean and broad bean [22,24]. Other authors have also confirmed positive effects of legume fertilization with sulphur on the content and uptake of nitrogen. The effect of nitrogen application on the yield and quality of soybean crops is also a frequent subject of research [14,28–30]. Most of the research on the reaction of soybean to nitrogen and sulfur fertilization was conducted in countries with a warm climate, i.e., Iran [31,32] India [33–37]. There are few studies conducted in temperate climate conditions especially with newer not genetically modified cultivars—on the effect of nitrogen and sulphur application on the yield and chemical composition of soybean seeds [12]. In view of the above, a study was carried out to test the effect of nitrogen and sulphur application on the yield of soybean and the content of protein, fat and selected macroelements in the seeds. The main objective of the research was to determine how much (and when) N and S fertilization should be used to obtain Ethe best production characteristics in soybean cultivation in the climatic conditions of south-eastern Poland.

2. Materials and Methods

2.1. Site Description, Experimental Design and Field Management

A 3-year field experiment was carried out in 2015–2017 under rainfed conditions on a private farm in Zamość District, in south-eastern Poland (50°43′34″ N, 23°39′11″ E).

The experimental field was located on soil with the granulometric composition of clayey silt, slightly acidic, with moderate content of phosphorus and potassium (Table 1). The conventional (not genetically modified) soybean cultivar Amandine, a medium-early cultivar (maturity group MG 000) with high yield potential, was planted in the experiment.

Table 1. Selected soil properties (0–30 cm depth) in experimental field.

Soil Characteristic	Value
Texture class [1]	Clayey silt
Sand (2–0.05 mm)	21%
Silt (0.05–0.002)	70%
Clay (<0.002)	9%
pH_{KCl}	6.6
Total nitrogen (g kg^{-1})	1.3
Organic carbon (g kg^{-1})	19.1
Available forms (mg kg^{-1})	
Phosphorus	175
Potassium	201
Magnesium	54
Copper	43.2
Manganese	187.2
Iron	942
Zinc	9.6

[1] According to [38].

The following factors were analyzed in the experiment:

I. Nitrogen application rate: 0, 30 and 60 kg ha^{-1} applied at different times (before sowing and/or at the start of the seed filling stage BBCH 73–75)

II. Sulphur application rate: 0 and 40 kg ha^{-1} applied in two portions: half during the development of lateral shoots (BBCH 209–210) and half at the start of flowering (BBCH 60–61)

The phonological stages (BBCH) of soybean were encoding according to [39]. The nitrogen dose resulted from fertilizer recommendations for legumes in Poland and previous own research [8]. The sulphur dose was determined based on the recommendation of other authors [14,18,22].

Thus, the following 14 fertilizer combinations were obtained (Table 2).

The experiment was set up in a randomized split-plot design in three replicates. The area of the plots was 12.5 m^2 (2.5 m × 5 m). Nitrogen application rates has been designated as the main plot and sulphur application rate as the subplot.

Soybean was sown between 30 April and 2 May, at a rate of 70 seeds m^{-2}, and the row spacing was 20 cm. Soybean seeds were prepared for sowing in FIX FERTIG technology. In this process the seeds are coated with rhizobia together with a polymer, which acts as a preservative and also protects against solar radiation. Phosphorus (in the form of Fos Dar fertilizer) and potassium (60% potassium chloride) were applied once before sowing, at the same rates in all treatments: P—21.1 kg ha^{-1}, K—76.4 kg ha^{-1}. Nitrogen fertilizer was applied in the form of 34% ammonium nitrate, and sulphur in the form of magnesium sulphate heptahydrate. In each year of the study the precursor crop for soybean was winter wheat. After harvesting of the precursor crop, skimming was carried out (shallow ploughing at 8–10 cm), followed by harrowing and pre-winter ploughing to a medium depth (22–25 cm). Harrowing was carried out in early spring, followed by pre-sowing tillage with a tillage machine equipped with a roller. Sowing was carried out using a mechanical seeder for cereals. Plant protection was limited to chemical weed control. Directly after sowing herbicides were applied to the soil: Sencor Liquid 600 SC (biologically active substance—metribuzin) in the amount of 0.5 dm^3 ha^{-1} and Dual Gold 960 EC (biologically active substance—S-metolachlor) in the amount of 1.0 dm^3 ha^{-1}. Soybean was harvested with a combine at full maturity (BBCH 99) in the middle 10 days of September.

Table 2. Researched fertilizer combinations.

Fertilizer Combination	Dose and Time of Application	
	Nitrogen	Sulphur
N0-S0	without nitrogen	without sulphur
N0-S40	without nitrogen	40 kg S ha^{-1}
N30 (30:0)-S0	30 kg N ha^{-1} before sowing	without sulphur
N30 (30:0)-S40	30 kg N ha^{-1} before sowing	40 kg S ha^{-1}
N30 (15:15)-S0	15 kg N ha^{-1} before sowing + 15 kg N ha^{-1} at the start of the seed filling	without sulphur
N30 (15:15)-S40	15 kg N ha^{-1} before sowing + 15 kg N ha^{-1} at the start of the seed filling	40 kg S ha^{-1}
N30 (0:30)-S0	30 kg N ha^{-1} at the start of the seed filling	without sulphur
N30 (0:30)-S40	30 kg N ha^{-1} at the start of the seed filling	40 kg S ha^{-1}
N60 (15:45)-S0	15 kg N ha^{-1} before sowing + 45 kg N ha^{-1} at the start of the seed filling	without sulphur
N60 (15:45)-S40	15 kg N ha^{-1} before sowing + 45 kg N ha^{-1} at the start of the seed filling	40 kg S ha^{-1}
N60 (30:30)-S0	30 kg N ha^{-1} before sowing + 30 kg N ha^{-1} at the start of the seed filling	without sulphur
N60 (30:30)-S40	30 kg N ha^{-1} before sowing + 30 kg N ha^{-1} at the start of the seed filling	40 kg S ha^{-1}
N60 (45:15-S0	45 kg N ha^{-1} before sowing + 15 kg N ha^{-1} at the start of the seed filling	without sulphur
N60 (45:15)-S40	45 kg N ha^{-1} before sowing + 15 kg N ha^{-1} at the start of the seed filling	40 kg S ha^{-1}

2.2. *Features of the Yield Components and Laboratory Analysis*

After harvest, at the full maturity stage (BBCH 99) of entire plot with a combine, the seed yield was determined and expressed per hectare for a moisture level of 15%. After harvesting, seed samples (0.5 kg) were taken from every plot for chemical analysis. Prior to harvest of soybean, 10 plants were randomly selected from each plot to determine elements of the yield components: plant height, pod number per plant, seed number per pod, and thousand seed weight.

The soybean seeds were analyzed for:

- crude protein by Kjeldahl's method (according to CLA/PSO/13),
- crude fat by the Soxhlet extraction-weighing method (according to CLA/PSO/10);
- total sulphur by the Bradley–Lancaster nephelometric method (following wet mineralization using concentrated sulphuric acid with 30% perhydrol),
- phosphorus by spectrophotometry (according to CLA/PLC/28),
- potassium, magnesium and calcium by Atomic Absorption Spectrometry with excitation in the air-acetylene flame (according to CLA/ASA/2).

The analyses were carried out at the Central Laboratory of Agroecology of the University of Life Sciences in Lublin. The total protein content in the seeds was calculated as 6.26 × total N.

2.3. *Statistical Analysis*

Statistical analysis of the results was performed by analysis of variance (ANOVA) in STATISTICA 13 PL software (Tulsa, OK, USA). Three-way ANOVA was carried out to determine the effect of the year, nitrogen fertilization, and sulphur fertilization on soybean yield, its components, and the chemical composition of the seeds. The effect of year, nitrogen fertilization, cultivar, and their interactions were analyzed using a split–split-plot design with the year being designed as whole plots, nitrogen fertilization as subplots, and sulphur fertilization as sub–subplots. The differences between means were determined using Tukey's post-hoc test at $p < 0.05$. As no significant interaction of the factors and years of the study was confirmed, the results presented are averages from three years. The effect of the interaction of factors is presented. If the interaction of factors was not significant, the main effect was presented.

3. Results

3.1. Meteorological Conditions

The experiment was carried out under rainfed conditions The weather conditions during the study period are presented in Figure 1. Meteorological data come from the Lublin-Felin automatic meteorological station, located on the premises of the Department of Plant Production Technology and Commodity Science of the University of Life Sciences in Lublin. The station is located approximately 90 km from the site of the field experiment the meteorological data were used to calculate Selyaninov's hydrothermal coefficient (Figure 2) according to the following formula: $k = (p \times 10)/\Sigma t$ where p is the sum of ten-day monthly precipitation (mm) and Σt is the sum of average daily temperatures from a tenday period/month (°C). Ranges of values for the coefficients were designated according to the scale developed by Skowera et al. [40].

In 2015 and 2016 the precipitation totals in the period from April to September were similar (330–335 mm) and higher than the long-term average. In 2017 the precipitation total during the growing season was about 358 mm, which was close to the long-term average. Particularly heavy rainfall was noted in May and April 2015 and in July 2016 and 2017. In 2015 very low rainfall was noted in June and in August.

Selyaninov's coefficient (Figure 2) indicated that the period from May to June in 2015 and 2016 was dry or very dry. In 2017 highly unfavorable temperature and moisture conditions were noted in the period from June to August.

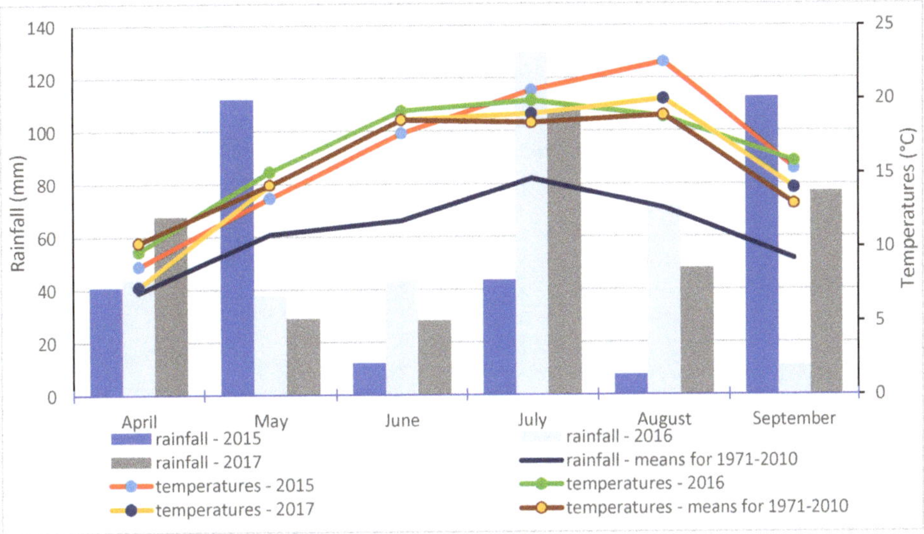

Figure 1. Air temperature (each point represents the average temperature of each month) and rainfall distribution during the growing period of April–September as compared to the long-term means (1971–2010).

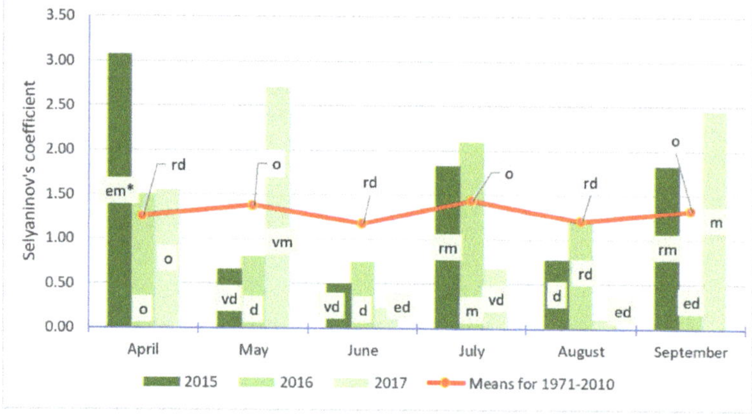

Figure 2. Selyaninov's coefficient during the growing period as compared to the long-term means (1971–2010). * Interpretation of the value of the Sielianinov coefficient according to [40]: ed—extremely dry, vd—very dry, d—dry, rd—rather dry, o—optimal, rm—rather moist; m—moist, vm—very moist, em—extremely moist.

3.2. Yield and Elements of the Yield Structure

The lowest yield of soybeans seeds, at a level of 20.9 dt ha^{-1}, was obtained from the control treatment, in which no fertilizer was applied (Figure 3). Fertilization with nitrogen and sulphur significantly increased the yield of soybean seeds. The most beneficial effect was obtained in the treatments in which 60 kg N was applied 1/2 before sowing +1/2 after emergence and 3/4 before sowing +1/4 after emergence. In these combinations, the seed yield was 8–10 dt ha^{-1} higher than in the combination without fertilizer, while sulphur application had no significant effect on yield. In the remaining treatments, sulphur application significantly increased the yield of soybean seeds (Figure 3). It should be noted that the greatest differences were noted when sulfur was applied in the combination without nitrogen fertilizer, were the yield of soybean seed increased by 18%. In the combinations with 30 kg N ha^{-1}, sulphur application increased yield by 6–7%.

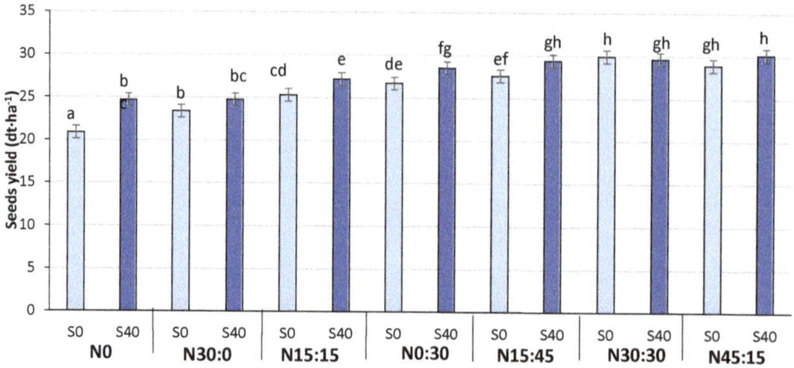

Figure 3. Effects of nitrogen × sulphur application on the yield of soybean seeds. Means marked with the same letter do not differ significantly at p = 0.05, bars mean SE. S0—without sulfur, S40—40 kg S ha^{-1}, N0—without nitrogen, N30:0—30 kg N ha^{-1} before sowing, N15:15—15 kg N ha^{-1} before sowing + 15 kg N ha^{-1} at the start of the seed filling, N:0:30—30 kg N ha^{-1} at the start of the seed filling, N15:45—15 kg N ha^{-1} before sowing + 30 kg N ha^{-1} at the start of the seed filling, N30:30—30 kg N ha^{-1} before sowing + 30 kg N ha^{-1} at the start of the seed filling, N45:15—45 kg N ha^{-1} before sowing + 15 kg N ha^{-1} at the start of the seed filling.

The fertilizer combinations affected the elements of the yield structure. Plant height increased significantly relative to the control following application of nitrogen at 30 kg in its entirety before sowing and application of 60 kg irrespective of the means of application. The highest plants were obtained from the combination with application of 45 kg N before sowing followed by foliar application of 15 kg. Plant height was also increased by sulphur application, but these were minor differences amounting to 4 cm (Figure 4). The interaction of the factors was not shown to significantly influence the height of the soybean plants.

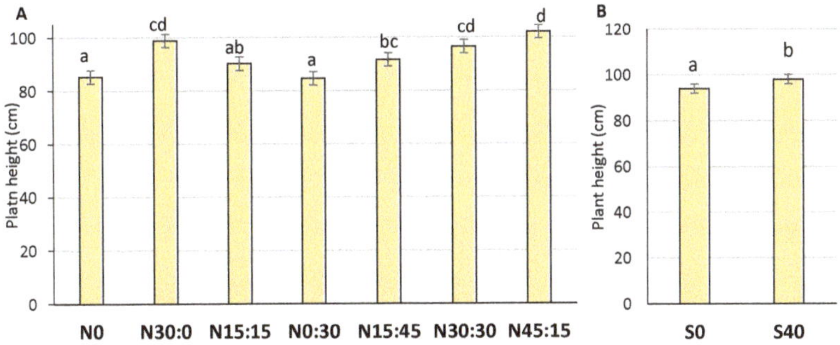

Figure 4. Effects of nitrogen (**A**), sulphur (**B**) application on the height of soybean plants. Means marked with the same letter do not differ significantly at $p = 0.05$, bars mean SE. S0—without sulfur, S40—40 kg S ha^{-1}, N0—without nitrogen, N30:0—30 kg N ha^{-1} before sowing, N15:15—15 kg N ha^{-1} before sowing + 15 kg N ha^{-1} at the start of the seed filling, N:0:30—30 kg N ha^{-1} at the start of the seed filling, N15:45—15 kg N ha^{-1} before sowing + 30 kg N ha^{-1} at the start of the seed filling, N30:30—30 kg N ha^{-1} before sowing + 30 kg N ha^{-1} at the start of the seed filling, N45:15—45 kg N ha^{-1} before sowing + 15 kg N ha^{-1} at the start of the seed filling.

All of the nitrogen fertilizer combinations significantly increased the number of pods per plant in comparison to the control. Sulphur application also increased the pod number per plant, especially in the treatment without nitrogen fertilizer, and in the combination with 30 kg applied 1/2 before sowing +1/2 after emergence or applied in its entirety after emergence (Figure 5). Nitrogen application significantly increased the seed number per pod, except in the case of 30 kg applied 1/2 before sowing +1/2 after emergence. The differences between the other nitrogen combinations were not significant (Figure 6). Sulphur application was not shown to significantly influence the seed number per pod. There was also no significant interaction between application of nitrogen and sulphur.

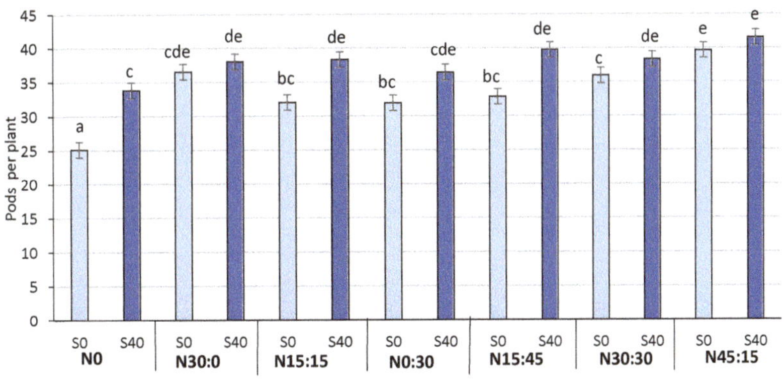

Figure 5. Effects of nitrogen × sulphur application on the number of pods per plant. Means marked with the same letter do not differ significantly at $p = 0.05$, bars mean SE. S0—without sulfur, S40-40 kg S ha^{-1}, N0—without nitrogen, N30:0—30 kg N ha^{-1} before sowing, N15:15—15 kg N ha^{-1}

before sowing + 15 kg N ha^{-1} at the start of the seed filling, N:0:30—30 kg N ha^{-1} at the start of the seed filling, N15:45—15 kg N ha^{-1} before sowing + 30 kg N ha^{-1} at the start of the seed filling, N30:30—30 kg N ha^{-1} before sowing + 30 kg N ha^{-1} at the start of the seed filling, N45:15—45 kg N ha^{-1} before sowing + 15 kg N ha^{-1} at the start of the seed filling.

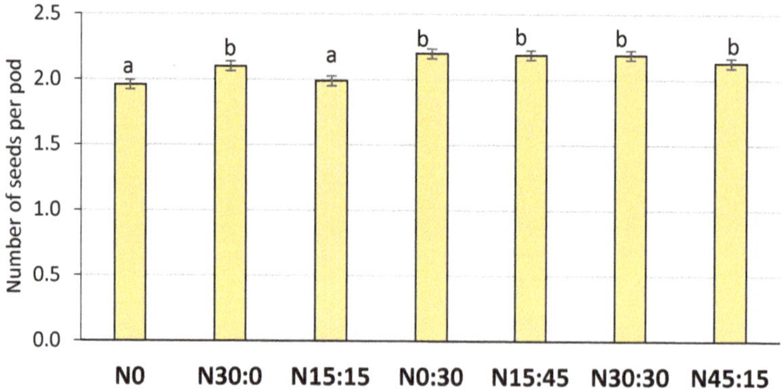

Figure 6. Effects of nitrogen application on the seed number per pod. Means marked with the same letter do not differ significantly at $p = 0.05$, bars mean SE N0—without nitrogen, N30:0—30 kg N ha^{-1} before sowing, N15:15—15 kg N ha^{-1} before sowing + 15 kg N ha^{-1} at the start of the seed filling, N:0:30—30 kg N ha^{-1} at the start of the seed filling, N15:45—15 kg N ha^{-1} before sowing + 30 kg N ha^{-1} at the start of the seed filling, N30:30—30 kg N ha^{-1} before sowing + 30 kg N ha^{-1} at the start of the seed filling, N45:15—45 kg N ha^{-1} before sowing + 15 kg N ha^{-1} at the start of the seed filling.

The nitrogen fertilizer used in the experiment increased the thousand seed weight as well, except in the case of 30 kg applied 1/2 before sowing +1/2 after emergence. The highest value was obtained following application of 60 kg with 1/4 applied before sowing +3/4 after emergence or 1/2 before sowing +1/2 after emergence (Figure 7). Seed yield was positively correlated with the number of pods per plant and the seed number per pod (Table 3).

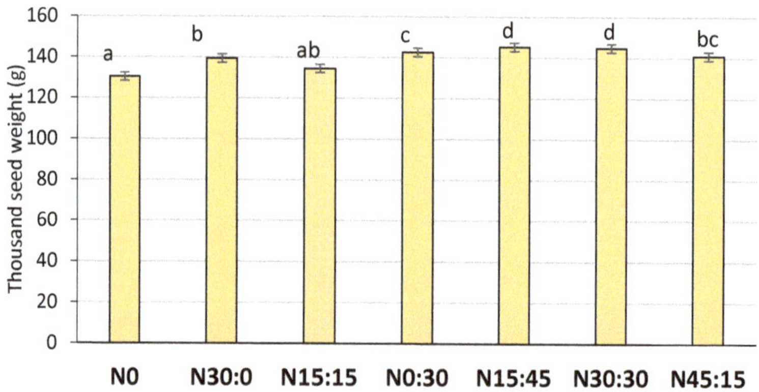

Figure 7. Effects of nitrogen application on the thousand seed weight. Means marked with the same letter do not differ significantly at $p = 0.05$, bars mean SE. N0—without nitrogen, N30:0—30 kg N ha^{-1} before sowing, N15:15–15 kg N ha^{-1} before sowing + 15 kg N ha^{-1} at the start of the seed filling, N:0:30–30 kg N ha^{-1} at the start of the seed filling, N15:45—15 kg N ha^{-1} before sowing + 30 kg N ha^{-1} at the start of the seed filling, N30:30—30 kg N ha^{-1} before sowing + 30 kg N ha^{-1} at the start of the seed filling, N45:15—45 kg N ha^{-1} before sowing + 15 kg N ha^{-1} at the start of the seed filling.

Table 3. Pearson's correlation coefficients between seed yield and selected elements of the yield structure.

Features	Number of Pods per Plant	Seed Number per Pod	Thousand Seed Weight
Yield	0.437 *	0.409 *	0.188

* Significant at $p = 0.05$.

3.3. Content and Yield of Crude Protein and Crude Fat

Nitrogen application increased the crude protein content in the soybean seeds relative to the control, except for 30 kg applied 1/2 before sowing +1/2 after emergence or applied in its entirety after emergence. The differences between the remaining variants of nitrogen application were not significant. Sulphur application increased the content of crude protein only in the treatment without nitrogen and the treatments with lower application of nitrogen, either entirely foliar or 1/2 before sowing +1/2 after emergence. All fertilizer combinations significantly increased the yield of crude protein per hectare relative to the control (Figure 8).

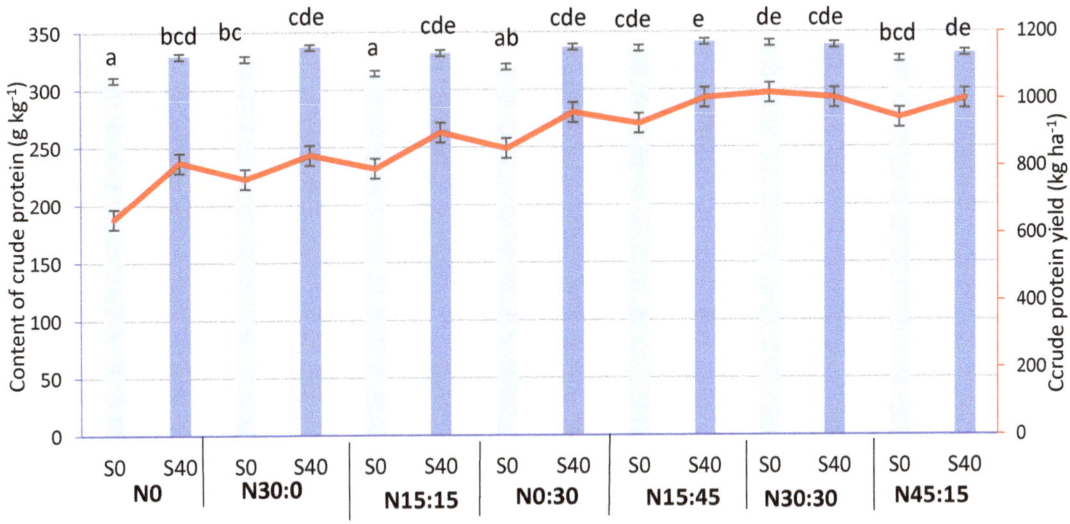

Figure 8. Effects of nitrogen × sulphur application on the content of crude protein in soybean seeds and protein yield. Means marked with the same letter do not differ significantly at $p = 0.05$, bars mean SE, red line–protein yield. S0–without sulfur, S40—40 kg S ha^{-1}, N0–without nitrogen, N30:0—30 kg N ha^{-1} before sowing, N15:15—15 kg N ha^{-1} before sowing + 15 kg N ha^{-1} at the start of the seed filling, N:0:30—30 kg N ha^{-1} at the start of the seed filling, N15:45—15 kg N ha^{-1} before sowing + 30 kg N ha^{-1} at the start of the seed filling, N30:30—30 kg N ha^{-1} before sowing + 30 kg N ha^{-1} at the start of the seed filling, N45:15—45 kg N ha^{-1} before sowing + 15 kg N ha^{-1} at the start of the seed filling.

The lowest crude fat content, 200 g kg^{-1}, was obtained in the soybean seeds from the control treatment. Application of nitrogen and sulphur increased the crude fat content in the soybean seeds. The most beneficial effect on crude fat content in the seeds was obtained in the case of higher nitrogen application (60 kg ha^{-1}), either 1/2 before sowing +1/2 after emergence or 3/4 before sowing +1/4 after emergence, in combination with sulphur application (Figure 9).

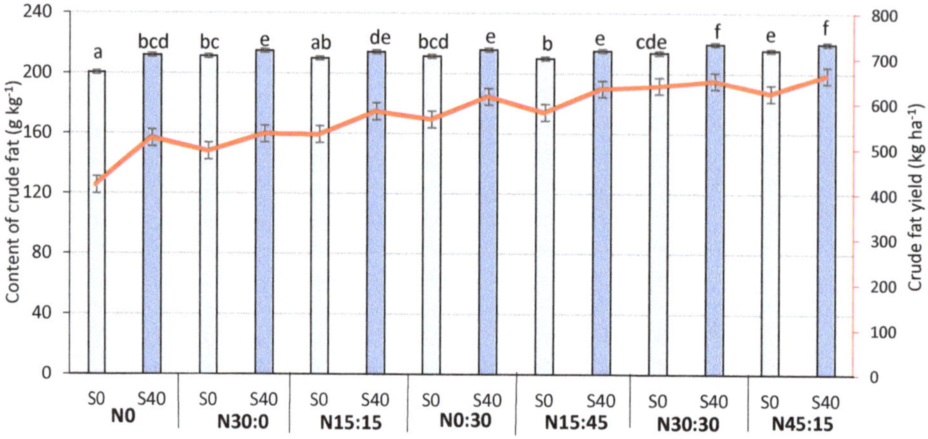

Figure 9. Effects of nitrogen × sulphur application on the content of crude fat in soybean seeds and fat yield. Means marked with the same letter do not differ significantly at $p = 0.05$, bars mean SE, red line–protein yield. S0—without sulfur, S40—40 kg S ha^{-1}, N0–without nitrogen, N30:0—30 kg N ha^{-1} before sowing, N15:15—15 kg N ha^{-1} before sowing + 15 kg N ha^{-1} at the start of the seed filling, N:0:30—30 kg N ha^{-1} at the start of the seed filling, N15:45—15 kg N ha^{-1} before sowing + 30 kg N ha^{-1} at the start of the seed filling, N30:30–30 kg N ha^{-1} before sowing + 30 kg N ha^{-1} at the start of the seed filling, N45:15—45 kg N ha^{-1} before sowing + 15 kg N ha^{-1} at the start of the seed filling.

3.4. Content of Macroelements

Nitrogen application generally reduced the phosphorus content in the soybean seeds relative to the control, except for 30 kg applied 1/2 before sowing +1/2 after emergence. Potassium content in the soybean seeds ranged from 12.40 to 15.42 g kg^{-1} and was highest in the control treatment. Top dressing application of nitrogen in the amount of 30 kg as well as 60 kg applied 1/2 before sowing + 1/2 after emergence significantly reduced the content of potassium, and sulphur application in these combinations had no significant effect. Sulphur application reduced potassium content in the soybean seeds when applied in the following nitrogen combinations: N30 (15:15), N60 (15:45) and N60 (45:15)—Table 4.

Table 4. Effect of nitrogen × sulphur application on the content of macroelements in soybean seeds.

Fertilization		Macroelements (g kg^{-1})				
Nitrogen	Sulphur	P	K	Ca	Mg	S
0	0	8.65 e	15.42 i	1.10 a	2.70 a	2.57 a
	40	8.96 f	14.70 g	1.51 d	2.83 a	3.06 c
N30 (30:0)	0	7.96 b	13.68 ef	1.98 g	3.94 fg	2.67 ab
	40	8.12 bc	12.80 c	2.17 h	3.79 de	3.19 cd
N30 (15:15)	0	8.43 de	14.81 gh	1.38 c	3.00 c	2.55 a
	40	8.53 de	12.40 b	1.95 g	2.93 bc	3.31 de
N30 (0:30)	0	7.96 b	14.73 g	1.14 a	3.05 c	2.67 ab
	40	7.87 b	13.90 f	1.70 e	2.98 bc	3.25 cd
N60 (15:45)	0	7.94 b	15.10 hi	1.42 c	3.92 efg	2.70 ab
	40	8.44 de	13.41 de	1.83 f	3.79 def	3.48 e
N60 (30:30)	0	7.58 a	12.44 b	1.86 f	4.04 g	2.64 ab
	40	8.01 b	11.94 a	1.87 f	3.79 def	3.79 f
N60 (45:15)	0	8.38 cd	14.99 gh	1.27 b	3.95 g	2.80 b
	40	8.44 de	13.10 cd	1.55 d	3.71 d	3.93 f

Means in columns marked with the same letter do not differ significantly at $p = 0.05$.

Nitrogen in the amount of 30 kg applied entirely after emergence or in the amount of 60 kg applied 1/2 before sowing +1/2 after emergence or 3/4 before sowing +1/4 after emergence increased the calcium content in the soybean seeds. In these combinations, sulphur application had no significant effect. In the remaining combinations of nitrogen fertilizer and in the control without nitrogen, sulphur application significantly increased the calcium content in the soybean seeds by 28–47%.

Nitrogen application in the amount of 30 kg, either in its entirety before sowing or 1/2 before sowing +1/2 after emergence, significantly increased the content of magnesium in the soybean seeds by 9.8–13.0% compared to the control. The remaining variants of nitrogen fertilization, N30 (0:30) and N60 irrespective of the means of application, significantly increased the magnesium content in the soybean seeds relative to the control, by 31.0–41.3%, as well as in comparison with 30 kg applied before sowing or in two equal portions, before sowing and after emergence. Sulphur application did not significantly influence magnesium content in the soybean seeds, except in the case of 60 kg N applied 1/2 before sowing +1/2 after emergence or 3/4 before sowing +1/4 after emergence. In these cases, sulphur application decreased magnesium content by 6.0–6.4%.

Sulphur content in the soybean seeds ranged from 2.55 to 3.93 g kg^{-1}. Nitrogen fertilization did not significantly increase sulphur content, except for 60 kg applied 3/4 before sowing +1/4 after emergence. Sulphur application significantly increased the content of this element in the soybeans, both in the control and in every combination of nitrogen application. The highest increase in sulphur content following application of this microelement was observed for the combinations with 60 kg N applied 1/2 before sowing +1/2 after emergence and 3/4 before sowing +1/4 after emergence—by 1.55 g kg^{-1} and 1.13 g kg^{-1}, respectively, i.e., by 43.6% and 40.4%.

4. Discussion

Soybean meets its demand for nitrogen through biological fixation of atmospheric nitrogen as well as from mineral fertilizers. According to [7–9,12] starter nitrogen fertilizer has been shown to increase soybean seed yield. This is confirmed in the present study as well. Nitrogen application before sowing significantly increased seed yield compared to the unfertilized control. The increase in seed yield following application of nitrogen before sowing can be ascribed to the increase in the activity of the root system, rate of photosynthesis, and leaf area index [31,41,42]. Our results are in agreement with the work of Osborne and Riedell [30], who pointed out that soybean grain yield at 16 kg N ha^{-1} starter nitrogen rate, was significantly higher (by 6%) than that at no N treatment with no difference in grain N or oil concentration.

Soybean has a relatively high demand for N, especially during the seed-filling stage. Supply of N to soybean plants during peak demand for seed filling can supplement existing N resources, thereby preventing premature ageing of plants and increasing seed yield [32]. This is confirmed by the present study. The highest seed yield was obtained in the combinations in which 60 kg N ha^{-1} was applied 1/2 before sowing +1/2 after emergence and 3/4 before sowing +1/4 after emergence. In these combinations the seed yield was 8–10 dt ha^{-1} higher than in the combinations without nitrogen application and 5.5–6.5 dt ha^{-1} higher than following application of 30 kg it its entirety before sowing or in two portions, before sowing followed by foliar application. These results are in agreement with research by Zainab et al. [43], who obtained the highest seed yield by applying 60 kg N ha^{-1} in two portions—one quarter before sowing, and the rest at the start of seed-filling. According to Schweiger et al. [44], only 40% to 52% of total nitrogen uptake by soybean originates in symbiotic nitrogen fixation, while the rest is nitrates taken up from the soil. Therefore if total N supply does not meet the needs of soybean, the plants remobilize N accumulated in the leaves to the seeds. This reduces photosynthesis and thus limits the yield potential of soybean [4]. But it is not always the rules. A study conducted in Iowa showed that nitrogen fertilizer applied early reproductive stage had no positive effect on plant DM, grain N concentration and removal, grain yield, or grain quality components at the time of planting had no significant effect on soybean leaf area or grain yield [29].

The increase in seed yield under the influence of nitrogen application was the effect of changes observed in elements of the yield structure, i.e., the pod number per plant and seed number per pod. This was confirmed by the correlation coefficients. Faligowska and Szukała [45] also report that fertilization of soybean with nitrogen has a beneficial effect on elements of the yield structure such as plant height, pod number per plant, and seed weight per plant.

Sulphur performs an important function in the synthesis of protein and essential sulphur-containing amino acids (methionine and cysteine) as well as vitamins and chlorophyll. It is needed for activation of certain enzymes and is an essential component of ferredoxin, which is involved in photosynthesis. Without adequate supply of sulphur, the plants cannot achieve their full yield potential or efficiently utilize applied N [33]. In the present study, application of 40 kg ha^{-1} of sulphur significantly increased the seed yield in the treatment without nitrogen application, by 3.78 dt ha^{-1} (18%), and the pod number per plant by 34.7%. The results are in agreement with those reported by Varun et al. [46] and Burkitbayev et al. [47], who found that seed yield, plant height, number of branches, pod number per plant, and 1000 seed weight of soybean were significantly higher following sulphur application. Similar results were obtained by Dhaker et al. [34], who reported that application of 40 kg S ha^{-1} significantly increased the number of pods per plant, pod length, 1000 seed weight, and seed yield. According to the authors, this may be explained by the fact that sulphur application modifies the physicochemical properties of the soil, thereby improving the availability of nutrients and promoting the growth and development of the plants. This in turn increases translocation of nutrients towards the generative organs and positively affects photosynthesis, and in consequence can substantially increase yields and elements of the yield structure.

Sulphur plays an important role in nitrogen metabolism in the plant, and its deficiency leads to a decrease in nitrogen utilization from fertilizers. According to Jamal et al. [35], S and N interact at the metabolic level in such a way that an imbalance in their availability reduces yields. Therefore S must be included in fertilizer recommendations in soil with S deficiency, and S and N should be administered in balanced quantities in order to obtain optimum yield. In the present study, sulphur application in the treatments in which nitrogen was applied in the amount of 30 kg increased seed yield. In contrast, in the treatments with higher nitrogen application of 60 kg ha^{-1}, sulphur application had no significant effect on seed yield or on most of the elements of the yield structure. This indirectly confirms that the yield-promoting effect of sulphur is mainly due to its effect on nitrogen metabolism. Sulphur performs a special function in nitrogen metabolism, increasing the rate of transformation of nitrogen taken up by the plant into protein. Nitrogen is the nutrient with the greatest yield-promoting effect, and therefore sulphur directly influences seed yield by influencing nitrogen metabolism [48]. In conditions of deficiency of one of these nutrients, the response to fertilization is poor; maximum yield can be achieved when there is an adequate amount of both elements [49]. Proper supply of plants with sulfur is important not only in terms of production, but also ecological. In conditions of sulfur deficiency, fertilizer nitrogen does not perform optimally. The effective use of nitrogen by plants is also important for the environment, because in conditions of sulfur deficiency, nitrogen may be lost as a result of nitrates penetrating into groundwater, as well as volatilization of gaseous forms into the atmosphere [50].

Soybean seeds have high content of protein and fat. The protein content in legume plants is influenced by genetic factors, i.e., the properties of the cultivar, but also by agricultural procedures, particularly the use of nitrogen fertilizer [4,9,11,12]. This was confirmed in the present study. Application of 60 kg of nitrogen significantly increased the protein content in the soybean seeds relative to the unfertilized control treatment. However, the effect of 30 kg of nitrogen was significant only when the full amount was applied before sowing. Both nitrogen and sulphur fertilization increased oil content in the seeds. Sulphur application also had a beneficial effect by increasing protein content in the seeds, but only in conditions without nitrogen fertilizer or the lower level of nitrogen applied in its entirety to the leaves or 1/2 before sowing + 1/2 after emergence. The beneficial effect of nitrogen application on protein in soybean seeds is confirmed by other authors [8,11]. However, Fecak et al. [51] found no significant changes in the content of total protein in soybean seeds after the application of differentiated nitrogen fertilization.

According to Kozłowska-Strawska and Kaczor [49], the main consequence of sulphur deficiency is reduced protein synthesis. Sulphur is an essential nutrient for the activity of enzymes taking part in nitrate reduction [52]. Therefore, plants grown in conditions without this element accumulate nitrogen in non-protein form (nitrates, amides and other compounds). Plants of the *Fabaceae* family, which are a valuable source of protein for people and animals, require adequate fertilization with sulphur [25]. Without this nutrient, they produce protein with much lower content of sulphur-containing amino acids, especially methionine, which is one of the most valuable amino acids determining the nutritional value of plants [22,36]. Sulphur is also involved in fat synthesis, and due to its metabolic functions it has an important influence on the quantity and quality of oil accumulated in the seeds. Sulphur compounds, such as vitamin H with coenzyme A, form an enzymatic system essential to synthesis of fatty acids. The results of research by other authors confirm that sulphur application increases the content of protein and fat in soybean seeds [35,37].

Soybean is an important source not only of protein and fats, but also of minerals such as potassium, magnesium, calcium, phosphorus, and iron [3,4,53]. In the present study, nitrogen application generally did not affect the content of sulphur and calcium in the soybean seeds, but it reduced the content of phosphorus and potassium (only in the case of foliar application of 30 kg and 60 kg (30:30) while increasing the content of magnesium, especially the higher rate of application. Sulphur application did not affect the content of phosphorus in the seeds but increased that of calcium and sulphur and decreased that of magnesium, although only in the combinations with higher nitrogen application. The findings of other authors are varied. According to Jarecki and Bobrecka-Jamro [4] and Shorabi et al. [32], nitrogen application did not significantly affect the content of phosphorus or potassium in soybean seeds. An increase in calcium content in the grain of cereals following sulphur application was reported by Brodowska and Kaczor [54]. Barczak et al. [14] observed a decrease in phosphorus content in response to application of various amounts of sulphur in an experiment on lupin. However, according to the authors, this may have been caused by 'dilution' of phosphorus as a result of the increased yield resulting from sulphur application, and not by the antagonistic interaction of phosphate ion (V) and sulphate ion (VI). A similar explanation of decreased content of minerals is given by Cakmak [55].

5. Conclusions

Application of nitrogen and sulphur significantly increased the seed yield of soybean. The most beneficial effect was obtained in the combinations with 60 kg N applied 1/2 before sowing +1/2 after emergence and 3/4 before sowing +1/2 after emergence. In these combinations, sulphur application did not significantly affect seed yield. In the remaining combinations of nitrogen application, sulphur application significantly increased the seed yield. Taking into account the yield and the chemical composition of the soybean seeds, fertilization with 60 kg N ha^{-1} in two portions can be recommended—1/2 or 3/4 before sowing and the remainder during the development of pods and seeds—in combination with sulphur application. In further research, it is worth paying attention to the form of sulfur introduced in fertilizers. An interesting issue is also the effect of sulfur fertilization on the composition of fatty acids and the biological value of protein.

Author Contributions: Conceptualization, A.G.; methodology, A.G., A.K.-D. and E.J.; investigation, A.G., E.J. and A.K.-D.; writing—original draft preparation, A.G., E.J., E.F.-O. and A.K.-D. All authors have read and agreed to the published version of the manuscript.

Funding: This research was funded by Minister of Education and Science of Poland, grant number RKS/S/55/22.

Institutional Review Board Statement: Not applicable.

Informed Consent Statement: Not applicable.

Data Availability Statement: Not applicable.

Conflicts of Interest: The authors declare no conflict of interest.

References

1. Kahraman, A. Nutritional value and foliar fertilization in soybean. *J. Elem.* **2017**, *22*, 55–66. [CrossRef]
2. Tyagi, S.D.; Khan, M.H.; Teixeira Da Silva, J.A. Yield stability of some soybean genotypes across diverse environments. *Int. J. Plant Breed.* **2011**, *5*, 37–41. [CrossRef]
3. Biel, W.; Gawęda, D.; Łysoń, E.; Hury, G. The effect of variety and agrotechnical factors on nutritive value of soybean seeds. *Acta Agrophs.* **2017**, *24*, 395–404.
4. Jarecki, W.; Bobrecka-Jamro, D. Effect of fertilization with nitrogen and seed inoculation with nitragina on seed quality of soya bean (*Glycine max* (L) Merrill). *Acta Sci. Pol. Agric.* **2015**, *14*, 51–59.
5. Salvagiotti, F.; Cassman, K.G.; Specht, J.E.; Walters, D.T.; Weiss, A.; Dobermann, A. Nitrogen uptake, fixation and response to fertilizer N in soybeans: A review. *Field Crops Res.* **2008**, *108*, 1–13. [CrossRef]
6. Saito, A.; Tanabata, S.; Tanabata, T.; Tajima, S.; Ueno, M.; Ishikawa, S.; Ohtake, N.; Sueyoshi, K.; Ohyama, T. Effect of nitrate on nodule and root growth of soybean (*Glycine max* (L.) Merr.). *Int. J. Mol. Sci.* **2014**, *15*, 4464–4480. [CrossRef] [PubMed]
7. Lorenc-Kozik, A.M.; Pisulewska, E. Effect of increasing levels of nitrogen fertilizer and microelements on seed yield of selected soybean cultivars. *Rośliny Oleiste-Oilseed Crops* **2003**, *24*, 131–142.
8. Szostak, B.; Głowacka, A.; Kasiczak, A.; Kiełtyka-Dadasiewicz, A.; Bąkowski, M. Nutritional value of soybeans and the yield of protein and fat depending on a cultivar and the level of nitrogen application. *J. Elem.* **2020**, *25*, 45–57. [CrossRef]
9. Bobrecka-Jamro, D.; Jarecki, W.; Buczek, J. Response of soya bean to different nitrogen fertilization levels. *J. Elem.* **2018**, *23*, 559–568. [CrossRef]
10. Pisulewska, E.; Lorenc-Kozik, A.; Borowiec, F. Effect of nitrogen fertilization level on seed yield, fat content and fatty acids composition of two soybean cultivars. *Rośliny Oleiste-Oilseed Crops* **1999**, *2*, 511–520.
11. Kulig, B.; Klimek-Kopyra, A. Sowing date and fertilization level are effective elements increasing soybean productivity in rainfall deficit conditions in Central Europe. *Agriculture* **2023**, *13*, 115. [CrossRef]
12. Lošák, T.; Ševčík, M.; Plchová, R.; von Bennewitz, E.; Hlušek, J.; Elbl, J.; Buňka, F.; Polášek, Z.; Antonkiewicz, J.; Varga, L.; et al. Nitrogen and sulphur fertilisation affecting soybean seed spermidine content. *J. Elem.* **2018**, *23*, 581–588. [CrossRef]
13. Caliskan, S.; Ozkaya, I.; Caliskan, M.E.; Arslan, M. Effects of nitrogen and iron fertilization on growth, yield and fertilizer use efficiency of soybean in a Mediterranean-type soil. *Field Crops Res.* **2008**, *108*, 126–132. [CrossRef]
14. Barczak, B.; Knapowski, T.; Kozera, W.; Ralcewicz, M. Effects of sulphur fertilisation on the content and uptake of macroelements in narrow-leaf lupin. *Rom. Agric. Res.* **2014**, *31*, 1–7.
15. Kozłowska-Strawska, J.; Badora, A. Selected problems of sulfur management in crops. *Pol. J. Nat. Sci.* **2013**, *28*, 309–316.
16. Zhao, F.J.; Wood, A.P.; McGrath, S.P. Sulphur nutrition of spring peas. *Asp. Appl. Biol.* **1999**, *56*, 189–194.
17. Aulakh, M.S. Crop responses to sulphur nutrition. In *Sulphur in Plants;* Abrol, Y.P., Ahmad, A., Eds.; Kluwer Academic Publishers: Dordrecht, The Netherlands, 2003; pp. 341–358.
18. Tiwari, K.N.; Gupta, B.R. Sulphur for Sustainable High Yield Agriculture in Uttah Pradesh. *Indian J. Fertil.* **2006**, *1*, 37–52.
19. Barczak, B.; Nowak, K.; Knapowski, T.; Ralcewicz, M.; Kozera, W. Reaction of narrow-leafed lupin (*Lupinus angustifolius* L.) to sulphur fertilization part I: Yield and selected yield structure components. *Fragm. Agron.* **2013**, *30*, 23–34.
20. Jamal, A.; Saleem Fazli, I.; Ahmad, S.; Abdin, M.Z.; Yun, S.J. Effect of nitrogen and sulphur application on nitrate reductase and ATP-sulphurylase activities in soybean. *Korean J. Crop Sci.* **2006**, *51*, 298–302.
21. Scherer, H.W.; Pacyna, S.; Spoth, K.R.; Schulz, M. Low levels of ferredoxin, ATP and leghemoglobin contribute to limited N_2 fixation of peas (*Pisum sativum* L.) and alfalfa (*Medicago sativa* L.) under S deficiency conditions. *Biol. Fertil. Soils* **2008**, *44*, 909–916. [CrossRef]
22. Głowacka, A.; Gruszecki, T.; Szostak, B.; Michałek, S. The response of common bean to sulphur and molybdenum fertilization. *Int. J. Agron.* **2019**, *2019*, 3830712. [CrossRef]
23. Tabatabai, M.A. Importance of sulphur in crop production. *Biogeochemistry* **1984**, *1*, 45–62. [CrossRef]
24. Pandurangan, S.; Sandercock, M.; Beyaert, R.; Conn, K.L.; Hou, A.; Marsolais, F. Differential response to sulfur nutrition of two common bean genotypes differing in storage protein composition. *Front. Plant Sci.* **2015**, *6*, 92. [CrossRef] [PubMed]
25. Zhao, F.J.; Evans, E.J.; Bilsborrow, P.E. Varietal differences in sulphur uptake and utilization in relation to glucosinolate accumulation in oilseed rape. In Proceedings of the 9th International Rapeseed Congress, Cambridge, UK, 4–7 July 1995; Volume 1, pp. 271–273.

26. Withers, P.J.A.; Evans, E.J.; Bilsborrow, P.E.; Milford, G.F.J.; McGrath, S.P.; Zhao, F.; Walker, K.C. Improving the prediction of sulphur deficiency in winter oilseed rape in the UK. In Proceedings of the 9th International Rapeseed Congress, Cambridge, UK, 4–7 July 1995; Volume 1, pp. 277–279.
27. Pacyna, S.; Schulz, M.; Scherer, H.W. Influence of sulphur supply on glucose and ATP concentrations of inoculated broad beans (*Vicia faba minor* L.). *Biol. Fertil. Soils* **2005**, *42*, 324–329. [CrossRef]
28. Islam, M.; Mohsan, S.; Ali, S. Effect of different phosphorus and sulfur levels on nitrogen fixation and uptake by chickpea (*Cicer arietinum* L.). *Agrociencia* **2012**, *46*, 1–13.
29. Barker, D.W.; Sawyer, J.E. Nitrogen Application to Soybean at Early Reproductive Development. *Agron. J.* **2005**, *97*, 615–619. [CrossRef]
30. Osborne, S.L.; Riedell, W.E. Starter Nitrogen Fertilizer Impact on Soybean Yield and Quality in the Northern Great Plains. *Agron. J.* **2006**, *98*, 1569–1574. [CrossRef]
31. Namvar, A.; Sharifi, R.S. Phenological and morphological response of chickpea (*Cicer arietinum* L.) to symbiotic and mineral nitrogen fertilization. *Zemdirbyste* **2011**, *98*, 121–130.
32. Sohrabi, Y.; Habibi, A.; Mohammadi, K.; Sohrabi, M.; Heidari, G.; Gholamreza, H.; Shiva, K.; Masoumeh, K. Effect of nitrogen (N) fertilizer and foliar-applied iron (Fe) fertilizer at various reproductive stages on yield, yield component and chemical composition of soybean (*Glycine max* L. Merr.) seed. *Afr. J. Biotechnol.* **2012**, *11*, 9599–9605. [CrossRef]
33. Kumar, S.; Wani, J.A.; Lone, B.A.; Fayaz, A.; Singh, P.; Qayoom, S.; Dar, Z.A.; Ahmed, N. Effect of phosphorus and sulphur on nutrient and amino acids content of soybean (*Glycine max* L. Merill) under 'Alfisols'. *J. Exp. Agric. Int.* **2017**, *16*, 1–7. [CrossRef]
34. Dhaker, S.C.; Mundra, S.L.; Nepalia, V. Effect of weed management and sulphur nutrition on productivity of soybean [*Glycine max* (L.) Merrill]. *Indian J. Weed Sci.* **2010**, *42*, 232–234. [CrossRef]
35. Jamal, A.; Fazli, I.S.; Ahmad, S.; Abdin, M.Z.; Yun, S.J. Effect of sulphur and nitrogen application on growth characteristics, seed and oil yields of soybean cultivars. *Korean J. Crop Sci.* **2005**, *50*, 340–345.
36. Sharma, A.; Sharma, S. Effect of nitrogen and sulphur nutrition on yield parameters and protein composition in soybean [*Glycine max* (L.) Merrill]. *J. Appl. Nat. Sci.* **2014**, *6*, 402–408. [CrossRef]
37. Devi, K.N.; Singh, L.N.K. Influence of sulphur and boron fertilization on yield, quality, nutrient uptake and economics of soybean (*Glycine max*) under upland conditions. *J. Agric. Sci.* **2012**, *4*, 1–10. [CrossRef]
38. IUSS Working Group WRB. *World Reference Base for Soil Resources 2014, Update 2015 International Soil Classification System for Naming Soils and Creating Legends for Soil Maps*; World Soil Resources Reports No. 106; FAO: Rome, Italy, 2015.
39. Lancashire, P.D.; Bleiholder, H.; Van den Boom, T.; Langelüddeke, P.; Strauss, R.; Weber, E.; Witzenberger, A. A uniform decimal code for growth stages of crops and weeds. *Ann. Appl. Biol.* **1991**, *119*, 561–601. [CrossRef]
40. Skowera, B.; Kopcinska, J.; Kopeć, B. Changes in thermal and precipitation conditions in Poland in 1971–2010. *Ann. Warsaw Univ. Life Sci. SGGW Land Reclam.* **2014**, *46*, 153–162. [CrossRef]
41. Gai, Z.; Zhang, J.; Li, C. Effects of starter nitrogen fertilizer on soybean root activity, leaf photosynthesis and grain yield. *PLoS ONE* **2017**, *12*, e0174841. [CrossRef]
42. Van Kessel, C.; Hartley, C. Agricultural management of grain legumes: Has it led to an increase in nitrogen fixation? *Field Crops Res.* **2000**, *65*, 165–181. [CrossRef]
43. Zainab, A.; Morteza, S.D.; Amir, A.M. Effect of different levels of nitrogen fertilizer on morphological traits and yield of soybean cultivar. *Adv. Environ. Biol.* **2014**, *8*, 334–337.
44. Schweiger, P.; Hofer, M.; Hartl, W.; Wanek, W.; Vollmann, J. N2 fixation by organically grown soybean in Central Europe: Method of quantification and agronomic effects. *Eur. J. Agron.* **2012**, *41*, 11–17. [CrossRef]
45. Faligowska, A.; Szukała, J. Influence of seed inoculation and nitrogen fertilization on morphological characters of legume crops. *Zesz. Probl. Postęp. Nauk Rol.* **2010**, *550*, 201–209.
46. Varun, B.K.; Patel, B.A.; Panchal, H.D.B.; Dalwadi, M.R.; Patel, J.C. Effect of different sources and levels of sulphur on yield and quality of soybean under inceptisol of middle Gujarat. *Asian J. Soil Sci.* **2011**, *6*, 85–87.
47. Burkitbayev, M.; Bachilova, N.; Kurmanbayeva, M.; Tolenova, K.; Yerezhepova, N.; Zhumagul, M.; Mamurova, A.; Turysbek, B.; Demeu, G. Effect of sulfur-containing agrochemicals on growth, yield, and protein content of soybeans (*Glycine max* (L.) Merr). *Saudi J. Biol. Sci.* **2021**, *28*, 891–900. [CrossRef] [PubMed]
48. Siddiqui, M.H.; Mohammad, F.; Khan, M.N.; Khan, M.M.A. Cumulative effect of soil and foliar application of nitrogen, phosphorus, and sulfur on growth, physico-biochemical parameters, yield attributes, and fatty acid composition in oil of erucic acid-free rapeseed-mustard genotypes. *J. Plant Nutr.* **2008**, *31*, 1284–1298. [CrossRef]
49. Kozłowska-Strawska, J.; Kaczor, A. Sulphur as a deficient element in agriculture—Its influence on yield and on the quality of plant materials. *Ecol. Chem. Eng. A* **2009**, *16*, 9–19.
50. . Schnug, E.; Haneklaus, S. Ecological aspects of plant sulphur supply. *Norw. J. Agric. Sci. Suppl.* **1994**, *15*, 149–156.
51. Fecák, P.; Šariková, D.; Černý, I. Influence of tillage system and starting N fertilization on seed yield and quality of soybean *Glycine max* (L. *) Merrill. Plant Soil Environ.* **2010**, *56*, 105–110. [CrossRef]
52. Sulieman, S.; Fischinger, S.A.; Gresshoff, P.M.; Schulzea, J. Asparagine as a major factor in the N-feedback regulation of N_2 fixation in *Medicago truncatula*. *Physiol. Plant.* **2013**, *140*, 21–31. [CrossRef]
53. Kraska, P.; Andruszczak, S.; Gierasimiuk, P.; Chojnacka, S. The effect of subsurface mineral fertilizer application on the yield and seed quality of soybean under no-tillage conditions. *Agron. Sci.* **2022**, *77*, 109–131. [CrossRef]

54. Brodowska, M.; Kaczor, A. The effect of various forms of sulphur and nitrogen on calcium and magnesium content and uptake in spring wheat (*Triticum aestivum* L.) and cocksfoot (*Dactylis glomerata* L.). *J. Elem.* **2009**, *14*, 641–647. [CrossRef]
55. Cakmak, I. Plant nutrition research: Priorities to meet human needs for food in sustainable ways. *Plant Soil* **2004**, *247*, 3–24. [CrossRef]

agronomy

MDPI

Article

Overexpression of *OsPHT1;4* Increases Phosphorus Utilization Efficiency and Improves the Agronomic Traits of Rice cv. Wuyunjing 7

Zhi Hu [1,†], Xu Huang [1,†], Xiaowen Wang [2], Huihuang Xia [1], Xiuli Liu [3], Yafei Sun [4], Shubin Sun [1], Yibing Hu [1,*] and Yue Cao [5,*]

1 College of Resources and Environmental Sciences, Nanjing Agricultural University, Nanjing 210095, China; 2019203047@njau.edu.cn (Z.H.); 2020203052@stu.njau.edu.cn (X.H.); 2019103121@njau.edu.cn (H.X.); sunshubin@njau.edu.cn (S.S.)
2 College of Horticulture, Nanjing Agricultural University, Nanjing 210095, China; wangxiaowen@njau.edu.cn
3 Institute of Virology and Biotechnology, Zhejiang Academy of Agricultural Sciences, Hangzhou 310004, China; liuxl@zaas.ac.cn
4 Institute of Eco-Environment and Plant Protection, Shanghai Academy of Agricultural Sciences, Shanghai 201803, China; sunsun_cool@126.com
5 Guangdong Provincial Key Laboratory for Environmental Pollution Control and Remediation Technology, School of Environmental Science and Engineering, Sun Yat-sen University, Guangzhou 510006, China
* Correspondence: huyb@njau.edu.cn (Y.H.); caoy85@mail.sysu.edu.cn (Y.C.)
† The authors contributed equally to this work.

Citation: Hu, Z.; Huang, X.; Wang, X.; Xia, H.; Liu, X.; Sun, Y.; Sun, S.; Hu, Y.; Cao, Y. Overexpression of *OsPHT1;4* Increases Phosphorus Utilization Efficiency and Improves the Agronomic Traits of Rice cv. Wuyunjing 7. *Agronomy* **2022**, *12*, 1332. https://doi.org/10.3390/agronomy12061332

Academic Editor: Antonio Lupini

Received: 29 December 2021
Accepted: 7 April 2022
Published: 31 May 2022

Abstract: Inorganic phosphate (Pi) is taken up by plant roots and translocated via phosphate transporters. Previously, we showed that phosphate transporter *OsPHT1;4* in the PHT1 family participates in phosphate acquisition and mobilization; it facilitates the embryo development of Japonica rice Nipponbare. This study investigated the potential of manipulating the expression of OsPHT1;4 to increase Pi acquisition efficiency and crop productivity in rice cv. Wuyunjing 7 (WYJ 7), a cultivar widely grown in Yangtze River Delta of China. The *OsPHT1;4* overexpression lines and wild-type WYJ 7 were treated under different Pi conditions in hydroponic and field experiments. Quantitative real-time RT-PCR analysis and the transgenic plants expressing GUS reporter gene indicate strong expression of *OsPHT1;4* in roots and leaf collars of cv. WYJ 7. The total P contents in shoots of the *OsPHT1;4*-overexpressing plants were significantly higher under Pi-deficient hydroponic conditions than the wild type under Pi sufficiency and deficiency. ^{33}Pi uptake and translocation assays confirmed the results. In the field condition, *OsPHT1;4* overexpression lines had a higher P concentration in tissues than the wild type control, and the panicle performance of the overexpression lines including the grain yield was improved as well. Taken together, our results show that OsPHT1;4 plays an important role in the acquisition and mobilization of Pi in WYJ 7, especially under Pi deficiency. The study highlights the importance of OsPHT1;4 in improving the agronomic traits of the widely grown rice cultivar in China.

Keywords: phosphate transporter; phosphurs nutrition; P uptake and transloation; grain yield; rice cultivar

1. Introduction

The essential macronutrient phosphorus (P) is a constituent of many biologically important molecules such as nucleic acids and phospholipids. It plays a pivotal role in plant growth and development and several metabolic pathways and energy transfer processes in plants [1]. However, inorganic phosphate (Pi), the major form of P taken up by the roots from the soil, is often a limiting nutrient in the rhizosphere [1]. The acquisition of Pi from the rhizosphere by the plant root and its translocation between cells is via Pi transporters (PTs) of the *PHT1* gene family [1,2]. The first genes encoding putative plant PTs were isolated from *Arabidopsis* [3]. Since then, a large number of PT genes have been identified from different plant families, including cereals, legumes, and Solanaceous species [4].

In rice (*Oryza sativa* L.), 13 genes encode proteins belonging to the PHT1 family [5,6]. Several of them have been functionally characterized. Generally, these PHT1 members display different characteristics either in response to Pi concentration or in physiological functions. *OsPHT1;2* and *OsPHT1;6* are responsive to Pi starvation. OsPHT1;2 is expressed throughout the stele in the primary and lateral roots under Pi deprivation; it is responsible for transporting Pi from roots to shoots, whereas OsPHT1;6 is expressed throughout the young primary and lateral roots under P deficiency. It plays an important role in both Pi uptake and translocation throughout the plant [7]. OsPHT1;1 is constitutively expressed in roots and shoots under both Pi-sufficient and -deficient conditions. It participates in Pi uptake and translocation under Pi-sufficient conditions [8]. Similarly, *OsPHT1;8* is expressed in various tissues independent of Pi supply, and overexpression of *OsPHT1;8* resulted in excessive Pi in both roots and shoots [9]. *OsPHT1;9* and *OsPHT1;10* are induced in root epidermis, root hairs, and lateral roots by Pi starvation [10]. *OsPHT1;3* mediates Pi uptake, translocation, and remobilization under extremely low Pi regimes [11], while *OsPHT1;11* and *OsPHT1;13* are specifically induced during mycorrhizal symbiosis [6,12].

We previously investigated the functions of *OsPHT1;4* in rice cv. *Nipponbare* [13]. Our findings suggest that OsPHT1;4 facilitates the acquisition and mobilization of Pi, and it plays an important role in the development of the embryo [13]. Since the Pi uptake efficiency and the development of the embryo are important events to crop production, we hypothesized that *OsPHT1;4* may play a major role in enhancing grain yield. Rice is one of the most important staple food crops for almost half of the world's population, of which about 90% is consumed in Asia alone. To meet the food demand of the rapidly growing world population, increasing crop yield through various biotechnologies including manipulating valuable gene expression proved to be promising [14]. Rice cv. Wuyunjing 7 (WYJ 7) is a japonica variety widely cultivated in China's Yangtze River Delta. Therefore, the study on the improvement of yield per plant for the crop has practical value.

In the present study, we investigated the functions of *OsPHT1;4* in WYJ 7 on phosphate uptake efficiency and its potential in yield improvement. We found that the uptake and mobilization of Pi significantly increased in all tissues of *OsPHT1;4*-overexpressing plants tested, and a pivotal role of OsPHT1;4 in the agronomic traits of the cultivar was identified. The results lay the foundation to crate rice plants that require less P fertilizer without sacrificing production by genetic engineering.

2. Materials and Methods

2.1. Plant Material and Growth Conditions

Rice [*Oryza sativa* L. ssp. *Japonica* cv. Wuyunjing 7 (WYJ 7)] was used as the wild-type and for transformation in this study.

Light-proofed plastic boxes of 7L were used for hydroponic cultivation. Seeds were surface-sterilized with diluted (1:3, *v*/*v*) $NaClO_4$ for 30 min, followed by rinsing with deionized water for 30 min. They were germinated in the dark at 25 °C for 3 days. Seedlings (10 days old) were transferred to the hydroponic boxes containing IRRI solution (1.25 mM NH_4NO_3, 1 mM each of $CaCl_2$ and $MgSO_4$, 0.5 mM Na_2SiO_3, 0.4 mM K_2SO_4, 0.2 mM KH_2PO_4, 20 μM Fe-EDTA, 20 μM H_3BO_3, 9.0 μM $MnCl_2$, 0.77 μM $ZnSO_4$, 0.39 μM Na_2MoO_4, and 0.32 μM $CuSO_4$). Each plastic box contains 20 rice seedlings, and rice seedlings were fixed with a sponge on a foam board covering the plastic boxes. +P treatment was implemented by following the recipe while −P treatment was carried out by reducing the KH_2PO_4 concentration of the recipe to 10 μM. Plants are grown hydroponically in an artificial greenhouse (16-h light, 30 °C/8-h dark, 22 °C; 70% relative humidity). Plant height and root length were measured after the plants had been treated in the solution for 21 days, and the shoots and roots were collected separately and dried in an oven at 70 °C for 1 week for phosphorus content determination. For qRT-PCR experiments, rice plants grown for 1 week under different treatments were put into liquid nitrogen before being stored at −80 °C.

For field experiments, seeds were first sterilized with 30% hydrogen peroxide for 30 min and then rinsed thoroughly with deionized water. Transgenic seeds were soaked in water containing 25 mg/L hygromycin, and wild-type seeds were soaked in water for 3 days, then the seedlings were sown in the Pailou Experimental Base of Nanjing Agricultural University for field growth. One month later, seedlings of the wild-type and OsPT4-Ox lines with the same growth vigor were selected and transplanted in the paddy fields at the interval of 21 cm between rows and 12 cm between plants. To follow the normal management, the paddy field was adjusted to contain 120 kg N/ha (urea), 70 kg P/ha (potassium dihydrogen phosphate), and 260 kg K/ha (potassium chloride). The rice plants mature after about 130 days of growth. After measuring agronomic traits such as plant height and tiller number, it was dissected into different parts and moved to a 70 °C oven to dry for a week, and then the samples were prepared to extract total phosphate. For qRT-PCR experiments, the fresh plant samples were put into liquid nitrogen before being stored at −80 °C after collection from the field.

2.2. RT-PCR and qRT-PCR Analyses

Total RNAs from various tissues of rice cv. WYJ 7 were isolated using the Trizol reagent (Invitrogen). Reverse Transcription-PCR (RT-PCR) was carried out by using gene-specific primers for *OsPHT1;4* (primers were listed in Supplementary Table S1). The PCR products were analyzed on an agarose gel (1%, *w*/*v*) and images were captured with a CCD camera. For qRT-PCR, first-strand cDNAs were synthesized from the total RNAs (HiScript II Q RT SuperMix for qPCR + gDNA wiper, Vazyme, Nanjing, China) according to the manufacturer's instructions followed by qRT-PCR (AceQ qPCR SYBR Green Master Mix, Vazyme on the StepOnePlus™ Real-Time PCR System, Applied Biosystems, Waltham, MA, USA) according to the manufacturer's instructions. *OsActin1* (LOC_Os03g50885) was used as an internal control for RT-PCR and qRT-PCR analyses.

2.3. Plasmids Construction and Plant Transformation

The full-length open reading frame (ORF) of *OsPHT1;4* was ligated under the control of a ubiquitin promoter in the pS1a-4 for *OsPHT1;4* overexpression (*OsPHT1;4*-Ox). The final constructs were transferred into *Agrobacterium tumefaciens* strain EHA105 by electroporation and subsequently transformed into rice WYJ 7 as described previously [15]. Three independent *OsPHT1;4*-Ox transgenic lines were used in this study. A list of primers used for the cloning is provided in Supplementary Table S2.

2.4. Southern Blot Analysis

The three independent *OsPHT1;4*-Ox transgenic lines were confirmed by Southern blot analysis. Genomic DNA (80–100 μg) was digested overnight with *Bam*HI and *Xho*I at 37 °C, separated by agarose gel (1%, *w*/*v*) electrophoresis, and blotted onto a nylon membrane (Hybond N+, Amersham, UK). The membrane was hybridized with a probe based on the hygromycin resistance gene fragment presented in the plasmid. The membrane was washed twice with a solution containing 1×SSC and SDS (0.1%, *w*/*v*) for 15 min at 65 °C. The signal on the washed membrane was captured (ScanMaker S260, Microtek, Budapest, Hungary).

2.5. Measurement of Total P Concentrations

Total P concentrations were determined after the plant samples had been dried at 70 °C for 3 days. The dried sample (50 mg) was predigested in glass tubes containing 1 mL of deionized water and 5 mL of H_2SO_4 overnight. Afterward, the tubes were heated at 180 °C for 20 min and then at 280 °C for 10 min. During the 30 min heating step, 30% hydrogen peroxide (50 μL) was added slowly every 10 min until the solution turned colorless [8,16]. After cooling to ambient temperature, the digested sample was diluted to 100 mL with distilled water. The total P concentration in the solution was assayed as described [16].

2.6. ^{33}P Uptake Assay

Ten-day-old wild-type and *OsPHT1;4-Ox* seedlings were transferred to +P medium (200 µM Pi) or −P medium (10 µM Pi) for a further 7 days, then transferred to medium containing 0.5 mM $CaCl_2$ and 2 mM MES (pH 5.5) for 10 min. Then plants were transferred to a 300 mL growth medium (as described above) containing 200 kBq of $H_3{}^{33}PO_4$ and incubated at 28 °C for 3 h. After labelling, roots were immersed in ice-cold desorption solution (0.5 mM $CaCl_2$, 100 µM unlabeled NaH_2PO_4, 2 mM MES, pH 5.5) for 10 min to remove the apoplastic ^{33}P. Roots were harvested and digested using $HClO_4$ and H_2O_2 at 70 °C for 2 h to 3 h with intermittent shaking until the solution was colorless. Following digestion, 1 mL TritonX-100 and 3 mL scintillation cocktail (Ultima Gold; Perkin-Elmer, Waltham, MA, USA) was added to the digested material and incubated for 24 h. ^{33}P radioactivity was determined by liquid scintillation counting (Tri-Carb 2100, Packard, Detroit, MI, USA). The sum of the ^{33}P cpm in the root and shoot portion of each plant was divided by the mass of the root to obtain the specific uptake of ^{33}P over the incubation period.

2.7. Statistical Analysis

Data were collected from two or three independent biological experiments and analyzed for significant differences (SPSS Statistics 20, IBM, Armonk, NY, USA). Duncan's multiple range test at $p < 0.05$ was carried out for all experiments to determine the significance of differences between the control and treatment plants.

3. Results

3.1. OsPHT1;4 Is Responsive to Pi Starvation and Is Abundantly Expressed in the Collar and Leaf Sheath of Rice cv WYJ 7

To determine the transcriptional expression of *OsPHT1;4* in response to Pi starvation and in different tissues of rice cv. WYJ 7, quantitative real-time PCR (qRT-PCR) analysis was employed (Figure 1). The transcript level of *OsPHT1;4* was higher in the root compared with the leaf blade under Pi-sufficient (+P) and -deficient (−P) conditions. Moreover, the expression of *OsPHT1;4* was significantly upregulated in response to Pi starvation in both root and leaf blade (Figure 1A). Previous studies have shown that the expression of *OsPHT1;4* was high in spikes [17], but we found different results in WYJ7. In the field growth conditions, the expression of this gene was highest in leaves and lowest in spike at the heading stage (Figure 1B), and *OsPHT1;4* was expressed significantly higher in leaf collar and leaf sheath compared to leaf blade at the mature stage (Figure 1C).

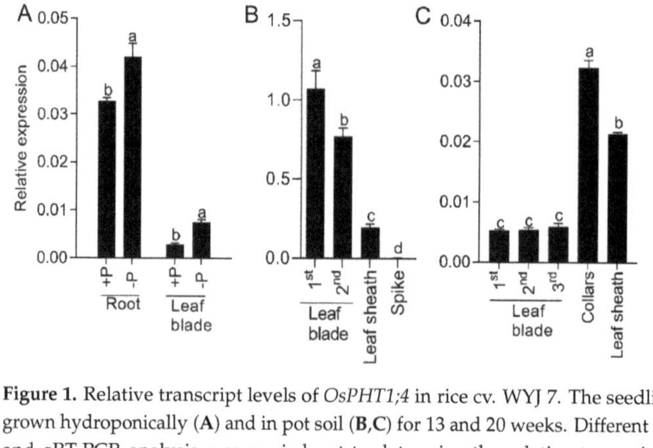

Figure 1. Relative transcript levels of *OsPHT1;4* in rice cv. WYJ 7. The seedlings (3 days old) were grown hydroponically (**A**) and in pot soil (**B**,**C**) for 13 and 20 weeks. Different tissues were harvested and qRT-PCR analysis was carried out to determine the relative transcript levels of *OsPHT1;4*. *OsActin1* was used as an internal control. Values are means ± SE (n = 3). Different letters above the bars indicate significant differences ($p < 0.05$, one-way ANOVA).

3.2. *Overexpression of OsPHT1;4 Promotes the Rice Growth under Both Pi-Sufficient and Pi-Deficient Conditions*

We created *OsPHT1;4*-overexpressing transgenic lines based on WYJ 7, hereafter referred to as *OsPHT1;4*-Ox lines, to test the gene's effect on plant performance. According to the results of RT-PCR, qRT-PCR, and Southern blot analysis (Supplementary Figure S1), we selected three independent homozygous *OsPHT1;4*-Ox lines, named Ox1, Ox2, and Ox3, for further study.

Wild-type and *OsPHT1;4*-Ox plants were grown in hydroponic culture and treated under Pi-sufficient (200 μM) and Pi-deficient (10 μM) conditions for 3 weeks. The primary root length of the *OsPHT1;4*-Ox lines was about 11.75% longer than that of the wild-type plants under the Pi-deficient condition, but no significant difference was observed under the Pi-sufficient condition (Figure 2A,B). Overexpression of *OsPHT1;4* resulted in a 7–12.6% increase in the shoot and root biomass independent of Pi supply conditions (Figure 2B–D). As a result, overexpression of *OsPHT1;4* in WYJ 7 promoted the growth of the Ox lines under both Pi-sufficient and Pi-deficient conditions by hydroponic cultivation.

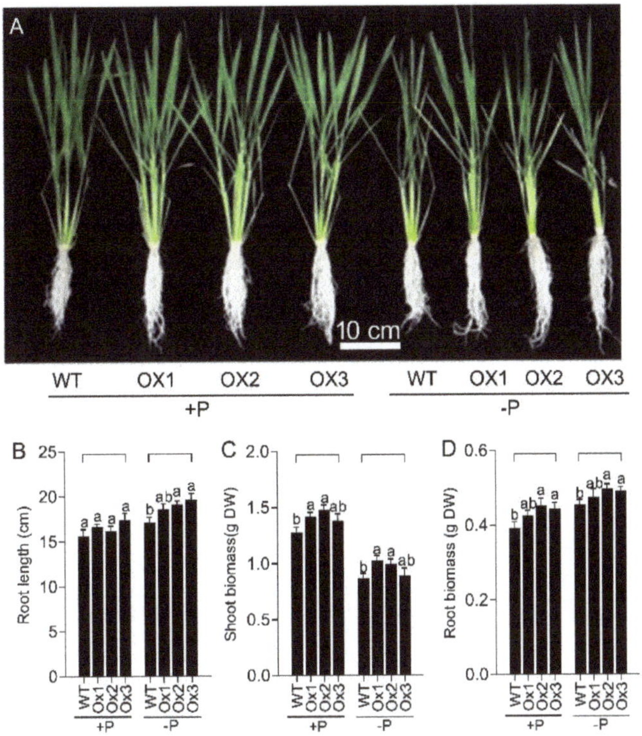

Figure 2. Overexpression of *OsPHT1;4* promotes rice vegetative growth. WT and *OsPHT1;4*-Ox lines (Ox1–Ox3) were grown hydroponically in (**A**) +P and −P media for 3 weeks. (**A**) Morphological comparison of WT and Ox1–Ox3 plants grown under +Pi and −Pi regimes. Comparisons of primary root length (**B**), shoot biomass (**C**), and root biomass (**D**) of the Ox plants and WT control. DW: Dry weight. Values are means ± SE ($n = 5$). Different letters above the bars indicate significant differences ($p < 0.05$, one-way ANOVA).

3.3. *Overexpression of OsPHT1;4 Increases Pi Uptake and Translocation Efficiency under the Hydroponic Condition*

To determine the effect of overexpressing *OsPHT1;4* on the maintenance of Pi status, we measured the total P concentration and content in shoots and roots of *OsPHT1;4*-Ox plants under both Pi-sufficient and Pi-deficient conditions. The total P contents in roots of *OsPHT1;4*-Ox plants were comparable to that of WT under both +P and −P. However, the *OsPHT1;4*-Ox plants had 3.37% and 42% higher total P content in shoots than the control under +P and −P, respectively (Figure 3A,B). The P uptake efficiency was calculated according to the modified method of the reference [18]. As shown in Figure 3C, the P uptake efficiency of the *OsPHT1;4*-overexpressing plants was 23% higher than that of the wild-type plants under the Pi-deficient conditions, although the concentrations were comparable between them under Pi-sufficiency (Figure 3C). We also calculated the root/shoot ratio of total P under different conditions, and the results showed that the overexpression tissues under the −P condition were significantly lower than that of the wild-type (Figure 3D). Based on these results, we conclude that overexpression of *OsPHT1;4* in WYJ 7 promoted the transport of P from roots to shoots and enhanced P uptake efficiency under Pi-deficient conditions.

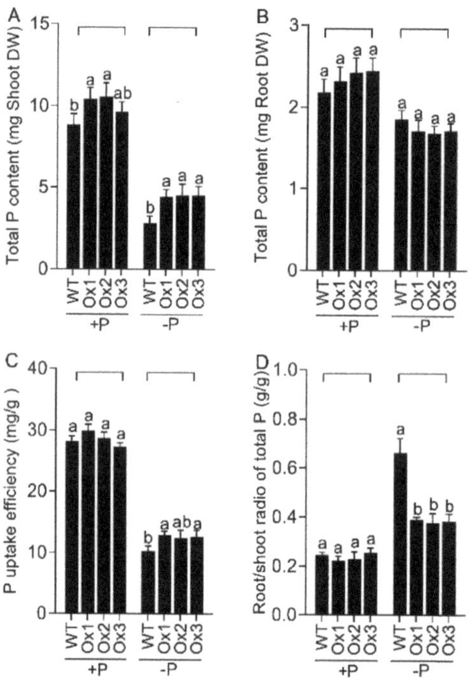

Figure 3. Overexpression of *OsPHT1;4* elevated total P concentration in Pi-deprived shoot. WT and Ox1–Ox3 seedlings (10-d-old) were grown hydroponically in +P and −P media for 3 weeks. Tissues were harvested and assayed for their total P concentrations. Comparisions of total P content of shoot (**A**), total P content of root (**B**), P uptake efficiency (**C**) and (**D**) root/shoot ratio of total P of WT and Ox1–Ox3 plants grown under +Pi and −Pi regimes. Values are means ± SE ($n = 5$). Different letters above the bars indicate significant differences ($p < 0.05$, one-way ANOVA).

To further determine the contribution of overexpression of *OsPHT1;4* in WYJ 7, Pi uptake assays were performed using isotope ^{33}Pi. The results showed that overexpression of the gene led to a significant increase of Pi uptake in roots under both Pi-sufficient and -deficient conditions. The ^{33}Pi uptake by roots of *OsPHT1;4*-Ox plants was 25–50% higher than that of the wild-type plants under Pi-sufficient conditions (Figure 4). Similarly, under Pi-deficient condition, ^{33}P uptake by roots of *OsPHT1;4*-Ox transgenic plants was about 15–35% higher compared with the wild-type plants (Figure 4). ^{33}Pi isotope experiment results further verified the roles of OsPHT1;4 in the uptake of Pi in roots.

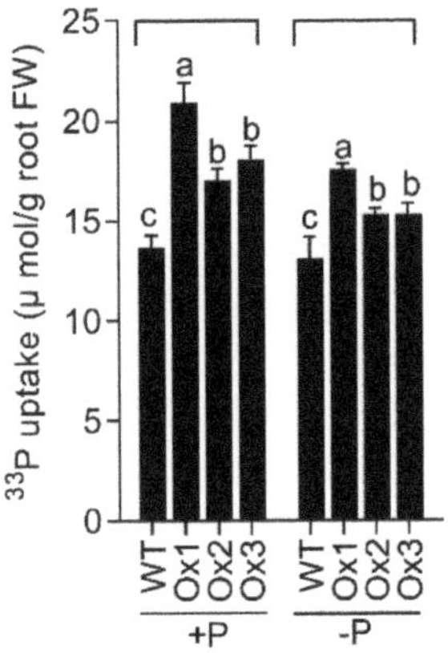

Figure 4. Overexpression of *OsPHT1;4* increases ^{33}Pi uptake rate. WT and Ox1–Ox3 seedlings (10 days old) were grown hydroponically in +P and −P media for 1 week and then transferred to the respective media containing ^{33}Pi to incubate for 3 h. Roots were harvested to determine the ^{33}Pi uptake. Values are means ± SE ($n = 5$). Different letters above the bars indicate significant differences ($p < 0.05$, one-way ANOVA). FW: Fresh weight.

The expression of *OsPHT1* family genes is a diversely regulated process, we thus examined the expression of other well-characterized *OsPHT1* family genes in the *OsPHT1;4*-overexpressing lines. The results showed that most genes did not change their expressions significantly in the lines, and only the expression of *OsPHT1;10* was significantly inhibited (Figure 5). This result shows that the overexpression of *OsPHT1;4* inhibits the expression of some other *OsPHT1* family genes under phosphorus-sufficient conditions, which is consistent with the previous investigation reported by Ye et al. [19].

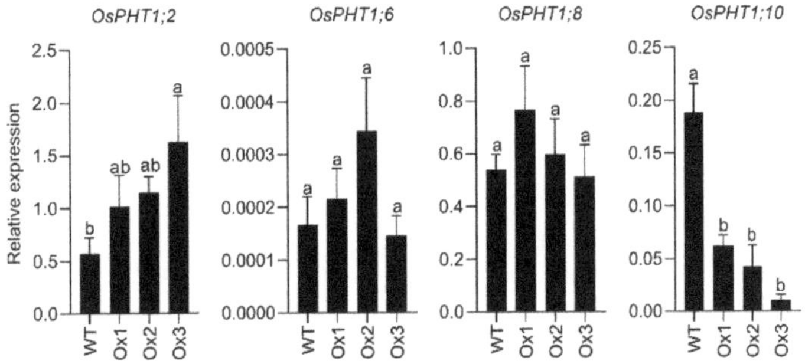

Figure 5. Expression of other *OsPHT1* family genes (*OsPHT1;2*, *OsPHT1;6*, *OsPHT1;8*, and *OsPHT1;10*) in the *OsPHT1;4*-Ox lines. WT and Ox1–Ox3 seedlings (10-d-old) were grown hydroponically in the +P media for 1 week. Values are means ± SE ($n = 3$). Different letters above the bars indicate significant differences ($p < 0.05$, one-way ANOVA).

3.4. Overexpression of OsPHT1;4 Enhances Translocation of P under Field Condition

To assess the impact of *OsPHT1;4* overexpression on Pi uptake and translocation during the entire rice growth cycle, we performed experiments on wild-type and *OsPHT1;4*-Ox plants grown in the paddy soil containing 25.25 mg kg^{-1} Pi. The total P concentrations in various tissues dissected from mature plants under the field were measured. They were significantly higher in all tissues of *OsPHT1;4*-Ox plants as compared to WT plants (Figure 6). The culm and leaf sheath from mature *OsPHT1;4*-Ox plants had 18% and 15% higher P concentrations, respectively, than those from the wild-type plants (Figure 6). Therefore, overexpression of *OsPHT1;4* in WYJ 7 significantly enhanced the accumulation of P in vegetative organs of the transgenic plants. Compared with that of the wild-type plants, the total P concentrations were higher in the panicle axis by 53%, in the unfilled rice hull by 58%, in the rice hull by 21%, and in brown rice by 14% (Figure 6), respectively. These results indicate that overexpression of *OsPHT1;4* was also effective in the accumulation of P in rice reproductive organs.

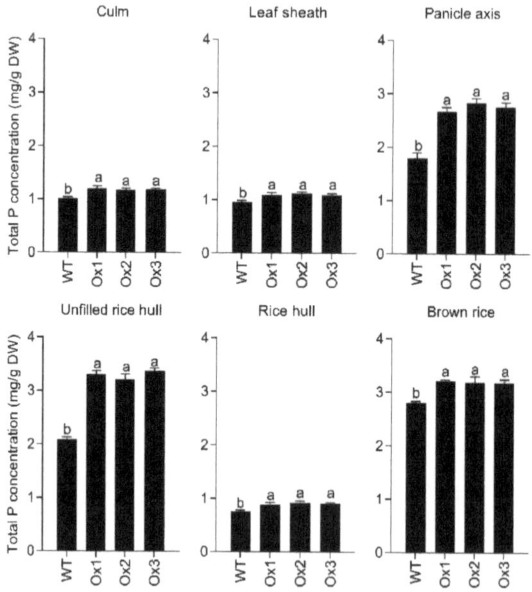

Figure 6. Overexpression of *OsPHT1;4* elevates total P concentration. WT and Ox1–Ox3 seedlings (10 days old) were grown to maturity in the field for 20 weeks. Various tissues indicated on the panel were harvested and the total P concentration was determined. Values are means ± SE (n = 5). Different letters above the bars indicate significant differences ($p < 0.05$, one-way ANOVA).

3.5. Enhanced Pi Utilization in OsPHT1;4-Ox Lines Promotes Grain Yield and Agronomic Traits of Rice cv. WYJ7 under Field Growth Conditions

The three *OsPHT1;4*-Ox lines were used for a field experiment in Nanjing, China, over three consecutive years. It was found that the grain density of each panicle of the *OsPHT1;4*-Ox plants was slightly bigger than those of the wild-type plants (Figure 7A). The grain yield per plant and 1000-grain weight of the *OsPHT1;4*-Ox plants were 5% and 7% greater than those of the wild-type, respectively (Figure 7B,C). However, the seed setting rate of the *OsPHT1;4*-Ox plants was not significantly different from that of the wild-type plants (Figure 7B). These results may be caused by the accumulation and translocation of P in the vegetative organs and its subsequent translocation to the reproductive organs. Since the grain yield per plant and 1000-grain weight of the *OsPHT1;4*-Ox plants increased without increasing the seed setting rate of the plants, the length and width of the wild-type and *OsPHT1;4*-Ox plant seeds were measured. As shown in Figure 7, the seeds from the *OsPHT1;4*-Ox plants were bigger, by 6% longer and 3% wider, than those from wild-type plants (Figure 7E–H). Taken together, the overexpression of *OsPHT1;4* promotes agronomic traits of rice under field growth conditions.

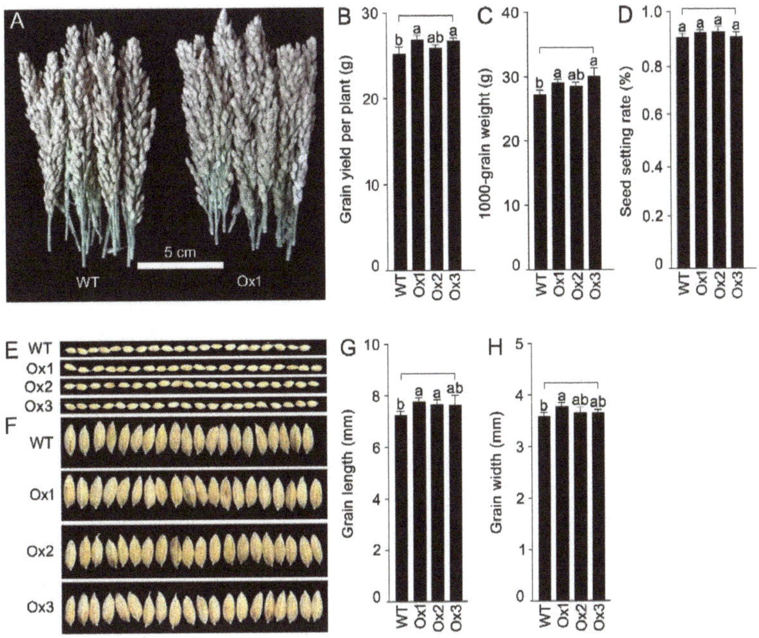

Figure 7. Overexpression of *OsPHT1;4* accentuates yield performance of WYJ 7. WT and Ox1–Ox3 plants were grown in the field as described in the legend of Figure 6. Photographs of material from the WT and Ox1 plants show the phenotypes of panicle (**A**), grain length (**E**), and grain width (**F**). Data are presented for grain yield per plant (**B**), percent seed setting rate (**C**), 1000-grain weight (**D**), grain length (**G**), and grain width (**H**). Values are means ± SE (n = 5). Different letters above the bars indicate significant differences ($p < 0.05$, one-way ANOVA).

4. Discussion

4.1. Overexpression of OsPHT1;4 Promotes the Growth and Grain Yield per Plant

The overexpression of *OsPHT1;4* in WYJ 7 promoted growth in the seedling stage. Both the shoot and root biomass of *OsPHT1;4*-Ox lines were greater than those of the wild type under both Pi-sufficient and Pi-deficient conditions. Moreover, under Pi-deficiency, the primary root length of the *OsPHT1;4*-Ox lines was longer than that of the wild-type plants. In previous work, *OsPht1;1*-Ox lines developed much longer and more dense root hairs, which led to an increase in Pi accumulation under P-replete conditions [8]. The overexpression of low-affinity phosphate transporter *OsPht1;2* in rice enhanced the absorption and translocation of P from roots to shoots, but with yellow leaves and short plants [20]. Although *OsPht1;8*-Ox lines improved the accumulation of P in plants, it hindered the growth of rice as the plant became shorter and displayed Pi toxicity symptoms [9]. The biomass of *OsPht1;3* overexpression lines was significantly decreased under both Pi-sufficient and -deficient conditions, and Pi toxicity symptoms could be observed under Pi-sufficiency [11]. We previously investigated the functions *OsPht1;6* in WYJ 7, the same rice cv. as we used in the present study, and found that the overexpression of *OsPht1;6* improved the absorption and utilization of P in plants. It also promoted the growth and development of the overexpression of plants [17]. In comparison with *OsPht1;6*, *OsPHT1;4* in WYJ 7 acts more effectively on the agronomic traits, such as grain size and yield per plant (Figure 7).

Improving rice grain yield is an important goal of agricultural production, as it is critical to ensure food security for nearly half of the world's population that are dependent on rice as a staple food. Our previous work demonstrated that the overexpression of *OsPht1;6* improved agronomic traits in rice WYJ 7, the main rice cultivar grown in China's Yangtze River Delta [17], by increasing the effective tiller number, grain weight per panicle, and grain yield per plant compared with the wild-type. In this study, our data showed that the overexpression of *OsPHT1;4* increases the 1000-grain weight and grain yield per plant compared with the wild-type of rice WYJ 7 (Figure 7B,C). These results indicate that both the overexpression of *OsPHT1;4* and *OsPht1;6* in rice WYJ 7 promote the growth of the plants and the grain yield, but the mechanisms of the gene's effects are different.

4.2. OsPHT1;4 Is Responsive to Pi Deficiency in WYJ 7 and Promotes P Uptake and Transport under Pi Deficiency

The transcript level of *OsPHT1;4* was higher in the roots than in the leaf blades of WYJ 7 under various Pi regimes (Figure 1A), and the transcript levels of *OsPHT1;4* were significantly upregulated by Pi starvation in both the roots and leaf blades (Figure 1A). We previously reported that the transcript level of *OsPHT1;4* was also higher in the root than in the leaf blade in rice cv. *Nipponbare* under Pi-sufficient (+P) and -deficient (−P) conditions [13]. However, the expression of *OsPHT1;4* was comparable in the roots under +P and −P conditions although the transcript level of *OsPHT1;4* was moderately induced by −P conditions in leaf blades [13]. These results show that their expression patterns of *OsPHT1;4* in rice cv. WYJ 7 and cv. *Nipponbare* are different.

The P concentrations of *OsPHT1;4*-overexpressing plants in rice cv. *Nipponbare* and *OsPht1;6*-overexpressing plants in rice cv. WYJ 7 are much higher than wild-type in both shoots and roots under Pi-sufficient (+P) and Pi-deficient (−P) conditions [13,17]. In the present work, the total P contents of *OsPHT1;4*-Ox plants in the shoots were significantly higher than that of the wild-type rice under the Pi-deficient conditions, while the total P contents in the roots were comparable to that of the wild type (Figure 3A,B,D). These data indicate that the overexpression of *OsPHT1;4* in WYJ 7 promotes the transport of P from roots to shoots, especially under Pi-deficient conditions. In addition, the P uptake efficiency of *OsPHT1;4*-overexpressing plants in WYJ 7 was significantly improved compared with the wide-type under Pi-deficient conditions (Figure 3C). The ratio of total P concentration in young leaves to that in old leaves of *OsPht1;6*-Ox plants was significantly higher than that of the wild-type plants under the two different Pi levels (40 and 80 mg fertilizer Pi kg^{-1} soil) tested in WYJ 7 [13]. These results indicate that both members in the rice PHT1 family can facilitate the uptake, translocation, and remobilization of Pi in rice [7,13,17]. Hopefully, the overexpression of *OsPHT1;4* in WYJ 7 is capable of maintaining normal growth in low-P soil or in the soil with less Pi fertilizer applied, which is one of the foci of agricultural production and environmental protection.

4.3. Overexpression of OsPHT1;4 Promotes Grain Yield Partly through Increasing Grain Size

Grain size is one of the key factors determining grain yield in crops [21,22]. However, the underlying molecular mechanism is largely unclear. OsMAPK6, a mitogen-activated protein kinase, influences rice grain size by influencing cell proliferation and reducing endogenous brassinosteroid levels of rice [23]. Similarly, mitogen-activated protein kinase 4 is a factor capable of affecting the grain size of rice [24]. In this study, we showed that OsPHT1;4-Ox plants increase the 1000-grain weight and grain yield per plant compared with wild-type WYJ 7 (Figure 7B,C). Furthermore, the lengths and widths of OsPHT1;4-Ox plants seeds increased significantly compared with the wild-type plants (Figure 7E–H). In our previous work in rice cv. Nipponbare, there were significant increases in 1000-grain weight and grain yield per plant in *OsPHT1;4*-Ox1 and -Ox2 compared with the WT [13]. A detailed investigation exhibited the genetic evidence for the role of *OsPHT1;4* during embryogenesis, and the embryo surface areas of Ox1 and Ox2 were 9–14% larger than that of the WT control [13]. These results provide a possible explanation for the increases of length and width of the grains in the *OsPHT1;4*-Ox plants (Figure 7), and *OsPHT1;4* may exert influence on the embryogenesis and seed development in rice cv. WYJ 7. Alternatively, the increase in grain size is the consequence of the enhanced uptake, translocation, and accumulation of P in above-ground parts of plants via the overexpression of *OsPHT1;4* in WYJ 7. In addition, *OsPht1;8*, another member of the rice PHIT1 family, demonstrated its effect on the allocation of Pi between the embryo and the endosperm [25]. This raised the possibility that a connection may exist between *OsPHT1;4* and *OsPht1;8* during embryogenesis and seed development, which is an interesting and important topic that deserves more detailed investigations.

The ultimate goal of agricultural research is to nurture more people with limited arable land on earth. How to improve yield per unit is critical to achieving the goal. Traditionally, different crop cultivars provide many valuable solutions for yield improvement. In rice, many cultivars including WYJ 7 show much better performance in agronomic traits than Nipponbare. This suggests that these practically used rice cultivars have their unique advantages in agriculture. In this study, our investigation shows that overexpression of *OsPHT1;4* in WYJ 7 increases phosphate utilization efficiency and improves the agronomic traits of the cultivar. It implies that the future improvement of rice yield by manipulating critical gene expression in optimized cultivar is still a feasible method in practical application.

Supplementary Materials: The following supporting information can be downloaded at: https://www.mdpi.com/article/10.3390/agronomy12061332/s1, Figure S1: Identification of *OsPHT1;4* Overexpressing Lines in WYJ 7. Table S1: List of primers used for semi-quantitative RT-PCR and quantitative real-time PCR analysis. Table S2. Primers used for vector construction.

Author Contributions: Conceptualization, Y.H., S.S. and Y.C.; methodology, Z.H., X.L. and Y.S.; formal analysis, X.H.; investigation, X.W., H.X., X.H. and Y.S.; data curation, Z.H., X.H., X.L., Y.S. and Y.C.; writing—original draft preparation, Z.H., Y.H. and Y.C.; writing—review and editing, Z.H., X.H., X.W., Y.H. and Y.C.; supervision, Y.H. and Y.C.; funding acquisition, Y.H., S.S. and Y.C. All authors have read and agreed to the published version of the manuscript.

Funding: This research was funded by the Guangdong Basic and Applied Basic Research Foundation, grant number 2021A1515011564; the National Natural Science Foundation of China, grant number 42107428; and the National Key Research and Development Program of China, grant number 2019YFD0900702.

Data Availability Statement: The data presented in this study are available on request from the corresponding author.

Conflicts of Interest: The authors declare no conflict of interest.

References

1. Raghothama, K.G. Phosphate Acquisition. *Annu. Rev. Plant Biol.* **1999**, *50*, 665–693. [CrossRef] [PubMed]
2. Mudge, S.R.; Rae, A.L.; Diatloff, E.; Smith, F.W. Expression analysis suggests novel roles for members of the Pht1 family of phosphate transporters in Arabidopsis. *Plant J.* **2002**, *31*, 341–353. [CrossRef] [PubMed]
3. Muchhal, U.S.; Pardo, J.M.; Raghothama, K.G. Phosphate transporters from the higher plant Arabidopsis thaliana. *Proc. Natl. Acad. Sci. USA* **1996**, *93*, 10519–10523. [CrossRef] [PubMed]
4. Gu, M.; Chen, A.; Sun, S.; Xu, G. Complex Regulation of Plant Phosphate Transporters and the Gap between Molecular Mechanisms and Practical Application: What Is Missing? *Mol. Plant* **2016**, *9*, 396–416. [CrossRef] [PubMed]
5. Goff, S.A.; Ricke, D.; Lan, T.-H.; Presting, G.; Wang, R.; Dunn, M.; Glazebrook, J.; Sessions, A.; Oeller, P.; Varma, H.; et al. A Draft Sequence of the Rice Genome (*Oryza sativa* L. ssp. *japonica*). *Science* **2002**, *296*, 92–100. [CrossRef]
6. Paszkowski, U.; Kroken, S.; Roux, C.; Briggs, S.P. Rice phosphate transporters include an evolutionarily divergent gene specifically activated in arbuscular mycorrhizal symbiosis. *Proc. Natl. Acad. Sci. USA* **2002**, *99*, 13324–13329. [CrossRef]
7. Ai, P.; Sun, S.; Zhao, J.; Fan, X.; Xin, W.; Guo, Q.; Yu, L.; Shen, Q.; Wu, P.; Miller, A.J.; et al. Two rice phosphate transporters, OsPht1;2 and OsPht1;6, have different functions and kinetic properties in uptake and translocation. *Plant J.* **2009**, *57*, 798–809. [CrossRef]

8. Sun, S.; Gu, M.; Cao, Y.; Huang, X.; Zhang, X.; Ai, P.; Zhao, J.; Fan, X.; Xu, G. A Constitutive Expressed Phosphate Transporter, OsPht1;1, Modulates Phosphate Uptake and Translocation in Phosphate-Replete Rice. *Plant Physiol.* **2012**, *159*, 1571–1581. [CrossRef]
9. Ye, Y.; Yuan, J.; Chang, X.; Yang, M.; Zhang, L.; Lu, K.; Lian, X. The Phosphate Transporter Gene OsPht1;4 Is Involved in Phosphate Homeostasis in Rice. *PLoS ONE* **2015**, *10*, e0126186. [CrossRef]
10. Wang, X.; Wang, Y.; Pineros, M.; Wang, Z.; Wang, W.; Li, C.; Wu, Z.; Kochian, L.; Wu, P. Phosphate transporters OsPHT1;9 and OsPHT1;10 are involved in phosphate uptake in rice. *Plant Cell Environ.* **2013**, *37*, 1159–1170. [CrossRef]
11. Chang, M.X.; Gu, M.; Xia, Y.W.; Dai, X.L.; Dai, C.R.; Zhang, J.; Wang, S.C.; Qu, H.Y.; Yamaji, N.; Ma, J.F.; et al. OsPHT1;3 Mediates Uptake, Translocation, and Remobilization of Phosphate under Extremely Low Phosphate Regimes. *Plant Physiol.* **2018**, *179*, 656–670. [CrossRef] [PubMed]
12. Glassop, D.; Godwin, R.M.; Smith, S.E.; Smith, F.W. Rice phosphate transporters associated with phosphate uptake in rice roots colonised with arbuscular mycorrhizal fungi. *Can. J. Bot.* **2007**, *85*, 644–651. [CrossRef]
13. Zhang, F.; Sun, Y.; Pei, W.; Jain, A.; Sun, R.; Cao, Y.; Wu, X.; Jiang, T.; Zhang, L.; Fan, X.; et al. Involvement ofOsPht1;4in phosphate acquisition and mobilization facilitates embryo development in rice. *Plant J.* **2015**, *82*, 556–569. [CrossRef] [PubMed]
14. Liu, K.M.; Wu, C.B.; Joung, S.J.; Tsai, W.P.; Su, K.Y. Multi-model approach on growth estimation and association with life history trait for elasmobranchs. *Front. Mar. Sci.* **2021**, *8*, 591692. [CrossRef]
15. Upadhyaya, N.M.; Surin, B.; Ramm, K.; Gaudron, J.; Schünmann, P.H.D.; Taylor, W.; Waterhouse, P.M.; Wang, M.-B. Agrobacterium-mediated transformation of Australian rice cultivars Jarrah and Amaroo using modified promoters and selectable markers. *Funct. Plant Biol.* **2000**, *27*, 201–210. [CrossRef]
16. Chen, A.; Hu, J.; Sun, S.; Xu, G. Conservation and divergence of both phosphate- and mycorrhiza-regulated physiological responses and expression patterns of phosphate transporters in solanaceous species. *New Phytol.* **2006**, *173*, 817–831. [CrossRef]
17. Zhang, F.; Wu, X.-N.; Zhou, H.-M.; Wang, D.-F.; Jiang, T.-T.; Sun, Y.-F.; Cao, Y.; Pei, W.-X.; Sun, S.-B.; Xu, G.-H. Overexpression of rice phosphate transporter gene OsPT6 enhances phosphate uptake and accumulation in transgenic rice plants. *Plant Soil* **2014**, *384*, 259–270. [CrossRef]
18. Elliott, G.C.; Läuchli, A. Phosphorus Efficiency and Phosphate–Iron Interaction in Maize 1. *Agron. J.* **1985**, *77*, 399–403. [CrossRef]
19. Ye, Y.; Li, P.; Xu, T.; Zeng, L.; Cheng, D.; Yang, M.; Luo, J.; Lian, X. OsPT4 Contributes to Arsenate Uptake and Transport in Rice. *Front. Plant Sci.* **2017**, *8*, 2197. [CrossRef]
20. Liu, F.; Wang, Z.; Ren, H.; Shen, C.; Li, Y.; Ling, H.Q.; Wu, C.; Lian, X.; Wu, P. OsSPX1 suppresses the fuction of OsPHR in the regulation of expression of OsPT2 and phosphate homeostasis in shoots of rice. *Plant J.* **2010**, *62*, 508–517. [CrossRef]
21. Ikeda, M.; Miura, K.; Aya, K.; Kitano, H.; Matsuoka, M. Genes offering the potential for designing yield-related traits in rice. *Curr. Opin. Plant Biol.* **2013**, *16*, 213–220. [CrossRef] [PubMed]
22. Xing, Y.; Zhang, Q. Genetic and Molecular Bases of Rice Yield. *Annu. Rev. Plant Biol.* **2010**, *61*, 421–442. [CrossRef] [PubMed]
23. Liu, S.; Hua, L.; Dong, S.; Chen, H.; Zhu, X.; Jiang, J.; Zhang, F.; Li, Y.; Fang, X.; Chen, F. Os MAPK 6, a mitogen-activated protein kinase, influences rice grain size and biomass production. *Plant J.* **2015**, *84*, 672–681. [CrossRef] [PubMed]
24. Duan, P.; Rao, Y.; Zeng, D.; Yang, Y.; Xu, R.; Zhang, B.; Dong, G.; Qian, Q.; Li, Y. SMALL GRAIN 1, which encodes a mitogen-activated protein kinase kinase 4, influences grain size in rice. *Plant J.* **2014**, *77*, 547–557. [CrossRef] [PubMed]
25. Li, Y.; Zhang, J.; Zhang, X.; Fan, H.; Gu, M.; Qu, H.; Xu, G. Phosphate transporter OsPht1;8 in rice plays an important role in phosphorus redistribution from source to sink organs and allocation between embryo and endosperm of seeds. *Plant Sci.* **2015**, *230*, 23–32. [CrossRef] [PubMed]

agronomy

MDPI

Article

Priestia sp. LWS1 Is a Selenium-Resistant Plant Growth-Promoting Bacterium That Can Enhance Plant Growth and Selenium Accumulation in *Oryza sativa* L.

Xiao-Rui Lin [1,2], Han-Bing Chen [3], Yi-Xi Li [1,2], Zhi-Hua Zhou [3], Jia-Bing Li [1,2], Yao-Qiang Wang [1,2], Hong Zhang [1,2], Yong Zhang [1,2], Yong-He Han [1,2,*] and Shan-Shan Wang [4,*]

1 College of Environmental Science and Engineering, Fujian Normal University, Fuzhou 350007, China; lamborghini186@126.com (X.-R.L.); yxli981112@163.com (Y.-X.L.); lijiabing@fjnu.edu.cn (J.-B.L.); wyq962237@163.com (Y.-Q.W.); zhanghong@fjnu.edu.cn (H.Z.); yongzhang@fjnu.edu.cn (Y.Z.)
2 Fujian Key Laboratory of Pollution Control and Resource Reuse, Fuzhou 350007, China
3 College of Life Science, Fujian Normal University, Fuzhou 350117, China; chenhanbing@fjnu.edu.cn (H.-B.C.); zzh-bio@fjnu.edu.cn (Z.-H.Z.)
4 College of Integrative Medicine, Fujian University of Traditional Chinese Medicine, Fuzhou 350122, China
* Correspondence: yhhan@fjnu.edu.cn (Y.-H.H.); sswang1208@163.com or 2020044@fjtcm.edu.cn (S.-S.W.)

Abstract: Selenium (Se) is essential for the basic functions of life, but the low daily intake of Se urges us to find reliable ways to increase food Se content. Plant-growth-promoting bacteria (PGPB) have shown potential in enhancing plant growth and Se accumulation. In this study, the soils collected from a Se tailing were used to isolate Se-tolerant PGPB. The results showed that a total of three strains were identified. Strain LWS1, belonging to *Priestia* sp., grew well in M9 medium and exhibited typical PGP characteristics by an IAA-production ability of 24.3 ± 1.37 mg·L^{-1}, siderophore-production ability of 0.23 ± 0.04 and phosphate-solubilizing ability of 87.5 ± 0.21 mg·L^{-1}. Moreover, LWS1 strain tolerated selenite (SeIV) up to 90 mM by a LC_{50} of 270.4 mg·L^{-1}. Further investigations demonstrated that the inoculation of strain LWS1 resulted in up to 19% higher biomass and 75% higher Se concentration in rice (*Oryza sativa* L.) than uninoculated treatments. Our study has provided evidence that microbial Se biofortification through inoculating with *Priestia* sp. strain LWS1 is an alternative way to improve Se uptake in crops and maintain human health.

Keywords: plant-growth-promoting bacteria; selenium; *Oryza sativa*; *Priestia* sp. LWS1; microbial biofortification

Citation: Lin, X.-R.; Chen, H.-B.; Li, Y.-X.; Zhou, Z.-H.; Li, J.-B.; Wang, Y.-Q.; Zhang, H.; Zhang, Y.; Han, Y.-H.; Wang, S.-S. *Priestia* sp. LWS1 Is a Selenium-Resistant Plant Growth-Promoting Bacterium That Can Enhance Plant Growth and Selenium Accumulation in *Oryza sativa* L.. *Agronomy* **2022**, *12*, 1301. https://doi.org/10.3390/agronomy12061301

Academic Editor: Pedro Palencia

Received: 29 April 2022
Accepted: 27 May 2022
Published: 29 May 2022

1. Introduction

Selenium (Se) is a naturally occurring metalloid with a low concentration and uneven distribution in the earth's crust. A typical concentration of soil Se ranges from 0.01 to 2 mg·kg^{-1}, with an average of 0.4 mg·kg^{-1} [1]. Se was considered a harmful element to human beings and animals for a long time after its discovery in 1817 [2]. The turning point came through the recognition of Se's role in preventing muscular dystrophy and liver cirrhosis in rats in 1957 [3]. We now know that the two-sided properties of Se are dependent on its intake content, and both deficiency and excessive intake are harmful to humans and animals [3]. Se deficiency has been linked to a range of health symptoms and diseases, including infertility in both men and women, miscarriage, cardiovascular disease, Keshan and Kashin–Beck disease, leukomyopathy, reduced immune function and an increased risk of various cancers [4–6]. Notably, excess Se intake can lead to hair loss and nervous system damage [7]. The recommended dietary allowance and tolerable upper limit of Se intake for an adult is 40 μg·day^{-1} and 400 μg·day^{-1}, respectively [8]. The German, Austrian and Swiss nutrition societies have revised the reference values for the intake of Se for men by 70 μg·day^{-1} and for women by 60 μg·day^{-1} [9]. Dietary Se speciation, including seleno-aminoacids, selenocysteine, and selenomethionine, all contribute to health improvement, probably due to the catalytic role of Se in several selenoproteins such as glutathione peroxidase, thioredoxin reductase, and iodothyronine-deiodinases [3,10]. Hence, increasing food Se content is a desirable way to increase Se intake and overcome the Se-deficiency problems.

Se source foods include organ meats and seafood, muscle meats, cereal and grains, agricultural crops, milk and dairy products, and fruits and vegetables [10]. Among which, rice (*Oryza sativa* L.) is a staple food that sustains half of the world's population [11], which can be an important source when increasing Se intake. Based on a daily rice consumption of 300 $g \cdot day^{-1}$, however, Williams et al. [12] revealed that 75% of the world's rice is not enough to provide 70% of the human body's Se needs. Se biofortification is considered to be a safe and effective method of increasing Se content in edible parts of crops, and thus increasing Se intake in the human body [13]. At present, Se biofortification in plants can be achieved by soaking seeds in Se solution, increasing soil Se concentration, foliar spray and other auxiliary methods, such as crop breeding and transgenic technology [5,14,15]. Although the application of exogenous Se can increase plant Se content, studies have shown that a high concentration of Se can inhibit plant growth. For example, when exposed to $\geq$5 $mg \cdot kg^{-1}$ Se, both the dry biomass and the antioxidant capacity of rice decreased [16]. Another study focusing on non-accumulator ryegrass also observed a negative relationship between solution Se and plant biomass [17]. Therefore, it is difficult to meet the requirements of biofortification with only the addition of exogenous Se.

Plant-growth-promoting bacteria (PGPB) are well-known microorganisms inhabiting the rhizosphere of plants, which can influence nutrient behaviors and plant growth via direct or indirect activities [18,19]. While biological nitrogen fixation, phosphate solubilization and phytohormone production, including indole-3-acetic acid (IAA), cytokinins and gibberellins contribute directly to plant growth, siderophore production, chitinase and glucanase production, antibiotic production, induced systematic resistance and the production of 1-aminocyclopropane-1-carboxylic acid deaminase are also important functions that regulate plant growth [19]. Of these, IAA production, siderophore production and phosphate solubilization are three indicators used to evaluate the PGP characteristics of PGPB [20]. Besides their roles in enhancing plant growth, PGPBs have also exhibited considerable effects regarding the remediation of organic- and heavy metal-contaminated soils [21,22]. For example, a recent study from Eze et al. [18] showed that the bioaugmentation with PGPB not only increased *Medicago sativa* biomass by 66%, but also resulted in a 91% removal of diesel hydrocarbons by an initial concentration of 4.59 $mg \cdot kg^{-1}$. The attractive applications of PGPB in promoting plant growth and heavy metal removal have also been widely documented [23,24].

As potential enhancers for Se biofortification, PGPBs have shown important benefits for Se uptake and accumulation in plants [25,26]. Among reported PGPBs, *Bacillus* spp. show great ability to enhance crop growth and Se accumulation. For example, by inoculating with *Bacillus* sp. YAM2, Yasin et al. [27] reported a promising Se biofortification strategy for wheat and potentially other crops. Given the wide use of arbuscular mycorrhizal fungi (AMF) in agricultural activities, co-inoculation of AMF with *Bacillus* sp. E5 also increased the Se content, chlorophyll and antioxidant enzyme content and plant biomass of lettuce [28]. Although Se is not an essential element for plants, it can benefit their growth [5,29]. As such, microbial Se fortification is a promising green way to elevate rice Se accumulation.

In the present study, we isolated a new Se-tolerant bacterial strain and evaluated its potential roles in promoting rice growth and Se accumulation. The full objectives were (1) identify the Se-tolerant ability of the newly isolated bacterial strain LWS1, (2) test the PGP characteristics of strain LWS1 and (3) evaluate the potential roles of strain LWS1 in plant growth and Se accumulation in rice. We believe that the results have provided a new bioresource, with potential use in promoting plant growth and Se accumulation in rice.

2. Materials and Methods

2.1. Microbial Screening and Identification

The soil used for bacterial isolation was collected from a Se-mine in Longyan, Fujian, China. Half of the collected soil was stored at −80 °C for 12 h and freeze-dried for 48 h till a balance weight was reached (Labconco FreeZone 6 plus, Kansas City, MI, USA). The soils were ground and sieved to <2 mm, and finally digested using the USEPA method 3050B [30]. The concentrations of typical heavy metal(loid)s were ($mg \cdot kg^{-1}$): 9.38 ± 0.17 (Se), 12.1 ± 0.21 (arsenic, As), 0.46 ± 0.06 (cadmium, Cd), 24.1 ± 0.53 (lead, Pb), 69.4 ± 1.39 (zinc, Zn) and 21.8 ± 2.16 (copper, Cu).

To isolate Se-tolerant bacteria, about 0.5 g of fresh soil was washed thrice with sterile phosphate-buffered solution (PBS, prepared by 19 mL 0.2 M NaH_2PO_4 and 81 mL 0.2 M Na_2HPO_4, pH 7.4) [31], followed by incubation in Luria–Bartani (LB) medium spiked with 100 $mg \cdot L^{-1}$ selenite (SeIV, Na_2SeO_3). The incubation conditions were 30 °C and 150 rpm for 12 h. The bacterial suspension was sub-cultured 3–5 times in a new medium to enrich the Se-tolerant bacteria. After a continuous gradient dilution, 50 μL of the established suspensions was spread on LB agar medium spiked with 100 $mg \cdot L^{-1}$ SeIV and incubated in an incubator at 30 °C until apparent colonies occurred. Each colony with a different morphology was picked out and re-incubated in LB and LB agar medium to obtain pure isolates. The suspension of each isolate was finally mixed with an equivalent volume of 50% glycerol and stored at −80 °C before further identification.

All isolates were re-incubated in LB medium at 30 °C and 150 rpm for 12 h. Aliquots of enriched suspensions were used to extract bacterial total DNA using a FastDNA® SPIN Kit for Soil (MP Biomedicals, LLC, 9 Goddard Irvine, CA, USA) according to the manufacturer's instructions [30]. In our study, the 16S rRNA gene for each isolate was amplified by PCR with the universal primers 5′-AGAGTTTGATCCTGGCTCAG-3′ (BACT27F) and 5′-ACGGCTACCTTGTTACGACTT-3′ (PROK1492R) on a T100 Thermocycler (Bio-Rad, 1000 Alfred Nobel Drive Hercules, CA, USA). The PCR mixtures consisted of 12.5 μL of 2 × Mix (Yifeixue Biotechnology, Nanjing, China), 1.5 μL of 10 μM mixed primer pair, 10 μL of PCR degrade water and 1 μL of DNA template with a normalized concentration of 50 $ng \cdot \mu L^{-1}$. The PCR programs included pre-denaturation (5 min at 94 °C), denaturation (35 s at 94 °C), annealing (30 s at 55 °C) and extension (1.5 min at 72 °C). The PCR cycles were 35 with a final extension at 72 °C for 10 min. Following an electrophoresis verification by 2% agarose gel, the PCR products were purified and sequenced by Allwegene Tech. (Beijing, China). The obtained sequences were analyzed by BLAST similarity search against the known sequences in NCBI database. Finally, the phylogenetic tree of 16S rRNA gene was constructed by MEGA 5.0 using the Neighbor-Joining algorithm. The nucleotide sequence of 16S rRNA gene of strain LWS1 was deposited in GenBank with the accession number of ON377339.

2.2. Analysis of Se Tolerance

To test Se tolerance, strain LWS1 was first incubated in LB medium at 30 °C and 150 rpm for 12 h. The established bacterial suspensions were centrifuged at 5000× *g* for 5 min to obtain biomass, which was then washed three times with PBS spiked with 0.9% NaCl. In our study, the Se-tolerant experiment was performed in M9 medium due to its clear components and lower interference in metabolism inside bacterial cells compared to LB medium. More details about M9 medium are available from Jouanneau et al. [32]. In brief, aliquots of the established bacterial suspensions were transferred to M9 medium to make the final OD_{600} of 0.01. The concentrations of solution SeIV were 0, 7.5, 15, 37.5, 75, 150, 375, 750, 1500, 3750 and 7500 $mg \cdot L^{-1}$. After 24 h of incubation, the OD_{600} of strain LWS1 was recorded to evaluate bacterial growth. To obtain a further evaluation of Se tolerance in strain LWS1, a nonlinear fitting based on the logistic model by using OriginPro 2021 (OriginLab Corporation, Northampton, MA, USA) was also performed. As a result, the median lethal dose (LC_{50}) of Se to bacterial strain was obtained.

2.3. Characterization of PGP Traits

The abilities of IAA biosynthesis, siderophore biosynthesis and phosphate solubilization can be regarded as typical indicators for PGP characteristics [20]. To evaluate the potential roles of strain LWS1 in regulating rice growth, the above-mentioned indexes were fully determined. Before characterization, the strain was incubated in LB medium at 30 °C and 150 rpm for 12 h. The biomass was collected by centrifugation at 5000× *g* for 5 min and re-suspended in sterile Milli-Q water to a uniform OD_{600} value. Aliquots of the resuspensions were transferred to following media for the analysis of PGP traits. The treatments without bacterial inoculation and with the inoculation of PGP strain *Pseudomonas* sp. PVR02 [33] were set as negative and positive controls, respectively.

To characterize the IAA biosynthesis ability of strain LWS1, 0.2 mL diluted suspension with OD_{600} = 1.5 was added into 20 mL sucrose minimal salts (SMS) medium [34]. The components of SMS medium include 10 $g \cdot L^{-1}$ sucrose, 2 $g \cdot L^{-1}$ K_2HPO_4, 1 $g \cdot L^{-1}$ $(NH_4)_2SO_4$, 0.5 $g \cdot L^{-1}$ $MgSO_4$, 0.5 $g \cdot L^{-1}$ yeast extract, 0.5 $g \cdot L^{-1}$ $CaCO_3$, 0.1 $g \cdot L^{-1}$ NaCl and 0.5 $g \cdot L^{-1}$ L-tryptophan (pH = 7.2). The strain was incubated at 30 °C and 150 rpm for 4 d, followed by centrifugation at 8000× *g* for 15 min. Finally, 1 mL of the supernatant was mixed with 2 mL Salkowski reagent consisting of 50 mL 35% $HClO_4$ and 1 mL 0.5 M $FeCl_3$. Following a stationary reaction for 40 min in the dark at room temperature, a pink color developed in the supernatant suspensions [34]. The absorbance of pink color was read at 530 nm. The IAA concentration was determined using a calibration curve of pure IAA as a standard following the linear regression analysis [23].

According to Schwyn and Neilands [35], the chrome azurol-S (CAS) analytical method was used to identify siderophore production ability. In briefy 0.2 mL bacterial suspension with OD_{600} = 1.5 was incubated in modified sucrose-aspartic acid medium, which contains 20 $g \cdot L^{-1}$ sucrose, 2 $g \cdot L^{-1}$ aspartic acid, 1 $g \cdot L^{-1}$ K_2HPO_4 and 0.5 $g \cdot L^{-1}$ $MgSO_4 \cdot 7H_2O$ with a pH of 7.0. The strain was incubated at 30 °C and 150 rpm for 2 d and centrifugated at 8000× *g* for 15 min. The supernatant was mixed with 1 mL CAS solution. A uninoculated medium was used as a blank control. After 3 h incubation in the dark at room temperature, the OD_{630} was measured and the siderophore production capacity was calculated as sample OD_{630}/control OD_{630} (λ/λ_0).

Regarding phosphate-solubilizing ability, the strain was incubated in National Botanical Research Institute's phosphate (NBRIP) growth medium [36]. The medium was composed of 5 $g \cdot L^{-1}$ glucose, 5 $g \cdot L^{-1}$ $MgCl_2 \cdot 6H_2O$, 0.25 $g \cdot L^{-1}$ $MgSO_4 \cdot 7H_2O$, 0.2 $g \cdot L^{-1}$ KCl, 0.1 $g \cdot L^{-1}$ $(NH_4)_2SO_4$ and 5 $g \cdot L^{-1}$ $Ca_3(PO_4)_2$ with a total volume of 20 mL and a pH of 7.0 ± 0.2. After an incubation at 30 °C and 150 rpm for 8 d, the suspensions were centrifuged at 8000× *g* for 15 min. The soluble phosphate was determined by the modified molybdenum blue method at 880 nm. In brief, 5 mL of the supernatants was collected and mixed with 0.5 mL of 10% (*m*/*v*) ascorbic acid and incubated for 30 s. A total of 1 mL of color developing agent (45 mL of 14% (*m*/*v*) ammonium molybdate solution + 5 mL of 3% (*m*/*v*) antimony potassium tartrate solution + 200 mL of 6 $mol \cdot L^{-1}$ H_2SO_4) was then added into the mixture and incubated for 15 min. Finally, the soluble phosphate concentration in the samples was calculated using a standard curve of pure phosphate solution following the linear regression analysis [37].

2.4. Evaluation of the Effects of Strain LWS1 on Plant Growth and Se Accumulation in Rice

Agricultural topsoil (0–15 cm) was collected from Fuzhou, China. The soils were air-dried and sieved to <2 mm before use. Different concentrations of SeIV-contaminated soils (0, 0.125, 0.175, 0.425, 3 and 6 $mg \cdot kg^{-1}$ of soil) were prepared by spiking 2.5 kg of soils with appropriate amounts of Na_2SeO_3. According to Eze et al. [38,39], the spiked soils were first mixed manually by hand wearing gloves, then thoroughly mixed using an IBC portable electric mixer (IBC H003). The mixing was performed for 15 min with a break and manual shaking after every 5 min to achieve complete homogeneity [39]. The homogenized soils were then aged for one month to make the soil system stable [40]. In brief, the soils were brought to 50% water-holding capacity and stored in a dark room to mimic the field drying–wetting cycles. Moreover, the plastic bags used to hold soils were partially opened for aeration.

The rice seeds of cultivar Yiyou 673 were soaked in sterile Milli-Q water for 1 h and disinfected by 75% ethanol for 30 s and 30% NaClO for 30 min. The absorbed chemical agents were removed following a thorough wash by sterile Milli-Q water. All seeds were geminated in a greenhouse under a light photoperiod of 16 h, a light density of 350 $\mu mol \cdot m^{-1} \cdot s^{-1}$, mean temperature of 25 °C and relative humidity of 70%, according to Xu et al. [41]. Rice plants with 2–3 leaves, of ~5 cm in height, were transferred to the aged soils in plastic pots, which contained 2.5 kg soils with a water depth of 3 cm. Before the experiment, strain LWS1 was activated in LB medium at 30 °C, 150 rpm for 12 h. Following centrifugation at 5000× *g* for 5 min, the obtained biomass was washed three times by sterile PBS and finally resuspended in sterile Milli-Q water to a uniform concentration. In each pot, 10 mL of bacterial suspension (1.2×10^7 $cfu \cdot mL^{-1}$) was added each month to the root base of rice plant. A treatment without bacterial inoculation was also set as control. The rice plants were grown in a greenhouse with the same conditions as described before.

After 4 months of growth, all rice plants were harvested and divided into root, husk, leaf and grain. The plant biomass was weighed and recorded. All plant tissues were freeze-dried for 48 h and digested according to USEPA method 3050B [30]. After centrifuging at 8000× *g* for 15 min, the supernatant was filtered by a 0.45 µm filter and diluted as required. In our study, the Se concentration was determined by graphite furnace atomic absorption spectroscopy (GFAAS). The instrument parameters used in the determination of Se were a wavelength of 196.0 nm, lamp current of 15 mA, slit width of 0.2 nm and sample volume of 20 µL, while the graphite furnace program for Se determination included 4 steps: drying at 120 °C for a ramp time of 55 s and hold time of 5 s, pyrolysis at 1400 °C for a ramp time of 10 s and hold time of 6 s, atomization at 2400 °C for a ramp time of 0 s and hold time of 2 s, and cleaning at 2700 °C for a ramp time of 1 s and hold time of 2 s [42]. Se stock solution (1000 $\mu g \cdot mL^{-1}$) purchased from the National Center for Standard Substances (Beijing, China) was diluted as necessary to establish the reference curve.

2.5. Statistical Analyses

All the experiments were conducted in four replicates. The data are presented as the mean value with standard error. Significant differences were determined according to two-way analysis of variance (ANOVA) by Tukey's multiple comparisons test at $p \leq 0.05$ using GraphPad Prism (Release 6.0, La Jolla, CA, USA).

3. Results and Discussion

3.1. Three Se-Tolerant Strains Were Isolated from Se Tailing Soil

Following a multiple enrichment in LB medium spiked with 100 $mg \cdot L^{-1}$ SeIV, a total of three culturable bacterial strains were isolated (Figure 1). While all strains formed a regular colony, only strain LWS1 had a flat colony structure (Figure 1). Moreover, the colony of strain LWS2-1 was yellow, but both LWS1 and LWS2-2 were of a white color (Figure 1). The colony morphology revealed that the three isolates belong to different microbial species. However, only strain LWS1 could grow well in M9 medium spiked with 100 $mg \cdot kg^{-1}$ Se. Considering that a flexible adaptability to complex inhabitants is an important trait for bacterial environmental applications [43], strain LWS1, which can grow in M9 medium with simple components, was further identified by 16S rRNA gene amplification and consequent evolutionary position analysis.

Figure 1. The colony morphology of the three bacterial strains isolated from Se tailing soil. The strains were grown on LB agar medium containing 100 mg·L^{-1} SeIV.

To further identify the taxonomic level of strain LWS1, its 16S rRNA gene sequence was aligned with the known sequences retrieved from the NCBI database. As shown in Figure 2, the 16S rRNA gene of strain LWS1 shared high similarity with several species of the genus *Bacillus*, with the highest BLAST scores (100%) observed in *Bacillus* sp. (in: Bacteria) IC-1C2 (MT649293.1). Compared with previous studies, the colony morphology of strain LWS1 showed the typical properties of *B. megaterium* [44], which was recently reclassified as *Priestia megaterium* [45]. Therefore, we preliminarily speculated that the bacterium was a new strain of *Priestia*, namely, *Priestia* sp. strain LWS1.

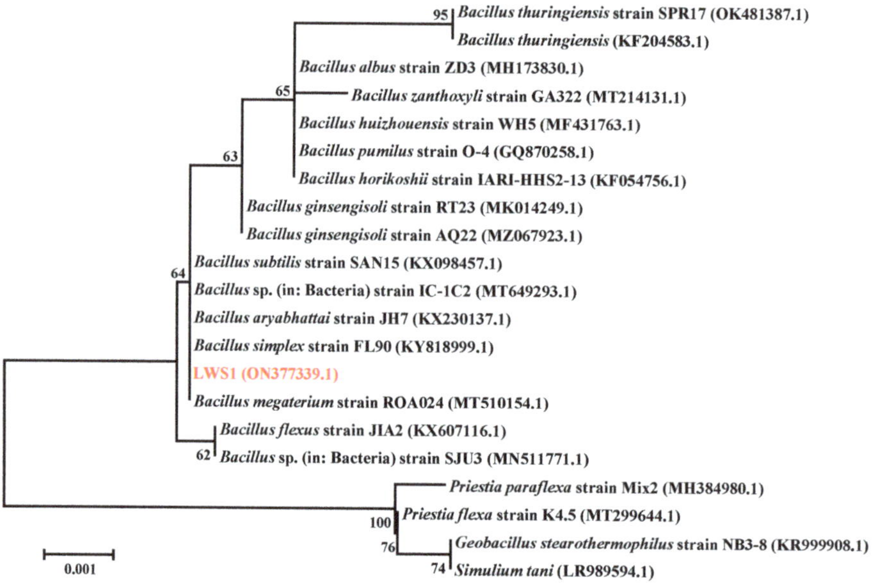

Figure 2. Neighbor-Joining phylogenetic tree of 16S rRNA sequence of strain LWS1 and reference sequences retrieved from NCBI database based on similarity analysis. The BLAST parameters included an expected threshold: 10, word size: 28, max matches in a query range: 0, match/mismatch scores: −1 and −2, gap costs: linear, filter: low-complexity regions and mask: lookup table only. The tree root was constructed with bootstrap values calculated from 1000 resamplings. The numbers at each node indicate the percentage of bootstrap supporting. The scale bar represents 10 substitutions per 100 bases. The sequence alignment and the phylogenic tree construction were performed by MEGA 6.0.

3.2. Strain LWS1 Showed Great Se Tolerance

To obtain an overview of the tolerance of *B. megaterium* to Se, strain LWS1 was exposed to different SeIV concentrations ranging from 0 to 7500 mg·L^{-1}. In general, strain LWS1 grew well in M9 medium spiked with ≤37.5 mg·L^{-1} Se, and the corresponding OD_{600} was between 1.938 and 2.215 (Figure 3A). The OD_{600} significantly decreased to 1.873 ($p < 0.05$) at 75 mg·L^{-1} Se (Figure 3A). When exposed to 375 mg·L^{-1} Se, the 57% growth in strain LWS1 was inhibited ($p < 0.01$; Figure 3A). In this study, the initial inoculating amount of strain LWS1 was recorded as OD_{600} = 0.01, and a considerable OD_{600} of 0.108 at a Se concentration of 7500 mg·L^{-1} suggested that strain LWS1 could tolerate a high concentration of Se (Figure 3A). A logistic fitting revealed a typical correlation between solution Se concentration and OD_{600} by a LC_{50} of 270.4 mg L^{-1} (R^2 = 0.9938, $p < 0.01$; Figure 3B).

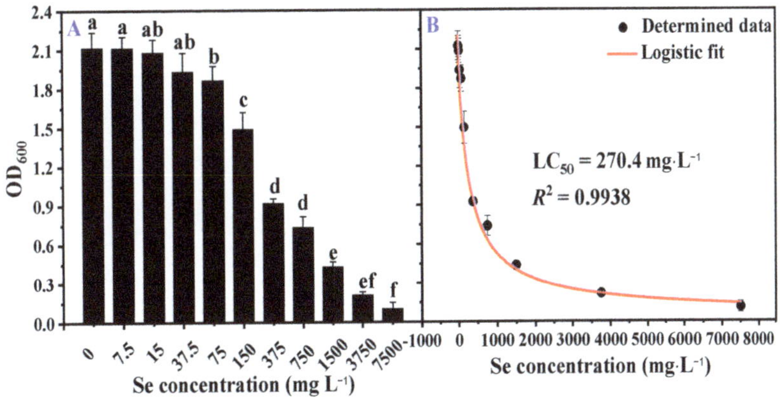

Figure 3. Se resistance of strain LWS1 grown in M9 medium containing SeIV for 24 h. Bacterial growth under different concentrations of Se from 0 to 7500 mg·L^{-1} (**A**) and logistic fitting to calculate the LC_{50} of Se to strain LWS1 (**B**). Different letters indicate significant differences between treatments based on two-way ANOVA by Tukey's multiple comparisons test at $p \leq 0.05$.

Microbes are key drivers influencing Se behaviors in the environment [46]. Bacterial tolerance to Se was widely reported in recent studies. *P. megaterium* (formerly *B. megaterium*) is a well-known bacteria, capable of great tolerance to Se. One possible process associated with Se tolerance in *P. megaterium* can be the reduction in SeIV to elemental Se [47], which can be supported by its potential use in the biosynthesis of Se nanoparticles [48,49]. Studies have also investigated the transformation of *P. megaterium* in the oxidation of elemental Se to SeIV [50]. However, there has been limited information regarding Se's inhibitory effect on *P. megaterium* and the response of *P. megaterium* to Se stress to date. Moreover, the reported strains of *P. megaterium* could tolerate both SeIV and selenate (SeVI), and the maximum inhibitory concentration (MIC) of SeIV was significantly lower than that observed in LWS1 [47,50]. Regarding other species in the genus *Bacillus*, *B. fusiformis* ARI 3, *Bacillus* sp. ARI 6 and *B. sphaericus* ARI 8 isolated from Se-contaminated soils had a MIC up to 450–600 μM for SeIV and 650–700 μM for SeVI [51]. However, several endobacteria, such as *Klebsiella* spp., *Bacillus* spp. and *Acinetobacter* sp., reported by Durán et al. [52] exhibited a surprising Se tolerance ability, with the MIC reaching 180 mM (SeIV). A similar finding was also found in Kuar et al. [53], where 80 mM Se only decreased the biomass of fungi *Fusarium equiseti* (SeF5) by about 30%. It is worth noting that, due to the lower toxicity of SeVI as compared to SeIV, microbes often show higher resistance to SeVI (Table 1). For example, *P. megaterium* CHP08, *Bacillus* sp. CHP07 and *Paenibacillus favisporus* CHP14 could tolerate a SeVI as high as 200–450 mM [54]. Our data suggested that strain LWS1 showed a great ability to tolerate Se stress like other reported Se-tolerant microbes.

Table 1. Se-tolerance ability and PGP characteristics of the representative microbes reported previously.

Microbes	Strain	Sources	Se Tolerance (MIC, mM)	IAA-Producing Ability (mg·L^{-1})	Siderophore-Producing Ability (λ/λ_0) [1]	Phosphate-Solubilizing Ability (mg·L^{-1})	Promoted Plants	Reference
Acinetobacter sp.	E6.2	Wheat stem	60	37.9	– [2]	–	Wheat	Durán et al. [52]
Priestia megaterium	CHP08	*Cardamine hupingshanensis*	50 (SeIV); 200 (SeVI)	11.5 ± 3.5	–	58.8 ± 24.1	*Brassica chinensis*	Li et al. [54]
Bacillus pichinotyi	YAM2	Sediment	–	12	n.m. [3]	–	Wheat (*Triticum aestivum* L.)	Yasin et al. [27]
Bacillus sp.	E6.1	Wheat stem	120	39.9	–	–	Wheat	Durán et al. [52]
Bacillus sp.	CHP07	*Cardamine hupingshanensis*	80 (SeIV); 450 (SeVI)	41.0 ± 11.9	–	–	*Brassica chinensis*	Li et al. [47]
Fusarium equiseti	SeF5	Soil	>80	64.28 ± 1.2	0.155	n.m.[c]	maize	Kaur et al. [53]
Klebsiella sp.	E2	Wheat stem	180	94.1	+	+	Wheat	Durán et al. [52]
Herbaspirillum camelliae	WT00C	Tea	n.m.	18.7	+	n.m.	Tea	Cheng et al. [43]
Paenibacillus favisporus	CHP14	*Cardamine hupingshanensis*	70 (SeIV); 400 (SeVI)	28.2 ± 4.5	+	27.6 ± 11.5	*Brassica chinensis*	Li et al. [47]
Pseudopestalotiopsis theae	SeF12	Soil	>80	52.89 ± 1.02	0.317	n.m.[c]	maize	Kaur et al. [53]
Priestia sp.	LWS1	Se tailing	>90	24.3 ± 1.37	0.23 ± 0.04	87.5 ± 0.21	Rice (*Oryza sativa* L.)	In this study

[1] 0.8–1.0, very weak; 0.6–0.8, weak; 0.4–0.6, moderate; 0.2–0.4, strong; 0–0.2, very strong. [2] not detected. [3] no mentioned.

3.3. Strain LWS1 Exhibited Typical PGP Characteristics

As previously noted, PGP characteristics are important in agricultural applications, especially in aiding plant growth and Se accumulation in crops [26,28,52]. As can be seen in Table 1, strain LWS1 was able to synthesize IAA and siderophore, and solubilize $Ca_3(PO_4)_2$ to sustain phosphorus acquisition. Specifically, strain LWS1 produced IAA by 24.3 ± 1.37 mg·L^{-1}, which is comparable to other reported bacteria (Table 1). In genus *Bacillus*, strain LWS1 showed a higher ability to produce IAA than strains CHP08 (11.5 ± 3.5 mg·L^{-1}) and YAM2 (12 mg·L^{-1}) [27,54]. It seems that microbial IAA production ability is associated with Se tolerance. For example, those microbes, such as *Bacillus* sp. E6.1, *F. equiseti* SeF5 and *Klebsiella* sp. E2, which tolerated Se up to 180 mM, also produced a higher IAA ranging from 39.9 to 94.1 mg·L^{-1} (Table 1). One explanation is that the production of IAA and the resistance to several heavy metals such as Se, tellurium and lead are probably relevant due to the associated genes often dispersing on the same plasmid [55]. Unlike IAA, the biosynthesis of siderophores is not a universal trait for Se-tolerant microbes (Table 1). To date, studies have found that bacterial siderophores can form stable complexes with lots of metals, including Fe, As, Al, Cd, Cu, Ga, In, Pb, and Zn, as well as the radionuclides including U and Np [20,56]. The siderophore-mediated Al uptake by *P. megaterium* ATCC 19213 was well-documented by Hu and Boyer [57]. Regarding *Priestia* sp. Strain LWS1, the siderophore-producing ability, present as λ/λ_0, was 0.23 ± 0.04 (Table 1). This indicated that strain LWS1 had a lower siderophore-producing ability than Se-tolerant fungi *F. equiseti* SeF5 by λ/λ_0 = 0.155, but this ability was higher than that observed in *Pseudopestaloti opsistheae* by 0.317 (Table 1). As noted by the criteria for siderophore-producing ability, strain LWS1 showed a strong ability to produce siderophores (Table 1).

It was interesting to note that strain LWS1 harbored the most efficient phosphate-solubilizing ability, i.e., 87.5 ± 0.21 mg·L^{-1}, in the representative microbial strains with a given Se tolerance (Table 1). The phosphate-solubilizing ability of *Priestia* sp. Strain LWS1 was higher than that of the *P. megaterium* strain HCP08 by 58.8 ± 24.1 mg·L^{-1} and *Paenibacillus favisporus* CHP14 by 27.6 ± 11.5 mg·L^{-1} isolated from the Se-hyperaccumulating plant *Cardamine hupingshanensis* [54]. In conclusion, strain LWS1 is a typical PGPB with IAA- and siderophore-production abilities and phosphate-solubilization abilities, which can be used as a potential enhancer to regulate plant growth.

3.4. Strain LWS1 Enhanced Plant Growth and Se Accumulation in Rice

After 4 months of growth, the dry biomass of different tissues of rice plant was recorded. In control treatments without bacterial inoculation, the root, husk, leaf and grain biomass of rice grown in the soils spiked with different Se concentrations ranged from 2.73 ± 0.45 g to 3.42 ± 0.30 g (Figure 4A), from 4.35 ± 0.43 g to 5.93 ± 0.59 g (Figure 4B), from 2.39 ± 0.18 g to 2.88 ± 0.25 g (Figure 4C) and from 4.16 ± 0.09 g to 4.92 ± 0.21 g (Figure 4D), respectively. However, the plant biomass in each treatment did not follow a typical trend. This can also be observed by the result for the total dry biomass of rice plant (Figure 4E). Our data were different from Dai et al. [16] in that the plant height, dry biomass, 1000-grain weight and rice yield all positively correlated to soil Se concentrations ranging from 0 to 5 mg·kg^{-1}. In fact, the role of Se in plant growth remains controversial and whether Se is essential for plants is yet to be established [5]. For the treatments inoculated with strain LWS1, however, the root, husk, leaf and total biomass of rice plants grown in Se-spiked soils obtained a larger biomass than the control (Figure 4). The highest difference could be observed for root biomass of 40% in group $Se_{0.425}$, for husk biomass of 12% in group $Se_{0.175}$, for leaf biomass of 41% in group $Se_{0.125}$ and for total biomass of 6% in group $Se_{0.175}$ (Figure 4). By comparing the two treatments with or without bacterial inoculation, the presence of strain LWS1 exhibited typical enhancements of plant growth. The significant increase in the root, husk, leaf, grain and total biomass of rice plant reached up to 47% ($Se_{0.425}$ in Figure 4A; $p < 0.01$), 51% ($Se_{0.175}$ in Figure 4B; $p < 0.01$), 63% ($Se_{0.125}$ in Figure 4C; $p < 0.01$), 17% (Se_6 in Figure 4D; $p < 0.05$) and 19% ($Se_{0.175}$ in Figure 4E; $p < 0.05$), respectively. It was apparent that both bacteria and Se contributed to rice growth, but Se-enhanced plant growth was dependent on the presence of strain LWS1.

To verify whether plant growth was associated with Se, we also tested Se concentration in different tissues of rice. As shown in Figure 5, unlike plant biomass, Se concentration in all tissues of rice without PGPB inoculation dramatically increased with the increase in soil Se. The root was the tissue accumulating most of the assimilated Se (Figure 5A), followed by husk (Figure 5B), with the lowest Se concentration observed in the leaf (Figure 5C) and grain (Figure 5D) of rice. The results were in accordance with previous research [16]. Similar to uninoculated treatments, the plant Se concentration in inoculated treatments also increased with the increase in soil Se, as did the Se distribution pattern in different rice tissues (Figure 5). The comparisons between uninoculated and inoculated treatments showed that bacterial inoculation increased plant Se by 12–106% in the root, 2.5–65% in the husk, 17–47% in the leaf and 16–47% in the grain (Figure 5). However, bacterial inoculation only significantly increased the root Se (1.92 ± 0.19 vs. 2.35 ± 0.24; $p < 0.01$) of rice when grown in the soils spiked with 6 mg·kg^{-1} Se (Figure 5A). Our data suggested that plant Se concentration was not correlated with plant biomass (Figures 4 and 5), suggesting the leading role of PGPB in promoting plant growth. In other words, the increase of Se accumulation in rice might be an indirect result of LWS1-mediated plant growth.

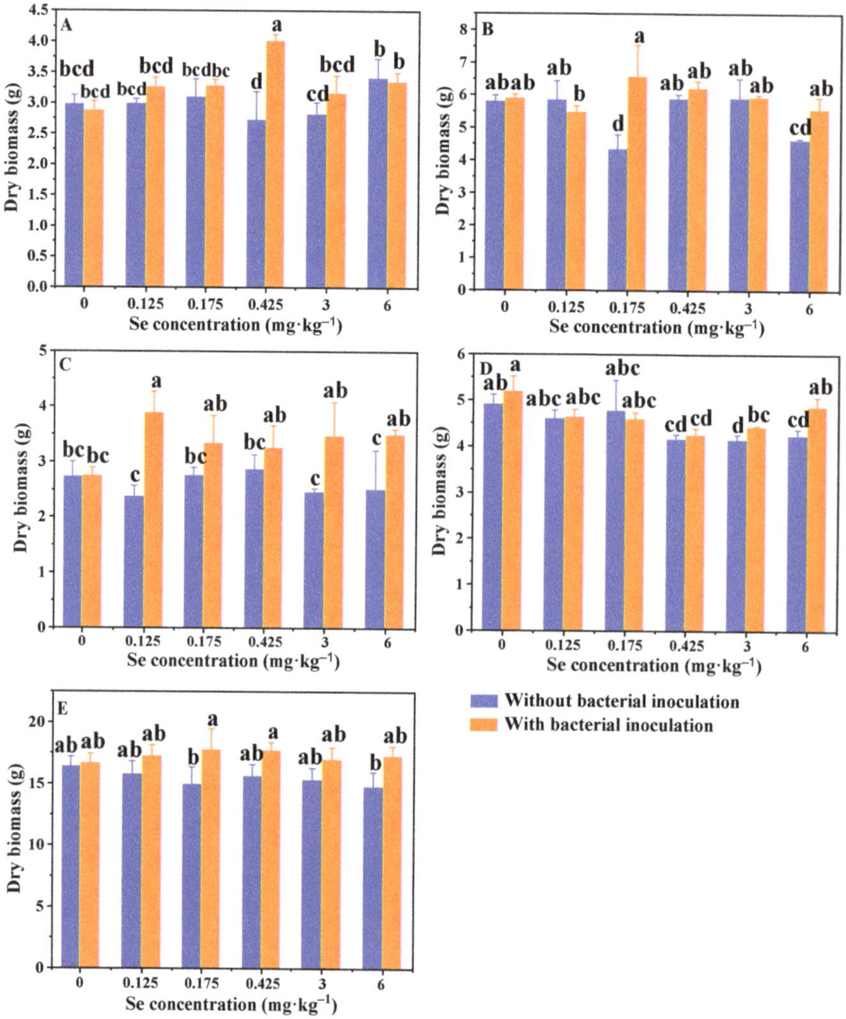

Figure 4. Dry weight of the root biomass (**A**), husk biomass (**B**), leaf biomass (**C**), grain biomass (**D**) and total biomass (**E**) of rice plant with or without the inoculation of strain LWS1 (n = 4). The plants were grown in soils spiked with 0–6 mg·kg^{-1} SeIV. Different letters indicate significant differences between treatments based on two-way ANOVA by Tukey's multiple comparisons test at $p \leq 0.05$.

In recent decades, several studies have focused on the use of *P. megaterium* to promote plant growth and remediation of heavy-metals-contaminated media, which is probably associated with its PGP characteristics and heavy-metal tolerance [58–60]. For example, a study from Rajkumar et al. [24] identified *P. megaterium* SR28C as a PGPB strain involved in the plant growth of *Brassica juncea*, *Luffa cylindrica* and *Sorghum halepense* and their tolerance to Ni stress. Similar to strain SR28C, endophytic *P. megaterium* BM18-2 has also shown dual roles in regulating Cd resistance and promoting plant growth [61]. Although *P. megaterium* is a well-known beneficial bacterium that has been industrially employed for more than 65 years [62], its application in the biofortification of plant growth and Se accumulation in rice is still limitedly documented. Our data suggested that PGPB and Se might play a synergistic role in enhancing plant growth and the involvement of beneficial PGPB was helpful for Se biofortification in rice [26]. However, bacterial inoculation had not significantly elevated Se accumulation in rice (Figure 5), indicating that other factors in the complex system (i.e., soil-microbe-plant system) also contributed to Se behaviors in the rhizosphere of rice and its translocation from the soils to the plants. As such, more details on the roles of *Priestia* sp. LWS1 in promoting plant growth and Se accumulation in rice still warrant future investigations.

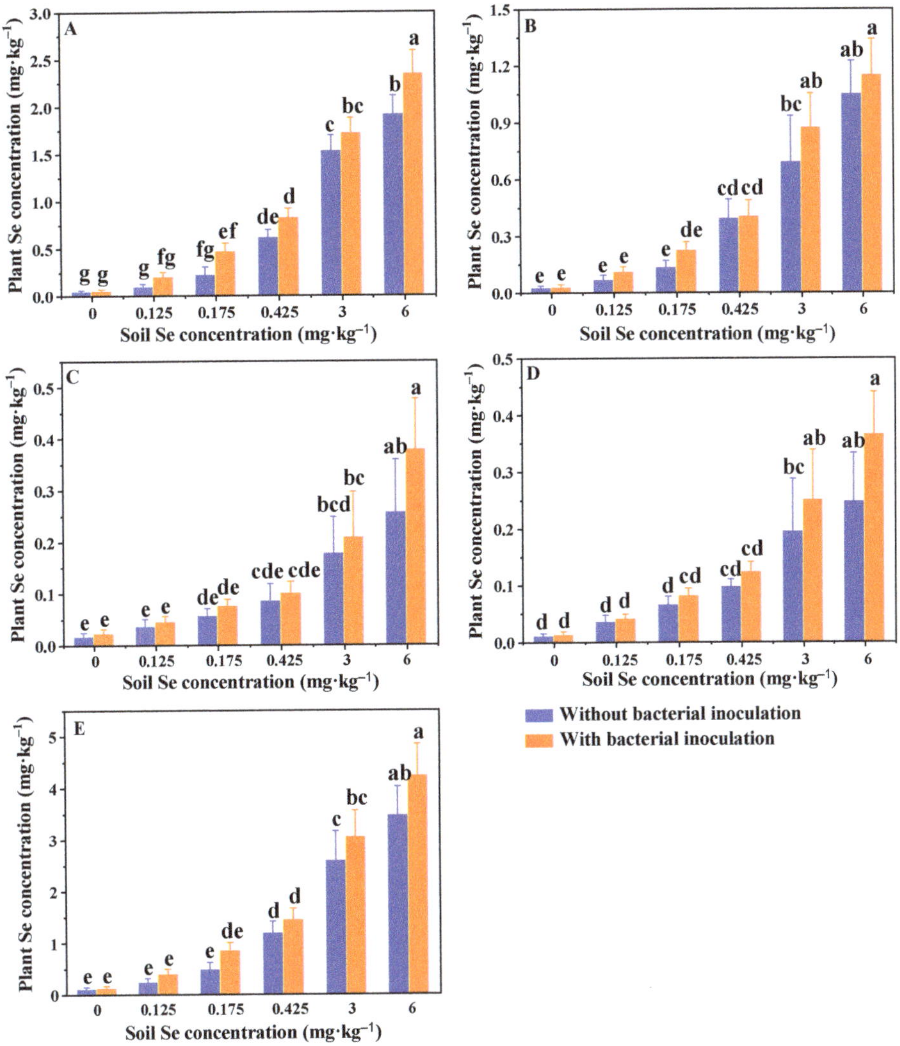

Figure 5. Se concentrations in the root biomass (**A**), husk biomass (**B**), leaf biomass (**C**), grain biomass (**D**) and total biomass (**E**) of rice plants with or without the inoculation of strain LWS1 ($n = 4$). The plants were grown in the soils spiked with 0–6 mg·kg^{-1} SeIV. Different letters indicate significant differences between treatments based on two-way ANOVA by Tukey's multiple comparisons test at $p \leq 0.05$.

4. Conclusions

In this study, a typical PGPB strain, namely *Priestia* sp. LWS1 was isolated from a Se tailing soil. Strain LWS1 could produce IAA by 24.3 ± 1.37 mg·L^{-1}, produce siderophore by 0.23 ± 0.04, solubilize phosphate by 87.5 ± 0.21 mg·L^{-1} and tolerate SeIV by up to 90 mM. The inoculation of strain LWS1 lead to the significant promotion of plant growth, but strain LWS1 might only one of the regulators in enhancing Se accumulation in rice. Our data shed light onto the potential use of *Priestia* sp. in Se biofortification in crops and improving Se-deficiency in human beings.

Author Contributions: Conceptualization, Y.-H.H. and S.-S.W.; Data curation, X.-R.L., H.-B.C., J.-B.L., Y.-H.H. and S.-S.W.; Funding acquisition, Y.-H.H.; Investigation, H.-B.C., Y.-H.H. and S.-S.W.; Methodology, X.-R.L., Y.-X.L., H.-B.C. and Y.-H.H.; Supervision, Y.-H.H. and S.-S.W.; Writing—original draft, X.-R.L., Y.-H.H. and S.-S.W.; Writing—review and editing, X.-R.L., H.-B.C., Y.-X.L., Z.-H.Z., J.-B.L., Y.-Q.W., H.Z., Y.Z., Y.-H.H. and S.-S.W. All authors have read and agreed to the published version of the manuscript.

Funding: This work was supported by the National Natural Science Foundation of China (41807110), the National Natural Science Foundation of Fujian Province (2021J01196), the Education Department Fund of Fujian Province (JAT170144), Undergraduate Education and Teaching Reform Research Project of Fujian Normal University and the Research Start-up Fund of Fujian Normal University (Y0720304x13).

Institutional Review Board Statement: Not applicable.

Informed Consent Statement: Not applicable.

Data Availability Statement: The data presented in this study are available upon request from the corresponding author.

Conflicts of Interest: The authors declare no conflict of interest.

References

1. Hartikainen, H. Biogeochemistry of selenium and its impact on food chain quality and human health. *J. Trace Elem. Med. Biol.* **2005**, *18*, 309–318. [CrossRef] [PubMed]
2. Lemly, A.D. Environmental implications of excessive selenium: A review. *Biomed. Environ. Sci.* **1997**, *10*, 415–435. [PubMed]
3. Gupta, M.; Gupta, S. An overview of selenium uptake, metabolism, and toxicity in plants. *Front. Plant Sci.* **2017**, *7*, 2074. [CrossRef] [PubMed]
4. Mistry, H.D.; Pipkin, F.B.; Redman, C.W.G.; Poston, L. Selenium in reproductive health. *Am. J. Obstet. Gynecol.* **2012**, *206*, 21–30. [CrossRef]
5. Sarwar, N.; Akhtar, M.; Kamran, M.A.; Imran, M.; Riaz, M.A.; Kamran, K.; Hussain, S. Selenium biofortification in food crops: Key mechanisms and future perspectives. *J. Food Compos. Anal.* **2020**, *93*, 103615. [CrossRef]
6. Ullah, H.; Liu, G.; Yousaf, B.; Ali, M.U.; Irshad, S.; Abbas, Q.; Ahmad, R. A comprehensive review on environmental transformation of selenium: Recent advances and research perspectives. *Environ. Geochem. Health* **2019**, *41*, 1003–1035. [CrossRef]
7. Vleet, J.F.V.; Ferrans, V.J. Etiologic factors and pathologic alterations in selenium-vitamin E deficiency and excess in animals and humans. *Biol. Trace Elem. Res.* **1992**, *33*, 1–21. [CrossRef]
8. Zhang, H.; Feng, X.; Chan, H.M.; Larssen, T. New insights into traditional health risk assessments of mercury exposure: Implications of selenium. *Environ. Sci. Technol.* **2014**, *48*, 1206–1212. [CrossRef]
9. Kipp, A.P.; Strohm, D.; Brigelius-Flohé, R.; Schomburg, L.; Bechthold, A.; Leschik-Bonnet, E.; Heseker, H.; German Nutrition Society (DGE). Revised reference values for selenium intake. *J. Trace Elem. Med. Bio.* **2015**, *32*, 195–199. [CrossRef]
10. Rayman, M.P. Selenium and human health. *Lancet* **2012**, *379*, 1256–1268. [CrossRef]
11. Chen, Y.; Han, Y.-H.; Cao, Y.; Zhu, Y.-G.; Rathinasabapathi, B.; Ma, L.Q. Arsenic transport in rice and biological solutions to reduce arsenic risk from rice. *Front. Plant Sci.* **2017**, *8*, 268. [CrossRef] [PubMed]
12. Williams, P.N.; Lombi, E.; Sun, G.-X.; Scheckel, K.; Zhu, Y.-G.; Feng, X.; Zhu, J.; Carey, A.-M.; Adomako, E.; Lawgali, Y.; et al. Selenium characterization in the global rice supply chain. *Environ. Sci. Technol.* **2009**, *43*, 6024–6030. [CrossRef] [PubMed]
13. Zhang, L.; Shi, W.; Wang, X. Difference in selenite absorption between high-and low-selenium rice cultivars and its mechanism. *Plant Soil* **2006**, *282*, 183–193. [CrossRef]
14. Zhu, Y.-G.; Pilon-Smits, E.A.H.; Zhao, F.-J.; Williams, P.N.; Meharg, A.A. Selenium in higher plants: Understanding mechanisms for biofortification and phytoremediation. *Trends Plant Sci.* **2009**, *14*, 436–442. [CrossRef] [PubMed]
15. Schiavon, M.; Nardi, S.; Vecchia, F.d.; Ertani, A. Selenium biofortification in the 21st century: Status and challenges for healthy human nutrition. *Plant Soil* **2020**, *453*, 245–270. [CrossRef]

16. Dai, Z.; Imtiaz, M.; Rizwan, M.; Yuan, Y.; Huang, H.; Tu, S. Dynamics of Selenium uptake, speciation, and antioxidant response in rice at different panicle initiation stages. *Sci. Total Environ.* **2019**, *691*, 827–834. [CrossRef]
17. Versini, A.; Di Tullo, P.; Aubry, E.; Bueno, M.; Thiry, Y.; Pannier, F.; Castrec-Rouelle, M. Influence of Se concentrations and species in hydroponic cultures on Se uptake, translocation and assimilation in non-accumulator ryegrass. *Plant Physiol. Bioch.* **2016**, *108*, 372–380. [CrossRef]
18. Eze, M.O.; Thiel, V.; Hose, G.C.; George, S.C.; Daniel, R. Enhancing rhizoremediation of petroleum hydrocarbons through bioaugmentation with a plant growth-promoting bacterial consortium. *Chemosphere* **2022**, *289*, 133143. [CrossRef]
19. Goswami, D.; Thakker, J.N.; Dhandhukia, P.C. Portraying mechanics of plant growth promoting rhizobacteria (PGPR): A review. *Cogent. Food Agric.* **2016**, *2*, 1127500. [CrossRef]
20. Rajkumar, M.; Ae, N.; Prasad, M.N.V.; Freitas, H. Potential of siderophore-producing bacteria for improving heavy metal phytoextraction. *Trends Biotechnol.* **2010**, *28*, 142–149. [CrossRef]
21. Grandlic, C.J.; Mendez, M.O.; Chorover, J.; Machado, B.; Maier, R.M. Plant growth-promoting bacteria for phytostabilization of mine tailings. *Environ. Sci. Technol.* **2008**, *42*, 2079–2084. [CrossRef] [PubMed]
22. Khan, S.; Afzal, M.; Iqbal, S.; Khan, Q.M. Plant–bacteria partnerships for the remediation of hydrocarbon contaminated soils. *Chemosphere* **2013**, *90*, 1317–1332. [CrossRef] [PubMed]
23. Jiang, C.-y.; Sheng, X.-f.; Qian, M.; Wang, Q.-y. Isolation and characterization of a heavy metal-resistant *Burkholderia* sp. from heavy metal-contaminated paddy field soil and its potential in promoting plant growth and heavy metal accumulation in metal-polluted soil. *Chemosphere* **2008**, *72*, 157–164. [CrossRef] [PubMed]
24. Rajkumar, M.; Ma, Y.; Freitas, H. Improvement of Ni phytostabilization by inoculation of Ni resistant *Bacillus megaterium* SR28C. *J. Environ. Manag.* **2013**, *128*, 973–980. [CrossRef] [PubMed]
25. Yang, D.; Hu, C.; Wang, X.; Shi, G.; Li, Y.; Fei, Y.; Song, Y.; Zhao, X. Microbes: A potential tool for selenium biofortification. *Metallomics* **2021**, *13*, mfab054. [CrossRef]
26. Ye, Y.; Qu, J.; Pu, Y.; Rao, S.; Xu, F.; Wu, C. Selenium biofortification of crop food by beneficial microorganisms. *J. Fungi* **2020**, *6*, 59. [CrossRef]
27. Yasin, M.; El-Mehdawi, A.F.; Anwar, A.; Pilon-Smits, E.A.H.; Faisal, M. Microbial-enhanced selenium and iron biofortification of wheat (*Triticum aestivum* L.)-Applications in phytoremediation and biofortification. *Int. J. Phytoremediat.* **2015**, *17*, 341–347. [CrossRef]
28. Durán, P.; Acuña, J.J.; Armada, E.; López-Castillo, O.M.; Cornejo, P.; Mora, M.L.; Azcón, R. Inoculation with selenobacteria and arbuscular mycorrhizal fungi to enhance selenium content in lettuce plants and improve tolerance against drought stress. *J. Soil Sci. Plant Nutr.* **2016**, *16*, 201–225. [CrossRef]
29. White, P.J. Selenium metabolism in plants. *BBA Gen. Subj.* **2018**, *1862*, 2333–2342. [CrossRef]
30. Han, Y.-H.; Yin, D.-X.; Jia, M.-R.; Wang, S.-S.; Chen, Y.; Rathinasabapathi, B.; Chen, D.-L.; Ma, L.Q. Arsenic-resistance mechanisms in bacterium *Leclercia adecarboxylata* strain As3-1: Biochemical and genomic analyses. *Sci. Total Environ.* **2019**, *690*, 1178–1189. [CrossRef]
31. Wang, S.-S.; Ye, S.-L.; Han, Y.-H.; Shi, X.-X.; Chen, D.-L.; Li, M. Biosorption and bioaccumulation of chromate from aqueous solution by a newly isolated *Bacillus mycoides* strain 200AsB1. *RSC Adv.* **2016**, *6*, 101153–101161. [CrossRef]
32. Jouanneau, Y.; Meyer, C.; Duraffourg, N. Dihydroxylation of four- and five-ring aromatic hydrocarbons by the naphthalene dioxygenase from *Sphingomonas* CHY-1. *Appl. Microbiol. Biotechnol.* **2016**, *100*, 1253–1263. [CrossRef] [PubMed]
33. Xu, J.-Y.; Han, Y.-H.; Chen, Y.; Zhu, L.-J.; Ma, L.Q. Arsenic transformation and plant growth promotion characteristics of As-resistant endophytic bacteria from As-hyperaccumulator *Pteris vittata*. *Chemosphere* **2016**, *144*, 1233–1240. [CrossRef] [PubMed]
34. Gordon, S.A.; Weber, R.P. Colorimetric estimation of indoleacetic acid. *Plant Physiol.* **1951**, *26*, 192–195. [CrossRef]
35. Schwyn, B.; Neilands, J. Universal chemical assay for the detection and determination of siderophores. *Anal. Biochem.* **1987**, *160*, 47–56. [CrossRef]
36. Nautiyal, C.S. An efficient microbiological growth medium for screening phosphate solubilizing microorganisms. *FEMS Microbiol. Lett.* **1999**, *170*, 265–270. [CrossRef]
37. Watanabe, F.S.; Olsen, S.R. Test of an ascorbic acid method for determining phosphorus in water and $NaHCO_3$ extracts from soil. *Soil Sci. Soc. Am. J.* **1965**, *29*, 677–678. [CrossRef]
38. Eze, M.O.; George, S.C.; Hose, G.C. Dose-response analysis of diesel fuel phytotoxicity on selected plant species. *Chemosphere* **2021**, *263*, 128382. [CrossRef]
39. Eze, M.O.; Hose, G.C.; George, S.C. Assessing the effect of diesel fuel on the seed viability and germination of *Medicago sativa* using the event-time model. *Plants* **2020**, *9*, 1062. [CrossRef]
40. Liang, S.; Guan, D.-X.; Ren, J.-H.; Zhang, M.; Luo, J.; Ma, L.Q. Effect of aging on arsenic and lead fractionation and availability in soils: Coupling sequential extractions with diffusive gradients in thin-films technique. *J. Hazard. Mater.* **2014**, *273*, 272–279. [CrossRef]
41. Xu, J.-Y.; Li, H.-B.; Liang, S.; Luo, J.; Ma, L.Q. Arsenic enhanced plant growth and altered rhizosphere characteristics of hyperaccumulator *Pteris vittata*. *Environ. Pollut.* **2014**, *194*, 105–111. [CrossRef] [PubMed]
42. Aleixo, P.C.; Nóbrega, J.A. Direct determination of iron and selenium in bovine milk by graphite furnace atomic absorption spectrometry. *Food Chem.* **2003**, *83*, 457–462. [CrossRef]

43. Bao, M.-T.; Wang, L.-N.; Sun, P.-Y.; Cao, L.-X.; Zou, J.; Li, Y.-M. Biodegradation of crude oil using an efficient microbial consortium in a simulated marine environment. *Mar. Pollut. Bull.* **2012**, *64*, 1177–1185. [CrossRef] [PubMed]
44. Zhou, J.; Ma, Q.; Yi, H.; Wang, L.; Song, H.; Yuan, Y.-J. Metabolome profiling reveals metabolic cooperation between *Bacillus megaterium* and *Ketogulonicigenium vulgare* during induced swarm motility. *Appl. Environ. Microbiol.* **2011**, *77*, 7023–7030. [CrossRef] [PubMed]
45. Hossain, T.J.; Das, M.; Ali, F.; Chowdhury, S.I.; Zedny, S.A. Substrate preferences, phylogenetic and biochemical properties of proteolytic bacteria present in the digestive tract of Nile tilapia (*Oreochromis niloticus*). *AIMS Microbiol.* **2021**, *7*, 528–545. [CrossRef] [PubMed]
46. Masscheleyn, P.H.; Patrick, W.H., Jr. Biogeochemical processes affecting selenium cycling in wetlands. *Environ. Toxicol. Chem.* **1993**, *12*, 2235–2243. [CrossRef]
47. Mishra, R.R.; Prajapati, u.; Das, J.; Dangar, T.K.; Das, N.; Thatoi, H. Reduction of selenite to red elemental selenium by moderately halotolerant *Bacillus megaterium* strains isolated from Bhitarkanika mangrove soil and characterization of reduced product. *Chemosphere* **2011**, *84*, 1231–1237. [CrossRef]
48. Hashem, A.H.; Abdelaziz, A.M.; Askar, A.A.; Fouda, H.M.; Khalil, A.M.A.; Abd-Elsalam, K.A.; Khaleil, M.M. *Bacillus megaterium*-mediated synthesis of selenium nanoparticles and their antifungal activity against *Rhizoctonia solani* in faba bean plants. *J. Fungi* **2021**, *7*, 195. [CrossRef]
49. Ramamurthy, C.; Sampath, K.S.; Arunkumar, P.; Kumar, M.S.; Sujatha, V.; Premkumar, K.; Thirunavukkarasu, C. Green synthesis and characterization of selenium nanoparticles and its augmented cytotoxicity with doxorubicin on cancer cells. *Bioproc. Biosyst. Eng.* **2013**, *36*, 1131–1139. [CrossRef]
50. Sarathchandra, S.U.; Watkinson, J.H. Oxidation of elemental selenium to selenite by *Bacillus megaterium*. *Science* **1981**, *211*, 600–601. [CrossRef]
51. Ghosh, A.; Mohod, A.M.; Paknikar, K.M.; Jain, R.K. Isolation and characterization of selenite- and selenate-tolerant microorganisms from selenium-contaminated sites. *World J. Microbiol. Biotechnol.* **2008**, *24*, 1607–1611. [CrossRef]
52. Durán, P.; Acuña, J.J.; Jorquera, M.A.; Azcón, R.; Paredes, C.; Rengel, Z.; de la Luz Mora, M. Endophytic bacteria from selenium-supplemented wheat plants could be useful for plant-growth promotion, biofortification and *Gaeumannomyces graminis* biocontrol in wheat production. *Biol. Fert. Soils* **2014**, *50*, 983–990. [CrossRef]
53. Kaur, T.; Vashisht, A.; Prakash, N.T.; Reddy, M.S. Role of selenium-tolerant fungi on plant growth promotion and selenium accumulation of maize plants grown in seleniferous soils. *Water Air Soil Pollut.* **2022**, *233*, 17. [CrossRef]
54. Li, Q.; Zhou, S.; Liu, N. Diversity of endophytic bacteria in *Cardamine hupingshanensis* and potential of culturable selenium-resistant endophytes to enhance seed germination under selenate stress. *Curr. Microbiol.* **2021**, *78*, 2091–2103. [CrossRef] [PubMed]
55. Huddedar, S.B.; Shete, A.M.; Tilekar, J.N.; Gore, S.D.; Dhavale, D.D.; Chopade, B.A. Isolation, characterization, and plasmid pUPI126-mediated indole-3-acetic acid production in *Acinetobacter* strains from rhizosphere of wheat. *Appl. Biochem. Biotechnol.* **2002**, *102*, 21–39. [CrossRef]
56. Ghosh, P.; Rathinasabapathi, B.; Teplitski, M.; Ma, L.Q. Bacterial ability in AsIII oxidation and AsV reduction: Relation to arsenic tolerance, P uptake, and siderophore production. *Chemosphere* **2015**, *138*, 995–1000. [CrossRef]
57. Hu, X.; Boyer, G.L. Siderophore-mediated Al uptake by *Bacillus megaterium* ATCC 19213. *Appl. Environ. Microbiol.* **1996**, *62*, 4044–4048. [CrossRef]
58. Njokua, K.L.; Akinyede, O.R.; Obidi, O.F. Microbial remediation of heavy metals contaminated media by *Bacillus megaterium* and *Rhizopus stolonifer*. *Sci. Afr.* **2020**, *10*, e00545. [CrossRef]
59. Zou, C.; Li, Z.; Yu, D. *Bacillus megaterium* strain XTBG34 promotes plant growth by producing 2-pentylfuran. *J. Microbiol.* **2010**, *48*, 460–466. [CrossRef]
60. Chakraborty, U.; Chakraborty, B.; Basnet, M. Plant growth promotion and induction of resistance in *Camellia sinensis* by *Bacillus megaterium*. *J. Basic Microbiol.* **2006**, *46*, 186–195. [CrossRef]
61. Wu, J.; Kamal, N.; Hao, H.; Qian, C.; Liu, Z.; Shao, Y.; Zhong, X.; Xu, B. Endophytic *Bacillus megaterium* BM18-2 mutated for cadmium accumulation and improving plant growth in Hybrid *Pennisetum*. *Biotechnol. Rep.* **2019**, *24*, e00374. [CrossRef] [PubMed]
62. Vary, P.S.; Biedendieck, R.; Fuerch, T.; Meinhardt, F.; Rohde, M.; Deckwer, W.-D.; Jahn, D. *Bacillus megaterium*—from simple soil bacterium to industrial protein production host. *Appl. Microbiol. Biotechnol.* **2007**, *76*, 957–967. [CrossRef] [PubMed]

agronomy

MDPI

Article

Regulation of Phosphorus Supply on Nodulation and Nitrogen Fixation in Soybean Plants with Dual-Root Systems

Hongyu Li, Xiangxiang Wang, Quanxi Liang, Xiaochen Lyu, Sha Li, Zhenping Gong, Shoukun Dong, Chao Yan and Chunmei Ma *

College of Agriculture, Northeast Agricultural University, Harbin 150030, China; lihongyu89103@126.com (H.L.); wxx15991204390@163.com (X.W.); l18730852087@163.com (Q.L.); xiaochenlyu@163.com (X.L.); risa_shasha@163.com (S.L.); gzpyx2004@163.com (Z.G.); shoukundong@163.com (S.D.); yanchao504@neau.edu.cn (C.Y.)
* Correspondence: chunmm@neau.edu.cn

Abstract: Phosphorus (P) is an important nutrient affecting nodulation and nitrogen fixation in soybeans. To further investigate the relationship of phosphorus with soybean nodulation and nitrogen fixation, the seedling grafting technique was applied in this study to prepare dual-root soybean systems for a sand culture experiment. From the unfolded cotyledon stage to the initial flowering stage, one side of each dual-root soybean system was irrigated with nutrient solution containing 1 mg/L, 31 mg/L, or 61 mg/L of phosphorus (phosphorus-application side), and the other side was irrigated with a phosphorus-free nutrient solution (phosphorus-free side), to study the effect of local phosphorus supply on nodulation and nitrogen fixation in soybean. The results are described as follows: (1) Increasing the phosphorus supply increased the nodules weight, nitrogenase activity, ureide content, number of bacteroids, number of infected cells, and relative expression levels of nodule nitrogen fixation key genes (*GmEXPB2*, *GmSPX5*, *nifH*, *nifD*, *nifK*, *GmALN1*, *GmACP1*, *GmUR5*, *GmPUR5*, and *GmHIUH5*) in root nodules on the phosphorus-application side. Although the phosphorus-application and phosphorus-free sides demonstrated similar changing trends, the phosphorus-induced increases were more prominent on the phosphorus-application side, which indicated that phosphorus supply systematically regulates nodulation and nitrogen fixation in soybean. (2) When the level of phosphorus supply was increased from 1 mg/L to 31 mg/L, the increase on the P– side root was significant, and nodule phosphorus content increased by 57.14–85.71% and 68.75–75.00%, respectively; ARA and SNA were 218.64–383.33% and 11.41–16.11%, respectively, while ureide content was 118.18–156.44%. When the level of phosphorus supply was increased from 31mg/L to 61mg/L, the increase in the regulation ability of root and nodule phosphorus content, ARA, SNA, and ureide content were low for roots, and the value for nodules was lower than when the phosphorus level increased from 1 mg/L to 31 mg/L. (3) A high-concentration phosphorus supply on one side of a dual-root soybean plant significantly increased the phosphorus content in the aboveground tissues, as well as the roots and nodules on both sides. In the roots on the phosphorus-free side, the nodules were prioritized for receiving the phosphorus transported from the aboveground tissues to maintain their phosphorus content and functionality.

Keywords: dual-root soybean; phosphorus; nodulation; nitrogen fixation; ultrastructure

Citation: Li, H.; Wang, X.; Liang, Q.; Lyu, X.; Li, S.; Gong, Z.; Dong, S.; Yan, C.; Ma, C. Regulation of Phosphorus Supply on Nodulation and Nitrogen Fixation in Soybean Plants with Dual-Root Systems. *Agronomy* **2021**, *11*, 2354. https://doi.org/10.3390/agronomy11112354

Academic Editor: Antonio Lupini

Received: 8 October 2021
Accepted: 17 November 2021
Published: 20 November 2021

1. Introduction

Phosphorus (P) is one of the three essential elements for plants and is the second most limiting element for plant growth. Currently, 40% of the world's cropland lacks phosphorus, resulting in lower yields [1]. Phosphorus is a very important nutritional element that affects soybean growth and nodule nitrogen fixation. Proper application of phosphorus fertilizer can regulate root nodules growth, nitrogenase activity, and metabolic pathways, as well as enhance the capacity of nitrogen-fixing root nodules -promoting nitrogen by phosphorus [2–9].

The effect of phosphorus on nitrogen fixation mainly includes soybean plant growth [4,10], nodule formation [6,9], and nodule metabolism [3]. The promotion of phosphorus in nitrogen fixation is achieved by stimulating the growth of the host plant rather than by promoting the growth of rhizobia or the formation and function of nodules [11–14]. Additionally, the optimal phosphorus environment for host plant growth and the phosphorus requirements for symbiotic nitrogen fixation are determined by the development and function of the root nodules [15]. Current studies on phosphorus metabolism in nodules of leguminous crops are mainly focused on the analysis of compounds that affect nodule nitrogen fixation and indirectly explain the mechanism of phosphorus inhibition of nodule nitrogen fixation [3]. Several studies have suggested that low phosphorus stress inhibits nitrogen-fixing enzyme activity in legume nodules resulting from reduced nodule ATP energy [16], leghemoglobin content [13], Fe ion content [17], and excessive secretion of organic acids [18]. Other studies found that legume crops can adapt to low phosphorus stress by increasing root nodule phytase and phosphatase activities [19,20].

To better understand the phosphorus regulation of growth and phosphorus uptake in leguminous plants, split-root, and dual-root experiments were performed. In Shu et al. [21], the growth and P uptake of the root system significantly increased on the phosphorus-application side and remained stable on the phosphorus-free side after P ($Ca_{10}(PO_4)6(OH)_2$, $FePO_4$, $C_6H_6O_{24}P_6Na_{12}$, or KH_2PO_4 was supplied to one side of the root system of *Lupinusalbus*. In Scott and Robson [22], supplying KH_2PO_4 to one side of the root system of subterranean clover significantly promoted P uptake by roots on the phosphorus-application side but not by roots on the phosphorus-free side and the shoots. However, Snapp and Lynch [23] found that a high-concentration of $NH_4H_2PO_4$ supply to the roots on one side of *Phaseolus vulgaris* L. promoted P uptake on both the phosphorus-application side and phosphorus-free side.

In existing studies, the uneven distribution of phosphorus fertilizer in the soil due to the application method results in a wide variation in the fertilizer available to the root system, with some roots receiving phosphorus fertilizer to meet their needs, while others do not. The existing studies on the regulatory effect of P supply on nodulation and nitrogen fixation in soybeans have focused mainly on its local effect under single-root conditions through direct contact and seldom on its systemic effect. Most of the research has mainly focused on the study of phosphorus in nodule-regulating substances but less on the mode of regulation. Split-root and dual-root experiments are effective methods to study the mode of regulation. In the above-described method of rooting, the root system of the legume crop plant was divided into two parts so that neither subsystem remained intact, which may affect the accuracy of the experimental results. In this study, a grafting approach of soybean seedlings [24] was used to prepare a dual-root system in soybean plants. Soybean seedlings were grafted to obtain a double-rooted soybean plant with complete roots on both sides, and the test results were more accurate. The dual-root system of soybean plants received different concentrations of P nutrient solution in the root system on one side and phosphorus-free nutrient solution on the other side [21,25]. Nodule weight, nitrogenase activity, ureide concentration, and the expression levels of key genes related to nodulation and nitrogen fixation were measured, and the nodule ultrastructure was observed. This study systematically investigated the effects of P supply on soybean nodulation and nitrogen fixation, providing a reference that clarifies the regulatory mechanisms of nodulation and nitrogen fixation in soybean.

2. Materials and Methods

This study was conducted at the Experimental Base of Northeast Agricultural University, Harbin city, Heilongjiang Province, China (126°43′ E, 45°44′ N) from 10 June to 25 July 2020. Sand-cultured dual-root soybean systems were used. The tested soybean variety was Kenfeng 16(Glycine max L. cv.) obtained from Heilongjiang Academy of Land Agricultural Reclamation Science, Heilongjiang, China).

2.1. Experimental Treatments

Dual-root soybean plants were prepared according to the grafting method described by Xia et al. [24] and they were potted with river sand. The method is described in detail in S1. Figure 1 shows the root morphology of a dual-root soybean plant under P31/0 treatment at the initial flowering stage, with the phosphorus-application side on the left and the phosphorus-free side on the right.

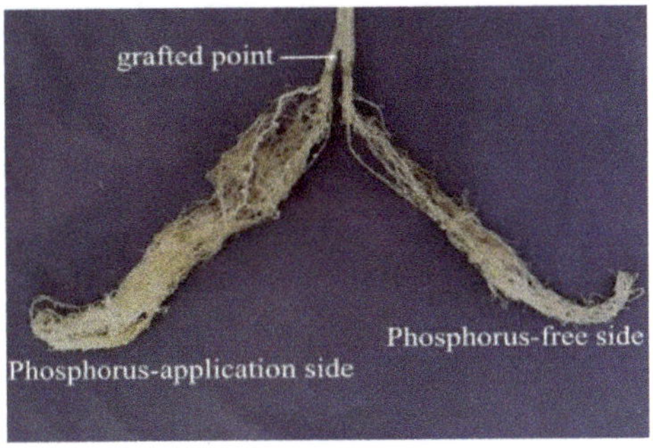

Figure 1. Roots of dual-root soybeans.

From the VC stage (unfolded cotyledon stage) to the R1 stage (initial flowering stage), the dual-root soybean plants were irrigated with a P-containing nutrient solution on one side (P+ side) and a P-free nutrient solution on the other side (P− side). On the P+ side, the nutrient solution applied contained P at three different concentrations, 1 mg/L (P1), 31 mg/L (P31), and 61 mg/L (P61), which were denoted as the P1/0, P31/0, and P61/0 treatments, respectively. Using KH_2PO_4 as the P source, the nutrient solution was prepared as proposed by Hoagland and Arnon [26] Yao et al. [27]. The potassium level in the P1/0 and P31/0 treatments was adjusted to the same level as that of P61/0. The types and concentrations of nutrient solution (mg/L) are shown in Tables S1 and S2.

From the VE stage (emergence stage) to the VC stage, the plants were irrigated with distilled water once a day at 250 mL on each side. Then, 250 mL of P-containing nutrient solution was applied on the P+ side, and 250 mL of P-free nutrient solution was applied to the P− side, once a day at 7:00–8:00 during the VC stage to V4 stage (fourth trifoliate leaf stage) and twice a day at 7:00–8:00 and 17:00–18:00 from the V4 stage to the R1 stage. Starting from the VC stage, field-grown soybean nodules cryopreserved from the previous year were ground and added to the nutrient solution (approximately 5 g/L), which was inoculated in the plants continuously for 5 days. The reproductive stage was recognized as described by Fehr et al. [28].

2.2. Sampling and Measurement

Samples were taken once in the V4 and R1 stages. The shoots were cut along the grafting site between 8:00 and 10:00 on a sunny day. The underground roots on both sides were washed with distilled water to remove the sand. For the V4 stage, the nodules of average size within 5 cm of the root base were harvested and fixed in formalin and then dehydrated with alcohol and glacial acetic acid (FFA) solution for ultrastructural observation. For the R1 stage, all the root nodules were harvested and stored in a freezer at −80 °C to measure the relative gene expression levels of *GmEXPB2*, *nifHDK*, *GmALN1*, etc. The intact roots in the V4 and R1 stages were used to measure the nitrogenase activity of the nodules. Subsequently, the nodules were removed and dried at 65°C to measure the dry weight of the P content of each part, as well as the nodule ureide content. Each measurement was repeated 4 times, 2 biological replicates, for a total of 8 plants.

Nodule nitrogenase activity was measured using the acetylene reduction activity (ARA) assay described by Gremaud and Harper [29].

For the ultrastructural observation of nodules, the removed nodules were halved longitudinally, and these halves were cut into 1 × 3 mm rectangular slices longitudinally with a blade, followed by FAA fixation and dehydration. Finally, ultrathin sections were cut with an LKB-V microtome. A Hitachi-600 transmission electron microscope was used for observation and imaging. The fixation and slide preparation procedures followed Goodchild and Bergersen) [30].

The P contents in plants were measured by the molybdenum antimony anti-colorimetric method with $CuSO_4$ and K_2SO_4 as catalysts and concentrated H_2SO_4 for digestion.

The nodule ureide content was measured following the method of Trijbels and Vogels [31].

For nodule RNA extraction and quantitative real-time reverse transcription-PCR (qRT-PCR) analysis, total RNA was extracted from root nodules using TRIzol reagent (Servicebio, Wuhan, China), and cDNA was synthesized using a RevertAid reverse transcription kit (Servicebio, Wuhan, China). The qRT-PCR analysis was performed by SYBR Green monitored qPCR (Servicebio, Wuhan, China), with the following reaction conditions: 95 °C for 30 s, 40 cycles of 95 °C for 5 s, 60 °C for 15 s, and 72 °C for 30 s. The primer sequences used for qRT-PCR amplification are shown in Table S3. Three biological replications were included. In this study, 16S rRNA was used as the reference for *nifD*, *nifH*, and *nifk* genes in the calculation of the qRT-PCR data, as previously described by Sulieman et al. [32], while 18S rRNA was used as the reference for *GmSPX5*, *GmACP1*, *GmUR5*, *GmPUR5*, and *GmHIUH5* genes in the calculation of the qRT-PCR data, as previously described by Carter et al. [33].

2.3. Statistical Analyses

All statistical analyses were performed using SPSS 21.0 (SPSS Inc., Chicago, Illinois). Before performing a one-way analysis of variance (ANOVA) on the data, all data were tested for normality, and Duncan's multirange test was used with a significance level of $p < 0.05$.

3. Results

3.1. Effects of P Supply on P Content and Dry Matter Accumulation in Soybean Plants with Dual-Root Systems

Table 1 shows that the P content was significantly higher in nodules than in the roots and shoots and higher on the P+ side than on the P− side at V4 and R1 for all treatments. At the V4 and R1 stages, the P content in the roots and shoots was similar between the P31/0 and P61/0 groups, but both groups had significantly higher P than the P1/0 group. At the V4 stage, the nodule P content increased significantly with the P concentration supplied. At the R1 stage, the nodule P content on the P+ side was significantly higher after the P31/0 and P61/0 treatments than after the P1/0 treatment but did not show any significant differences between P31/0 and P61/0 treatments. On the P− side, the nodule P content was similar between all treatments. At the V4 and R1 stage, the nodule P content on the P− side increased by 85.71% and 57.14%, respectively, under the P31/0 treatment when compared with P1/0, and by 0% and 9.09%, respectively, and under the P61/0 treatment when compared with P31/0. The root P content on the P– side increased by 68.75% and 75.00%, respectively, under the P31/0 treatment when compared with P1/0, and by 29.63% and 25.00%, respectively, under the P61/0 treatment when compared with P31/0. Under a low P supply (P1/0), the P contents in the roots and nodules were similar between the P+ side and P− side in the R1 stage, during which the difference in the nodule P content between the two sides reached an extremely significant level. Under the P31/0 and P61/0 treatments, the P content in the roots and nodules was higher on the P+ side than on the P− side, and all differences reached extremely significant levels at the V4 stage. In the R1 stage, the P content difference between the two sides was extremely significant in the roots under both the P31/0 and P61/0 treatments and was significant in the nodules under P31/0, but it was nonsignificant in the nodules of the P61/0-treated plants. The root P content ratio between the two sides (P−/P+ ratio) was 1.00 under P1/0; the P−/P+ ratio under the P31/0 and P61 treatments was 0.62 and 0.54 in the V4 stage and 0.55 and 0.63 in the R1 stage, respectively. The P−/P+ ratio of the nodule P content was greater than 0.80 under all treatments. This finding indicated that the root system prioritized the P− side nodules for distribution of P transported from the shoots to ensure normal functioning of the nodules.

Table 1. Effect of P supply on the P content of dual-root soybeans.

Treatments		Roots				Nodules				Shoot(%)
		P+(%)	P−(%)	IP–(%)	P−/P+	P+(%)	P−(%)	IP–(%)	P−/P+	
V4	P1/0	0.07 ± 0.00^{b}	0.07 ± 0.00^{b}	85.71	1.00	$0.20 \pm 0.01^{c**}$	0.16 ± 0.01^{c}	68.75	0.80	0.07 ± 0.01^{b}
	P31/0	$0.21 \pm 0.01^{a**}$	0.13 ± 0.01^{a}		0.62	$0.32 \pm 0.01^{b**}$	0.27 ± 0.01^{b}		0.84	0.16 ± 0.01^{a}
	P61/0	$0.24 \pm 0.02^{a**}$	0.13 ± 0.01^{a}	0.00	0.54	$0.44 \pm 0.01^{a**}$	0.35 ± 0.01^{a}	29.63	0.80	0.18 ± 0.01^{a}
R1	P1/0	0.07 ± 0.01^{b}	0.07 ± 0.00^{b}	57.14	1.00	0.16 ± 0.01^{b}	0.16 ± 0.01^{c}	75.00	1.00	0.09 ± 0.00^{b}
	P31/0	$0.20 \pm 0.01^{a**}$	0.11 ± 0.01^{a}		0.55	$0.34 \pm 0.01^{a*}$	0.28 ± 0.01^{b}		0.82	0.21 ± 0.01^{a}
	P61/0	$0.19 \pm 0.01^{a**}$	0.12 ± 0.01^{a}	9.09	0.63	0.36 ± 0.01^{a}	0.35 ± 0.02^{a}	25.00	0.97	0.22 ± 0.01^{a}

Note: All data in the table are expressed as the mean ± standard error (n = 4). Different lowercase letters indicate the differences between treatments at a significance level of 5%. Longitudinal comparison. * and ** denote a significant difference between the P+ side and the P− side at the 5% level and 1% level, respectively. IP– is the increment of P–.

Table 2 reveals that the dry weight showed an increasing trend in all parts of the soybean plants with increasing P supply on the P+ side. At the V4 stage, the dry weight of the shoots was significantly different between all treatments; at the R1 stage, although the difference was nonsignificant between P31/0 and P61/0, the dry weight of the aboveground tissues under these two treatments was significantly higher than that under the P1/0 treatment. On the P+ side, the dry weights of roots and nodules showed marked differences between treatments; on the P− side, the differences were nonsignificant between the P31/0 and P61/0 treatments. The root and nodule dry weights showed no significant differences between the P+ side and the P− side under a low P supply (P1/0); however, they were higher on the P+ side than on the P− side under the P31/0 and P61/0 treatments, demonstrating extreme differences in the plants supplied with higher concentrations of P. The P−/P+ ratio of root dry weight at V4 the stage was 0.91 under the P1/0 treatment, 0.49 under the P31/0 treatment, and 0.50 under the P61/0 treatment, and 0.41 under both the P31/0 and P61/0 treatment at R1ste stage. The P−/P+ ratio of nodule dry weight at the V4 stage was 1.00 under P1/0 and 0.22 under both P31/0 and P61/0, and it was 0.27 and 0.23 under the P31/0 and P61/0 treatments at the R1. These results show that the higher P supply increased the dry weights of all parts of the soybeans, and the increase was more prominent in the roots and nodules that directly contacted the supplied P. Combined with the data in Table 2, the distribution of P to the P− side was prioritized by absorption and transport from the P+ side, to support the growth of the root system, and the root P content was maintained at a reduced level to ensure root development; moreover, nodule growth was suppressed to reserve a certain level of P for nodule development and normal physiological functioning. The effect of phosphorus supply on the number of root nodules in double-rooted soybeans is shown in Table S4.

Table 2. Effect of P supply on the dry weight of dual-root soybeans.

	Treatments	Roots			Nodules			Shoots (g/Plant)
		P+(g/Plant)	P−(g/Plant)	P−/P+	P+(g/Plant)	P−(g/Plant)	P−/P+	
V4	P1/0	0.78 ± 0.03 c	0.71 ± 0.01 b	0.91	0.01 ± 0.000 c	0.01 ± 0.001 b	1.00	1.78 ± 0.11 c
	P31/0	1.72 ± 0.05 b**	0.84 ± 0.05 a	0.49	0.18 ± 0.006 b**	0.04 ± 0.001 a	0.22	3.99 ± 0.52 b
	P61/0	1.87 ± 0.01 a**	0.93 ± 0.03 a	0.50	0.23 ± 0.011 a**	0.05 ± 0.001 a	0.22	5.39 ± 0.40 a
R1	P1/0	0.92 ± 0.04 c	0.80 ± 0.03 b	0.87	0.02 ± 0.003 c	0.02 ± 0.003 b	1.00	2.09 ± 0.04 b
	P31/0	2.61 ± 0.01 b**	1.07 ± 0.03 a	0.41	0.22 ± 0.003 b**	0.06 ± 0.003 a	0.27	6.49 ± 0.71 a
	P61/0	2.73 ± 0.03 a**	1.11 ± 0.04 a	0.41	0.31 ± 0.012 a**	0.07 ± 0.003 a	0.23	7.20 ± 0.26 a

Note: All data in the table are expressed as the mean ± standard error (n = 4). Different lowercase letters indicate the differences between treatments at a significance level of 5%. Longitudinal comparison. ** denotes a significant difference between the P+ side and the P− side at the 1% level, respectively.

3.2. Effect of P Supply on the Activity of Nitrogenase and the Nodule Ureide Content in Soybean Plants with Dual-Root Systems

Table 3 shows the specific nitrogenase activity (SNA) and acetylene reduction activity (ARA) on both sides of the dual-root soybean plants under different levels of P supply at the V4 and R1 stages. As Table 3 shows, both ARA and SNA increased with the P supply and showed significant differences between treatments on both sides of the dual-root system of soybean plants at the V4 and R1 stages. At the V4 stage, ARA and SNA on the P+ side increased by 2288.00% and 29.00%, respectively, under the P31/0 treatment when compared with P1/0, and by 52.43% and 19.44%, respectively, under the P61/0 treatment when compared with P31/0. On the P− side, P31/0 increased ARA and SNA by 383.33% and 16.11%, respectively, compared with P1/0, and P61/0 increased them by 36.21% and 10.46%, respectively, when compared with P31/0. At the R1 stage, ARA and SNA on the P+ side increased by 1568.33% and 56.12%, respectively, under P31/0 treatment, compared with P1/0, and by 60.64% and 10.23%, respectively, under P61/0, compared with P31/0. On the P− side, P31/0 increased ARA and SNA by 218.64% and 11.44%, respectively, compared with P1/0, and P61/0 increased them by 50.00% and 15.78%, respectively, compared with P31/0. The change in SNA with the enhancement in P supply was less prominent than that of ARA, suggesting that the decrease in ARA mainly resulted from the decreases in nodule dry weight and number. In addition, ARA and SNA did not show significant differences between the P+ and P− sides under the P1/0 treatment, but they were markedly higher on the P+ side than on the P− side, and the differences were extremely significant under the P31/0 and P61/0 treatments. The P−/P+ ratio of ARA was 0.96-0.98 under the P1/0 treatment and only 0.17–0.19 under the P31/0 and P61 treatments. The P−/P+ ratio of SNA was greater than 0.75 under all treatments. These results indicate that P deficiency could significantly inhibit ARA and suppress SNA less markedly in soybean.

Table 3. Effects of P supply on ARA and SNA in dual-root soybeans.

	Treatments	ARA				SNA			
		P+(C_2H_4 μmol^{-1}h^{-1}Plant^{-1})	P-(C_2H_4 μmol^{-1}h^{-1}Plant^{-1})	IP−	P−/P+	P+(C_2H_4 μmol^{-1} g^{-1}h^{-1})	P-(C_2H_4 μmol^{-1} g^{-1}h^{-1})	IP−	P−/P+
V4	P1/0	0.25 ± 0.01 c	0.24 ± 0.01 c		0.96	25.76 ± 0.87 c	24.96 ± 0.47 c		0.97
	P31/0	5.97 ± 0.04 b**	1.16 ± 0.06 b	383.33	0.19	33.23 ± 0.31 b**	28.98 ± 0.54 b	16.11	0.87
	P61/0	9.10 ± 0.35 a**	1.58 ± 0.01 a	36.21	0.17	39.69 ± 1.27 a**	32.01 ± 0.17 a	10.46	0.81
R1	P1/0	0.60 ± 0.01 c	0.59 ± 0.02 c		0.98	29.81 ± 0.49 c	29.81 ± 1.18 c		1.00
	P31/0	10.01 ± 0.17 b**	1.88 ± 0.07 b	218.64	0.19	46.54 ± 0.34 b**	33.21 ± 0.82 b	11.41	0.75
	P61/0	16.08 ± 0.76 a**	2.82 ± 0.16 a	50.00	0.18	51.30 ± 0.90 a**	38.45 ± 0.65 a	15.78	0.75

Note: All data in the table are expressed as the mean ± standard error (n = 4). Different lowercase letters indicate the differences between treatments at a significance level of 5%. Longitudinal comparison. ** denotes a significant difference between the P+ side and the P− side at the 1% level, respectively. IP– is the increment of P–.

Table 4 shows the ureide content of root nodules at the V4 and R1 stages in soybean plants supplied with different levels of P. Table 4 also reveals that nodule ureide content at the V4 and R1 stages showed a changing trend similar to those of SNA and ARA, shown in Table 3—namely, they increased with P supply and varied significantly between treatments on both sides. At the V4 and R1 stages, the ureide content on the P+ side increased by 234.31% and 281.03%, respectively, when the P supply was increased from P1/0 to P31/0. The ureide content on the P+ side increased by 7.92% from P31/0 to P61/0 at the V4 stage and decreased by 0.90% at the R1 stage. On the P− side, the ureide content at the V4 and R1 stages increased by 118.18% and 156.44% from P1/0 to P31/0 and by 6.02% and 9.27% from P31/0 to P61/0, respectively. In addition, the ureide content did not show any significant difference between the P+ and P− sides under the P1/0 treatment, but it was extremely significantly higher on the P+ side than on the P− side under the P31/0 and P61/0 treatments. The P−/P+ ratio of the ureide content was 0.97-0.87 under the P1/0 treatment and only 0.59-0.65 under the P31/0 and P61 treatments. The results indicate that P deficiency can severely inhibit the synthesis of ureides in soybean and more markedly increase the number of nodules that directly contact P.

Table 4. Effect of P supply on the ureide content in nodules of dual-root soybeans.

	Treatments	Ureide Content (mg/g DW)		IP−(%)	P−/P+
		P+	P−		
V4	P1/0	1.02 ± 0.02 [c]	0.99 ± 0.04 [b]	118.18	0.97
	P31/0	3.41 ± 0.08 [b**]	2.16 ± 0.08 [a]		0.63
	P61/0	3.68 ± 0.11 [a**]	2.29 ± 0.05 [a]	6.02	0.62
R1	P1/0	1.16 ± 0.01 [b]	1.01 ± 0.01 [b]	156.44	0.87
	P31/0	4.42 ± 0.09 [a**]	2.59 ± 0.13 [a]		0.59
	P61/0	4.38 ± 0.03 [a**]	2.83 ± 0.11 [a]	9.27	0.65

Note: All data in the table are expressed as the mean ± standard error (n = 4). Different lowercase letters indicate the differences between treatments at a significance level of 5%. Longitudinal comparison. ** denotes a significant difference between the P+ side and the P− side at the 1% level, respectively. IP– is the increment of P–.

3.3. Effect of P Supply on the Ultrastructure of Soybean Plants with Dual-Root Systems Root Nodules

Rhizobial infection of host cells is the first step in nodule development. Host cells can be classified as infected cells (ICs) and uninfected cells (UCs). Figure 2A shows a cross-section of the nodule ultrastructure under a transmission electron microscope (4000×) at the V4 stage on both sides of the dual-root soybean plants with different levels of P supply. A treatment comparison revealed that the number of ICs was largest under the P31/0 treatment and smallest under P1/0. There was a very slight difference in the area of nodule infection between P31/0 and P61/0. In contrast, the number of UCs showed the opposite changing pattern with the increase in P supply: it was largest under P1/0 and smallest under P31/0, and the number was slightly larger under P61/0 than under P31/0. In addition, there were more ICs on the P+ side than on the P− side. Rhizobia enter the host cell through an infection thread to form a bacteroid (Bt), and the rhizobia that have successfully invaded the host cells can divide to form multiple new Bts. Figure 2B shows that the number of Bts at the V4 stage was counted on the cross sections of nodules under a transmission electron microscope (20,000×) on both sides of the dual-root soybean plants with different levels of P supply. The number of Bts in root nodules was smallest under the P1/0 treatment and largest under the P31/0 treatment, and it was slightly smaller under P61/0 than under P31/0. In addition, the nodules on the P+ side had more Bts than those on the P− side. The results show that the numbers of ICs and Bts on both sides of the soybean root systems can be significantly affected by supplying P to only one side, and the changes increase with the increase in P supply. The effect of P supply on the number of IC, UC, and Bt nodules in double-root soybeans is shown in Table S5.

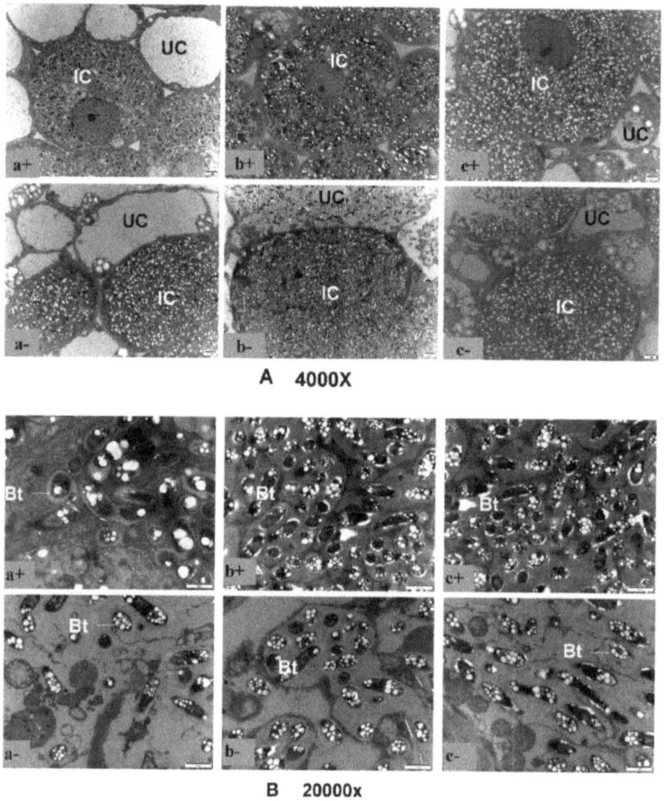

Figure 2. Ultrastructure of nodules on both sides of a dual-root soybean plant at V4 stage: (**A**) is under a transmission electron microscope 4000×, (**B**) is under a transmission electron microscope 20,000×; a+: P1/0 P+; a−: P1/0 P−; b+: P31/0 P+; b−: P31/0 P−; c+: P61/0 P+; c−: P61/0 P−; IC: infected cell; UC: uninfected cell; Bt: bacteroid.

3.4. Effect of P Supply on the Expression of Key Genes Related to Nodulation and Nitrogen Fixation in Soybean Plants with Dual-Root Systems

Figure 3 shows the relative expression levels of key genes related to nodulation and nitrogen fixation in nodules on both sides of the dual-root soybean plants with different levels of P supply. Figure 3A shows that with P31/0 as a control, the expression levels of *GmEXPB2*, *nifD*, *nifH*, and *GmPUR5* on the P+ side increased significantly with an enhanced P supply; *GmSPX5nifK*, *GmALN1*, and *GmACP1* expressions were higher under P31/0 than under P1/0, and the difference in P31/0 vs. P61/0 was not significant. *GmHIUH5* expression was significantly higher for P61/0 than P31/0, and the difference between P31/0 and P1/0 was not significant; *GmUR5* was significantly higher for P31 than P61/0, and P61/0 was significantly higher than P1/0. In Figure 3B, with P31/0 as a control, the gene expression levels are presented.

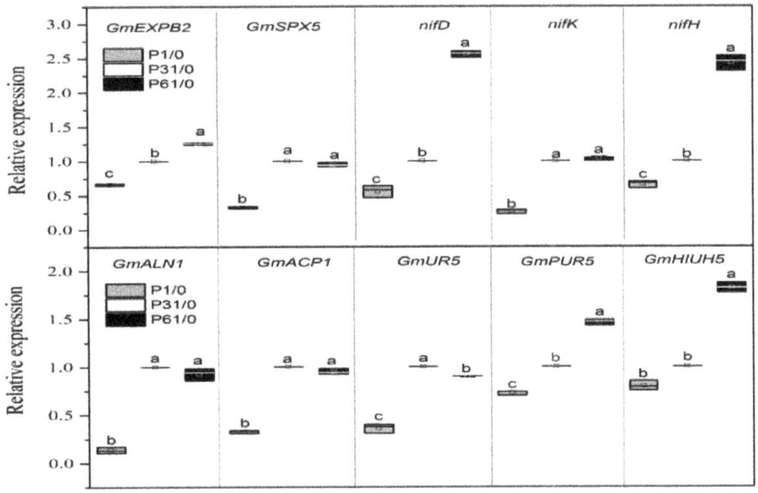

(A)

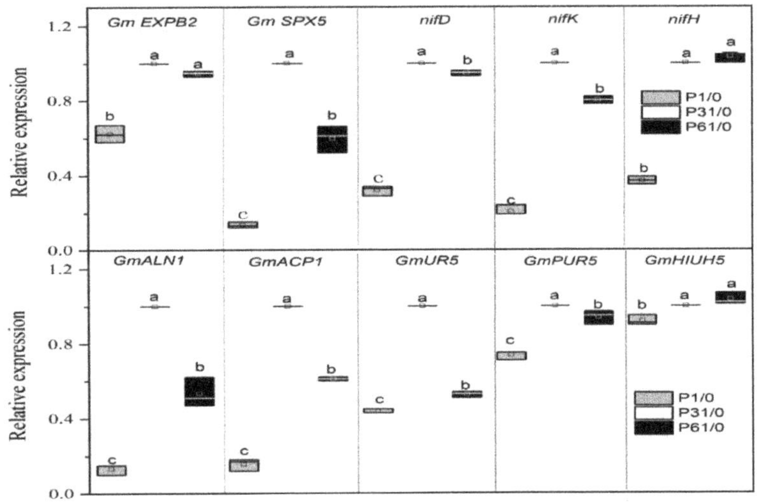

(B)

Figure 3. *Cont.*

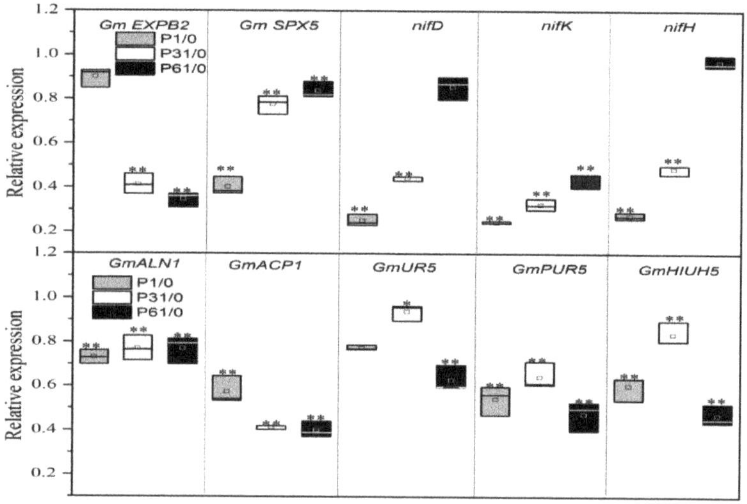

(C)

Figure 3. (**A**) Relative gene expression on the phosphorus-application side (the P31/0 as control). (**B**) Relative gene expression on the phosphorus-free side(the P31/0 as control). (**C**) Relative expression of phosphorus-free genes side (phosphorus-application side as a control). Effects of P supply on the relative expression levels of key genes related to nodulation and nitrogen fixation on both sides of the dual-root soybean. In (**A**,**B**), different lowercase letters indicate the significant differences between treatments at the 5% level. In (**C**), *,** denotes a significant difference between the P+ side and the P− side, with P** indicating a significant difference at the 1% level; *GmEXPB2* is nodulation-related key gene; *nifD*, *nifH*, and *nifK* are regulatory key genes of nitrogen fixation; *GmALN1*, *GmACP1*, *GmUR5*, *GmPUR5*, *GmHIUH5*.

The expression levels of *GmEXPB2*, *GmSPX5*, *nifD*, *nifK*, *GmALN1*, *GmACP1*, *GmUR5*, and *GmPUR5* on the P− side showed a trend of first increasing and then decreasing with the increase in P supply. All three gene expression levels increased significantly when the P supply increased from 1 mg/L to 31 mg/L; however, after the P supply reached 31 mg/L, the *GmEXPB2*, *nifH*, and *GmHIUH5* expression levels changed nonsignificantly, and *GmSPX5*, *nifD*, *nifK*, *GmALN1*, *GmACP1*, *GmUR5*, and *GmPUR5* expression levels markedly decreased with the increase in P supply. In Figure 3C, compared with the P+ side (1.0), gene expression levels on the P− side were all low. In addition, the relative *GmEXPB2* and *GmACP1* expression decreased with the increase in P supply, showing an extremely significant difference between the two sides under P31/0 and P61/0 and no significant difference under P1/0. *GmACP1* showed highly significant differences at all phosphorus supply levels. All expression levels of *GmSPX5*, *nifD*, *nifK*, and *nifH* increased with the increase in P supply. *nifD* and *nifH* showed extremely significant differences between the two sides under P1/0 and P31/0 but no significant difference under P61/0, while *GmSPX5* and *nifK* showed an extremely significant difference between the two sides under all levels of P supply. Gene expressions of *GmUR5*, *GmPUR5*, and *GmHIUH5* tended to increase and then decrease with increasing phosphorus supply levels. *GmEXPB2* showed highly significant differences at the phosphorus supply levels of P31/0 and P61/0, and no significant differences at P1/0, while *GmPUR5* and *GmHIUH5* showed highly significant differences at all phosphorus supply levels. One-sided P supply significantly affected the relative expression levels of genes regulating nodulation and nitrogen fixation in soybean, and the demand for P to meet the needs of gene expression on the P+ side was greater than that on the P−side.

4. Discussion

4.1. Effects of P Supply on Nodulation and Nitrogen Fixation in Soybean Plants

Increasing a soybean's P supply can significantly increase its dry weight, the number of nodules [10,13,34], and nodule nitrogenase activity [5,35,36]. In this study, we applied the grafting approach for soybean seedlings developed by Xia et al. [24] to prepare a dual-root system of soybean plants, which received different concentrations of P nutrient solution at the root system on the phosphorus-application side and phosphorus-free nutrient solution on the phosphorus-free side. We found that the increase in P supply significantly enhanced the dry weight, ARA, and SNA of nodules on the phosphorus-application side and indirectly affected these parameters on the phosphorus-free side of the dual-root soybeans, and the increases were more marked in the roots and nodules on the side in direct contact with a P supply. Le Roux et al. [18] found that low P treatment significantly reduced the content of ureides in soybean nodules. Magadlela et al. [37] found that the ureide content decreased in the nodules but increased in the roots of the legume tree after low P treatment. The present study found that increasing the P supply on only one side increased the ureide content in nodules on both sides of the dual-root soybeans, further confirming that P supply systemically affects nitrogen fixation in soybean nodules.

Isidra-Arellano et al. [38] reported that low P treatment reduced the number of curled root hairs and reduced the relative gene expression levels of *PvNSP2*, *PvNIN*, and *PvFLOT2*, which control the formation of infection threads in common beans. Gentili et al. [39] found that P at a moderate concentration most significantly stimulated cell divisions in the cortex, nodule primordia emergence, and initial nodule emergence in the roots of *Alnusincana*. In the present study, by observing the nodule ultrastructure of dual-root soybeans at the V4 stage, we found that as the P supply increased, the formation of nodule Bts and the number of ICs both increased. *GmEXPB2* and *GmSPX5* are associated with the formation and extension of cell walls during nodule formation and development in soybean [5,40]. Li et al. [40] found that high P treatment increased the relative expression of the *GmEXPB2* gene in the nodules, but this effect was reversed after 14 days of treatment. Zhou et al. [41] found that P supply increased the relative gene expression level of *GmEXPB2* in roots and leaves of transgenic soybean. *GmSPX1* has been documented to exhibit no response to Pi starvation in soybean roots [42]. In the present study, the relative *GmEXPB2* and *GmSPX5* expression levels in soybean nodules were increased by a high P supply and decreased by a low P supply, which is consistent with previous studies. Nitrogenase is encoded by the nitrogen fixation gene *nif*. In a nitrogenase system, *nifD* and *nifK* are the structural genes encoding the MoFe protein subunit of the nitrogenase complex, and *nifH* encodes ferritin [43,44]. Nasr Esfahani et al. (2016) [45] stated that low P treatment reduces the relative expression levels of *nifH* and *nifK* and suppresses nitrogenase activity in nodules. Sulieman et al. [32] reported that low P treatment reduced the number and dry weight of root nodules and increased the relative expression levels of *nifH*, *nifD*, and *nifK* in soybean nodules. Our finding that low P supply reduced the relative expression levels of *nifH*, *nifD*, and *nifK* in soybean nodules contrasts with the findings of Sulieman et al. [32]. The inconsistency might be attributed to the fact that in this study, low P supply decreased ARA and SNA and, thus, suppressed the expression of genes regulating nodulation and nitrogen fixation in nodules. *GmALN1 (Allantoinase 1)*, *GmACP1 (Acid Phosphatase)*, *GmUR5 (AIRS synthetase)*, *GmPUR5(AIRS synthetase)*, and *GmHIUH5* (hydroxyisourate hydrolase) are the key genes for ureide synthesis, and they are significantly expressed in soybean nodules [33]. Alamillo et al. [46] found that ALN gene expression also increased in *Phaseolus vulgaris* L. roots and in shoots in response to drought. Díaz-Leal et al. [47] found that nitrate stress reduced the ALN gene expression in roots, stems, and leaves of *Phaseolus vulgaris* L. The present study found that increasing the P supply on only one side increased the key genes for ureide synthesis in nodules on both sides of dual-root soybeans. It is worth mentioning that increasing the P supplied to only one side of the dual-root soybeans promoted the formation of nodule Bts, increased the number of ICs, and elevated the relative expression of regulatory genes for nodulation and nitrogen fixation (*GmEXPB2*, GmSPX5, *nifH*, *nifD*, *nifK*, *GmALN1*, *GmACP1*, *GmUR5*, *GmPUR5*, and *GmHIUH5*) on the phosphorus-free side in our study. From the perspectives of nodule structure and relative key gene expression, our findings further confirm that the local P supply can systemically regulate nodulation and nitrogen fixation in soybean.

4.2. Effects of P Supply Level on P Absorption and Transport in Soybean Plants

The level of P supply can affect the P absorption of soybean plants [35,48,49]. The P contents in the aboveground tissues and nodules of soybean increase with the increase in P supply [50]. In a study that supplied soybean plants with 0.005 and 0.025 mM KH_2PO_4, Georgiev and Tsvetkova [51] found that treatment with 0.025 mM KH_2PO_4 increased the P content in the root system but reduced the P content in the shoots of the plants. In the present study, applying high concentrations of P to only one side significantly increased the P content in the shoots, roots, and nodules on both sides of dual-root soybean plants. Under low P stress, nodules are prioritized for receiving P to ensure their growth [4,52,53]. In a study conducted by Qin et al. [8], where 5 or 250 μM KH_2PO_4 was applied to soybean roots under hydroponic conditions and ^{33}P-labeled KH_2PO_4 was applied to the bottom of the root area without nodules, 5 μM KH_2PO_4 treatment resulted in a significantly higher ^{33}P content in soybean nodules than 250 μM KH_2PO_4 treatment. Al-Niemi et al. [54] applied 0 and 5 mM KH_2PO_4 to common bean plants that had been supplied with 0.75 mM KH_2PO_4 until the V2 stage and then applied ^{33}P-labeled KH_2PO_4 to the bottom area of the roots without nodules or to the entire root system. They found that P-free treatment promoted the transport of ^{33}P from the roots and nodular surface into the nodules. In the present study, we also found that P transported from the aboveground tissues to the root system on the P− side might be prioritized for root nodules to maintain their function.

Using the root separation method to apply P to only one side of subterranean clover and *Lupinusalbus*, Scott et al. [22] Shu et al. [21] found that P uptake in the roots was enhanced on only the P-supply side and remained stable on the non-P-supply side. Burleigh et al. [55], Shen et al. [25], and Wouterlood et al. [56] applied high concentrations of P to one side and a low concentration or no P to the other side of chickpea, *Lupinusalbus* and *truncatula* plants, respectively, using the root separation method, and they found that the P uptake by the roots was enhanced on both sides. Compared with both sides of the roots supplied with phosphorus, the treatment with no phosphorus supply on one side of the roots inhibited *Phaseolus vulgaris* root growth on the phosphorus supply side (Bonser et al. [57]), while low levels of phosphorus on one side of the roots and normal levels of phosphorus on the other side inhibited normal phosphorus supply soybean root nodule growth on the low-level phosphorus supply side; however, the relative expression levels of genes PvNIN, PvRIC1, and PvRIC2, which regulate the formation of nodules by rhizobial symbiosis, were only upregulated on the low-phosphorus side, according to Isidra-Arellano et al. [58]. In our study, after supplying different concentrations of P for a long time on one side of the dual-root soybeans, the P content in the roots and nodules on the P− side significantly increased with the increase in P supply to the P+ side, which is different from the results of Scoot et al. [22] and Shu et al. [21]. Our experiment was conducted under sand culture conditions to eliminate soil phosphorus interference, which may account for the differences in the results.

5. Conclusions

1. In the dual-root soybean plants supplied with P on only one side, the nodule weight, ARA, SNA, ureide content, the number of Bts, the number of ICs, and relative expression levels of key genes related to nodulation and nitrogen fixation (*GmEXPB2*, GmSPX5, *nifH*, *nifD*, *nifK*, *GmALN1*, *GmACP1*, *GmUR5*, *GmPUR5*, and *GmHIUH5*) on the P+ side increased with the increase in P supply during the VC-R1 period. Those on the P− side showed the same pattern as those on the P+ side but less prominently. Therefore, P supply regulates soybean nodulation and nitrogen fixation.

2. When the level of phosphorus supply was increased from 1 mg/L to 31 mg/L, the increase on the P− side root was significant, and nodule phosphorus content increased by 57.14–85.71% and 68.75–75.00%; ARA and SNA were 218.64–383.33% and 11.41–16.11%, respectively, and ureide content was 118.18–156.44%. When the level of phosphorus supply was increased from 31 mg/L to 61 mg/L, the increases in the regulation ability of the root and nodule phosphorus content were only 0.00–9.09% and 25.00–29.63%, respectively; ARA and SNA were 36.21–50.00% and 10.46–15.78%, respectively, and ureide content was 6.02–9.27%.
3. When the dual-root soybean plants were supplied with different concentrations of P on one side, the P content in the roots and nodules on the P− side increased with the increase in P supply. These findings show that P can be transported to the roots and nodules on the non-supplied side through the shoots, and the nodules are prioritized to receive P.

Supplementary Materials: The following are available online at https://www.mdpi.com/article/10.3390/agronomy11112354/s1, Table S1: The elements of P or K in different P level treatments, Table S2: Concentrations of elements in nutrient medium of the sand culture, Table S3: Primer sequences of key genes for nodulation and nitrogen fixation in soybean, Table S4: Effect of phosphorus supply on the number of nodules in double-root soybeans, Table S5: Effect of P supply on the number of IC, UC, and Bt in nodules of double-root soybeans.

Author Contributions: Conceptualization, H.L., C.M., and Z.G.; data curation, H.L.; funding acquisition, X.L. and Z.G.; investigation, X.W., Q.L., and S.L.; methodology, H.L. and S.D.; resources, C.M. and C.Y.; software, H.L.; writing—original draft preparation, H.L.; writing—review and editing, H.L. and C.M.; All authors have read and agreed to the published version of the manuscript.

Funding: We are grateful for the support from the National Key Research and Development Programme, Integration and Demonstration of High Quality, Simple, and Efficient Cultivation Techniques for Soybean (Grant Number 2020YFD1000903).

Data Availability Statement: Not Applicable.

Acknowledgments: We are grateful to the College of Agriculture of Northeast Agricultural University for providing laboratory space. We thank the American Journal Experts for their linguistic assistance during the preparation of this manuscript.

Conflicts of Interest: The authors declare no conflict of interest. The funders had no role in the design of the study; in the collection, analyses, or interpretation of data; in the writing of the manuscript, or in the decision to publish the results.

References

1. Vance, C.P.; Uhde-Stone, C.; Allan, D.L. Phosphorus acquisition and use: Critical adaptations by plantsfor securing a nonrenewable resource. *New Phytol.* **2010**, *157*, 423–447. [CrossRef]
2. Jakobsen, I. The role of phosphorus m nitrogen fixation by young pea plants (Pisum sativum). *Physiol. Plant.* **1985**, *64*, 190–196. [CrossRef]
3. Chen, Z.; Cui, Q.; Liang, C.; Sun, L.; Tian, J.; Liao, H. Identification of differentially expressed proteins in soybean nodules under phosphorus deficiency through proteomic analysis. *Proteomics* **2011**, *11*, 4648–4659. [CrossRef]
4. Sulieman, S.; Ha, C.V.; Schulze, J.; Tran, L.S. Growth and nodulation of symbiotic Medicago truncatula at different levels of phosphorus availability. *J. Exp. Bot.* **2013**, *64*, 2701–2712. [CrossRef]
5. Xue, Y.; Zhuang, Q.; Zhu, S.; Xiao, B.; Liang, C.; Liao, H.; Tian, J. Genome Wide Transcriptome Analysis Reveals Complex Regulatory Mechanisms Underlying Phosphate Homeostasis in Soybean Nodules. *Int. J. Mol. Sci.* **2018**, *19*, 2924. [CrossRef]
6. Taliman, N.; Dong, Q.; Echigo, K.; Raboy, V.; Saneoka, H. Effect of Phosphorus Fertilization on the Growth, Photosynthesis, Nitrogen Fixation, Mineral Accumulation, Seed Yield, and Seed Quality of a Soybean Low-Phytate Line. *Plants* **2019**, *8*, 119. [CrossRef] [PubMed]
7. Bulgarelli, R.G.; De Oliveira, V.H.; de Andrade, S.A.L. Arbuscular mycorrhizal symbiosis alters the expression of PHT1 phosphate transporters in roots and nodules of P-starved soybean plants. *Theor. Exp. Plant Physiol.* **2020**, *32*, 243–253. [CrossRef]
8. Qin, L.; Zhao, J.; Tian, J.; Chen, L.; Sun, Z.; Guo, Y.; Lu, X. The High-Affinity Phosphate Transporter GmPT5 Regulates Phosphate Transport to Nodules and Nodulation in Soybean. *Plant Physiol.* **2012**, *159*, 1634–1643. [CrossRef] [PubMed]

9. Lu, M.; Cheng, Z.; Zhang, X.-M.; Huang, P.; Fan, C.; Yu, G.; Chen, F.; Xu, K.; Chen, Q.; Miao, Y.; et al. Spatial Divergence of PHR-PHT1 Modules Maintains Phosphorus Homeostasis in Soybean Nodules. *Plant Physiol.* **2020**, *184*, 01209–02019. [CrossRef]
10. Israel, D.W. Investigation of the role of phosphorus in symbiotic dinitrogen fixation. *Plant Physiol.* **1987**, *84*, 835–840. [CrossRef]
11. Israel, D.W. Symbiotic dinitrogen fixation and host-plant growth during development of and recovery from phosphorus deficiency. *Physiol. Plant.* **1993**, *88*, 294–300. [CrossRef]
12. Tsvetkova, G.; Georgiev, G. Effect of phosphorus nutrition on the nodulation, nitrogen fixation and nutrient-use efficiency of Bradyrhizobiumjaponicun soybean (Glycine max L. Merr.) symbiosis. *Bulg. J. Plant Physiol.* **2003**, *3*, 315–335.
13. Miao, S.; Qiao, Y.; Han, X.-Z.; An, M. Nodule Formation and Development in Soybeans (Glycine max L.) in Response to Phosphorus Supply in Solution Culture1. *Pedosphere* **2007**, *17*, 36–43. [CrossRef]
14. Dong, Q.; Echigo, K.; Raboy, V.; Saneoka, H. Seedling growth, physiological characteristics, nitrogen fixation, and root and nodule phytase and phosphatase activity of a low-phytate soybean line. *Plant Physiol. Biochem.* **2020**, *149*, 225–232. [CrossRef]
15. Cassman, K.; Whitney, A.; Stockinger, K. Root Growth and Dry Matter Distribution of Soybean as Affected by Phosphorus Stress, Nodulation, and Nitrogen Source1. *Crop. Sci.* **1980**, *20*, 239–244. [CrossRef]
16. Sa, T.M.; Israel, D.W. Energy status and functioning of phosphorus-deficient soybean nodules. *Plant Physiol.* **1991**, *97*, 928–935. [CrossRef] [PubMed]
17. Tang, C.; Hinsinger, P.; Drevon, J.J.; Jaillard, B. Phosphorus Deficiency Impairs Early Nodule Functioning and Enhances Proton Release in Roots of Medicago truncatula L. *Ann. Bot.* **2001**, *88*, 131–138. [CrossRef]
18. Le Roux, M.; Khan, S.; Valentine, A. Nitrogen and carbon costs of soybean and lupin root systems during phosphate starvation. *Symbiosis* **2009**, *48*, 102–109. [CrossRef]
19. Mandri, B.; Drevon, J.J.; Adnane, B.; Oufdou, K.; Faghire, M.; Plassard, C.; Payre, H.; Ghoulam, C. Interactions between common bean genotypes and rhizobia strains isolated from moroccan soils for growth, phosphatase and phytase activities under phosphorus deficiency conditions. *J. Plant Nutr.* **2012**, *35*, 1477–1490. [CrossRef]
20. Araújo, A.P.; Plassard, C.; Drevon, J.J. Phosphatase and phytase activities in nodules of common bean genotypes at different levels of phosphorus supply. *Plant Soil* **2008**, *312*, 129. [CrossRef]
21. Shu, L.; Shen, J.; Rengel, Z.; Tang, C.; Zhang, F.; Cawthray, G. Formation of cluster roots and citrate exudation by Lupinusalbus in response to localized application of different phosphorus sources. *Plant Sci.* **2007**, *172*, 1017–1024. [CrossRef]
22. Scott, B.; Robson, A. The Distribution of Mg, P and K in the Split Roots of Subterranean Clover. *Ann. Bot.* **1991**, *67*, 251–256. [CrossRef]
23. Snapp, S.; Lynch, J. Phosphorus Distribution and Remobilization in Bean Plants as Influenced by Phosphorus Nutrition. *Crop. Sci.* **1996**, *36*, 929–935. [CrossRef]
24. Xia, X.; Ma, C.; Dong, S.; Xu, Y.; Gong, Z. Effects of nitrogen concentrations on nodulation and nitrogenase activity in dual root systems of soybean plants. *Soil Sci. Plant Nutr.* **2017**, *63*, 470–482. [CrossRef]
25. Shen, J.; Li, H.; Neumann, G.; Zhang, F. Nutrient uptake, cluster root formation and exudation of protons and citrate in Lupinusalbus as affected by localized supply of phosphorus in a split-root system. *Plant Sci.* **2005**, *168*, 837–845. [CrossRef]
26. Hoagland, D.R.; Arnon, D.I. The water culture method for growing plants without soil. *Calif. Agric. Exp. Stn. Circ.* **1950**, *347*, 33.
27. Yao, Y.; Wu, D.; Gong, Z.; Zhao, J.; Ma, C. Variation of nitrogen accumulation and yield in response to phosphorus nutrition of soybean (Glycine max L. Merr.). *J. Plant Nutr.* **2018**, *41*, 1138–1147. [CrossRef]
28. Fehr, W.R.; Caviness, C.E.; Burmood, D.T.; Pennington, J.S. Stage of Development Descriptions for Soybeans, Glycine Max (L.) Merrill1. *Crop. Sci.* **1971**, *11*, 929–931. [CrossRef]
29. Gremaud, M.F.; Harper, J.E. Selection and initial characterization of partially nitrate tolerant nodulation mutants of soybean. *Plant Physiol.* **1989**, *89*, 169–173. [CrossRef]
30. Goodchild, D.J.; Bergersen, F.J. Electron microscopy of the infection and subsequent development of soybean nodule cells. *J. Bacteriol* **1966**, *92*, 204–213. [CrossRef]
31. Trijbels, F.; Vogels, G.D. Degradation of allantoin by Pseudomonas acidovorans. *Biochim. Acta* **1966**, *113*, 292–301. [CrossRef]
32. Sulieman, S.; Kusano, M.; Ha, C.; Watanabe, Y.; Abdalla, M.; Abdelrahman, M. Divergent metabolic adjustments in nodules are indispensable for efficient N2 fixation of soybean under phosphate stress. *Plant Sci.* **2019**, *289*, 110249. [CrossRef]
33. Carter, A.M.; Tegeder, M. Increasing Nitrogen Fixation and Seed Development in Soybean Requires Complex Adjustments of Nodule Nitrogen Metabolism and Partitioning Processes. *Curr. Biol.* **2016**, *26*, 2044–2051. [CrossRef]
34. Silva, R.; Gomes de Faria, A.; Neto, A.; Pereira, T.; Dias, J.; Martins dos Santos, A.; Junior, A.; Freitas, G.; Nascimento, V. High P Availability in Brazilian Cerrado Soils Reduces Nodulation and Grain Yield of Soybean. *Commun. Soil Sci. Plant Anal.* **2019**, *50*, 1267–1277. [CrossRef]
35. Rotaru, V.; Sinclair, T. Interactive influence of phosphorus and iron on nitrogen fixation by soybean. *Environ. Exp. Bot.* **2009**, *66*, 94–99. [CrossRef]
36. Wang, Y.; Yang, Z.; Youbin, K.; Li, X.; Li, W.; Du, H.; Zhang, C. GmPAP12 Is Required for Nodule Development and Nitrogen Fixation Under Phosphorus Starvation in Soybean. *Front. Plant Sci.* **2020**, *11*, 450. [CrossRef]
37. Magadlela, A.; Steenkamp, E.; Valentine, A. Variable P supply affects N metabolism in a legume tree, Virgiliadivaricata, from nutrient-poor Mediterranean-Type ecosystems. *Funct. Plant Biol.* **2015**, *43*, 287–297. [CrossRef]
38. Isidra-Arellano, M.; Reyero-SaavedraI, M.; Sánchez-Correa, M.; Pingault, L.; Sen, S.; Joshi, T. Phosphate Deficiency Negatively Affects Early Steps of the Symbiosis between Common Bean and Rhizobia. *Genes* **2018**, *9*, 498. [CrossRef] [PubMed]

39. Gentili, F.; Wall, L.G.; Huss-Danell, K. Effects of phosphorus and nitrogen on nodulation are seen already at the stage of early cortical cell divisions in Alnusincana. *Ann. Bot.* **2006**, *98*, 309–315. [CrossRef] [PubMed]
40. Li, X.; Zhao, J.; Tan, Z.; Zeng, R.; Liao, H. GmEXPB2, a Cell Wall b-Expansin, Affects Soybean Nodulation through Modifying Root Architecture and Promoting Nodule Formation and Development. *Plant Physiol.* **2015**, *169*, 2640–2653. [CrossRef]
41. Zhou, J.; Xie, J.; Liao, H.; Wang, X. Overexpression of β-expansin gene GmEXPB2 improves phosphorus efficiency in soybean. *Physiol. Plant.* **2013**, *150*, 194–204. [CrossRef]
42. Yao, Z.; Tian, J.; Liao, H. Comparative characterization of GmSPX members reveals that GmSPX3 is involved in phosphate homeostasis in soybean. *Ann. Bot.* **2014**, *114*, 477–488. [CrossRef] [PubMed]
43. Curatti, L.; Brown, C.S.; Ludden, P.W.; Rubio, L.M. Genes required for rapid expression of nitrogenase activity in Azotobacter-vinelandii. *Proc. Natl. Acad. Sci. USA* **2005**, *102*, 6291–6296. [CrossRef] [PubMed]
44. Esfahani, M.N.; Sulieman, S.; Schulze, J.; Yamaguchi-Shinozaki, K.; Shinozaki, K.; Tran, L. Approaches for enhancement of N2 fixation efficiency of chickpea (Cicer arietinum L.) under limiting nitrogen conditions. *Plant Biotechnol. J.* **2014**, *12*, 387–397. [CrossRef] [PubMed]
45. Nasr Esfahani, M.; Kusano, M.; Nguyen, K.H.; Watanabe, Y.; Ha, C.V.; Saito, K.; Sulieman, S.; Herrera-Estrella, L.; Tran, L.S. Adaptation of the symbiotic Mesorhizobium-chickpea relationship to phosphate deficiency relies on reprogramming of whole-plant metabolism. *Proc. Natl. Acad. Sci. USA* **2016**, *113*, 4610–4619. [CrossRef]
46. Alamillo, J.M.; Díaz-Leal, J.L.; Sánchez-Moran, M.V.; Pineda, M. Molec-ular analysis of ureide accumulation under drought stress in Phaseolus v-ulgaris L. *Plant Cell Env.* **2010**, *33*, 1828–1837. [CrossRef]
47. Díaz-Leal, J.L.; Gálvez-Valdivieso, G.; Fernández, J.; Pineda, M.; Alamillo, J.M. Developmental effects on ureide levels are mediated by tissue-specific regulation of allantoinase in Phaseolus vulgaris L. *J. Exp. Bot.* **2012**, *63*, 4095–4106. [CrossRef]
48. Vance, C.; Graham, P.; Allan, D. Biological Nitrogen Fixation: Phosphorus—A Critical Future Need? *Nitrogen Fixat. Mol. Crop. Product.* **2002**, *38*, 509–514. [CrossRef]
49. Zhang, Y.; Hongling, Q.I.; Fengxia, L.U.; Han, M.; Wang, P. Differences in Phosphorus Absorption and Utilization Efficiency of Soybean in Mature Period under Phosphorus Stress. *Agric. Biotechnol.* **2014**, *3*, 24–27.
50. Xue, A.; Guo, X.; Zhu, Q.; Zhang, H.; Wang, H. Effect of Phosphorus Fertilization to P Uptake and Dry Matter Accumulation in Soybean with Different P Efficiencies. *J. Integr. Agric.* **2014**, *13*, 326–334. [CrossRef]
51. Georgiev, G.I.; Tsvetkova, G.E. Changes in phosphate fractions growth r-ate, nodulation and nitrogn2 fixation of phosphorus starved soybean pla-nts. *J. Plant Nutr.* **2011**, *34*, 2055–2068. [CrossRef]
52. Thuynsma, R.; Valentine, A.; Kleinert, A. Phosphorus deficiency affects the allocation of below-ground resources to combined cluster roots and nodules in Lupinusalbus. *J. Plant Physiol.* **2014**, *171*, 285–291. [CrossRef] [PubMed]
53. Lazali, M.; Drevon, J.J. The nodule conductance to O_2 diffusion increases with phytase activity in N_2-fixing Phaseolus vulgaris L. *Plant Physiol. Biochem.* **2014**, *80*, 53–59. [CrossRef] [PubMed]
54. Al-Niemi, T.S.; Kahn, M.L.; Mc De Rmott, T.R. Phosphorus uptake by bean nodules. *Plant Soil* **1998**, *198*, 71–78. [CrossRef]
55. Burleigh, S.H.; Harrison, M.J. The down-regulation of Mt4-like genes by phosphate fertilization occurs systemically and involves phosphate translocation to the shoots. *Plant Physiol.* **1999**, *119*, 241–248. [CrossRef] [PubMed]
56. Wouterlood, M.; Lambers, H.; Veneklaas, E. Plant phosphorus status has a limited influence on the concentration of phosphorus-mobilising carboxylates in the rhizosphere of chickpea. *Funct. Plant Biol.* **2005**, *32*, 153–159. [CrossRef]
57. Bonser, A.; Lynch, J.; Snapp, S. Effect of phosphorus deficiency on growth angle of basal roots in Phaseolus vulgaris. *New Phytol.* **1996**, *132*, 281–288. [CrossRef]
58. Isidra-Arellano, M.C.; Pozas-Rodríguez, E.A.; RocíoReyero-Saavedra, M.; Arroyo-Canales, J.; Ferrer-Orgaz, S.; Socorro Sánchez-Correa, M.; Cardenas, L.; Covarrubias, A.A.; Valdés-López, O. Inhibition of legume nodulation by Pi deficiency is dependent on the autoregulation of nodulation (AON) pathway. *Plant J.* **2020**, *103*, 1125–1139. [CrossRef]

agronomy

MDPI

Article

Effect of Phosphorus Supply Levels on Nodule Nitrogen Fixation and Nitrogen Accumulation in Soybean (*Glycine max* L.)

Hongyu Li [1], Lihong Wang [2], Zuowei Zhang [2,*], Aizheng Yang [2] and Deping Liu [2]

[1] College of Agriculture, Northeast Agricultural University, Harbin 150030, China
[2] School of Water Conservancy and Civil Engineering, Northeast Agricultural University, Harbin 150030, China
* Correspondence: zzw@neau.edu.cn

Abstract: The specific mechanism by which phosphorus affects nodule nitrogen fixation and nitrogen absorption in soybeans remains inconclusive. To further quantitatively analyze the effect of phosphorus on nodule nitrogen fixation and nitrogen accumulation in soybeans, this experiment was carried out under sand culture conditions. The experiment consisted of six phosphorus supply levels (1 mg/L, 11 mg/L, 21 mg/L, 31 mg/L, 41 mg/L, 51 mg/L). The acetylene reduction method and ^{15}N tracer method (50 mg/L $(NH_4)_2SO_4$) were used to determine and analyze the nodule growth status, nodule nitrogenase activity, nitrogen content, and nodule nitrogen fixation rate at initial flowering (R1 stage), initial pod (R3 stage), seed filling (R5 stage) and maturity stages (R8 stage). The results are described as follows: 1. The nitrogen fixation of soybean nodules at different growth stages has different requirements for phosphorus supply levels. The initial flowering stage and seed-filling stage were 31 mg/L–41 mg/L, and the initial pod stage was 51 mg/L. 2. The nitrogen source in different parts of soybean showed different trends with different growth periods and phosphorus supply concentrations. Among them, from the initial flowering stage to the seed filling stage, the main body of the nitrogen supply of soybean shoots in the low phosphorus treatment (1 mg/L–31 mg/L) gradually changed from fertilizer nitrogen to nodule nitrogen fixation, while the main body of the nitrogen supply of soybean shoots in the high phosphorus treatment (41 mg/L–51 mg/L) always showed nodule nitrogen fixation and was transformed into fertilizer nitrogen at the mature stage. The main nitrogen supply to the roots of soybean at different levels of phosphorus supply from the initial flowering to the initial pods and maturity stage was fertilizer nitrogen, and the main nitrogen supply at the seed filling stage was nodule nitrogen fixation. The nitrogen supply to the main body of soybean nodules was constantly nodule nitrogen fixation. 3. Different phosphorus supply levels significantly affected the nitrogen fixation of soybean nodules ($R^2 \geq 0.803$), and both the acetylene reduction method and the ^{15}N tracer method could be used to determine the nitrogen fixation capacity of soybean nodules. This study indicated the optimal phosphorus supply level of nodules in different growth stages of soybean and clarified the main body of phosphorus supply in different parts of soybean at different growth stages, which pointed out the direction for further improving the utilization efficiency of soybean nitrogen and phosphorus fertilizer.

Keywords: phosphorus supply levels; nodulation; nitrogen accumulation; nitrogen fixation

Citation: Li, H.; Wang, L.; Zhang, Z.; Yang, A.; Liu, D. Effect of Phosphorus Supply Levels on Nodule Nitrogen Fixation and Nitrogen Accumulation in Soybean (*Glycine max* L.). *Agronomy* **2022**, *12*, 2802. https://doi.org/10.3390/agronomy12112802

Academic Editors: Gang Li, Dong-Xing Guan and Daniel Menezes-Blackburn

Received: 16 October 2022
Accepted: 3 November 2022
Published: 10 November 2022

1. Introduction

Phosphorus is one of the three basic elements required for crop growth, second only to nitrogen, which has a significant effect on the growth and nodulation process of leguminous crops [1–3]. According to statistics, 40% of the world's arable land grows crops whose yield is limited by phosphorus [4]. Studies have shown that the reasonable application of phosphorus fertilizer can regulate the growth and development of soybean nodules, enhance the nitrogen fixation capacity of nodules, and promote plant growth, thereby achieving the effect of phosphorus promoting nitrogen and increasing yield [5–11]. At present, the effects of phosphorus on nitrogen fixation in nodules mainly focus on the effect of different phosphorus levels on nodule growth and metabolism [12–14]. Some studies have suggested that low phosphorus concentration reduces the nitrogenase activity in nodules of legume crops due to the reduction of ATP energy in nodules [15], the decrease in leghemoglobin content [16], decreased Fe content [5], and excessive secretion of organic acids [17]. Other studies have found that leguminous crops adapt to low phosphorus concentrations by increasing phytase and phosphatase activity in nodules [18,19]. Some studies have also suggested that low phosphorus concentration inhibits the nitrogen fixation of leguminous nodules and the nitrogen absorption of plants [20,21]. Additional analysis of the physiological effects of phosphorus levels on nitrogen fixation and metabolism in nodules showed that the effect of phosphorus on nitrogen fixation in nodules was achieved by regulating nitrogen metabolism [22]. Characterized by low phosphorus conditions, the proportion of nodule nitrogen accumulated by legume crops decreased significantly, while the nitrogen content in plants and leaves increased, indicating that the assimilated nitrogen had exceeded the demand of plant growth and development for nitrogen, and the accumulation of asparagine in roots and nodules increased, further indicating that the effect of phosphorus on nodule nitrogen fixation was achieved by regulating nitrogen metabolism [12–14]. The above studies have shown that phosphorus levels have a significant effect on nitrogen fixation and nitrogen uptake by nodules of leguminous crops, but they have not been able to quantitatively characterize the main nitrogen supply of leguminous crops in each growth period under different phosphorus supply levels, nor have they affected different growth stages according to the growth process. An in-depth analysis of the effect of nitrogen fixation by soybean nodules under the phosphorus supply level is not conducive to improving the utilization efficiency of nitrogen and phosphorus fertilizers in soybean. Therefore, this study intended to quantitatively characterize the nodule growth status, nodule nitrogenase activity, nodule nitrogen fixation rate, and nitrogen accumulation of soybean plants with different phosphorus supply levels under sand culture conditions by the ^{15}N tracer method and provide a theoretical basis for analyzing the physiological mechanism of phosphorus application affecting soybean nodule nitrogen fixation and nitrogen absorption.

2. Materials and Methods

2.1. Test Site Overview

The study was conducted at the Experimental Base of Northeast Agricultural University in 2019, which is located in the Xiangfang District, Harbin City, Heilongjiang Province, China, with geographical coordinates of 126.43°43′ E, 45°44′ N. The annual precipitation is 500–550 mm, and the accumulated temperature is $\geq$10 °C and 2700 °C.

2.2. Test Materials

The soybean variety tested was Kenfeng 16 (*Glycine max* L.) (Heilongjiang Academy of Land Agricultural Reclamation Science, Heilongjiang, China). Labeled ammonium sulfate ^{15}N (3.36% abundance) was purchased from Shanghai Chemical Industry Research Institute, and the tested phosphorus source was KH_2PO_4.

2.3. Experimental Design

Six phosphorus levels were set up in the experiment: 1 mg/L, 11 mg/L, 21 mg/L, 31 mg/L, 41 mg/L, and 51 mg/L, expressed as P1, P11, P21, P31, P41, and P51, respectively. In each treatment, $(^{15}NH_4)_2SO_4$ with a ^{15}N abundance of 3.36% was used as a nitrogen source to analyze the nitrogen fixation ability of nodules. Phosphorus levels and the composition of the nutrient solution refer to Yao et al. [23] and Li et al. [24] using the nutrient solution formula, with a slight improvement. The specific components are shown in Tables S1 and S2.

2.4. Test Treatment

Potted cultivation with river sand was used in this study. Each plastic pot was 0.30 m in diameter and 0.28 m in height. Two drainage holes 1 cm in diameter were drilled at the bottom of the pot, with one on each side of the partition plate. Each pot was filled with 20 kg of washed sand. Before use, the sand was thoroughly washed with tap water and then rinsed twice with distilled water. The soybean seeds were sown in fine sand at a depth of 2 cm and cultured in an incubator at 30 °C for approximately 3 days. When the distance between the growing point of the cotyledon and the tip of the root reached approximately 7–10 cm, the root system of soybean seedlings was exposed by rinsing with distilled water; then, uniform seedlings were selected for transplanting, and two seedlings were kept per pot.

The inoculation of rhizobia was carried out when the opposite true leaves of soybean were fully expanded. The soybean nodules stored in the field in the previous year were ground into powder and then added to the nutrient solution, and each liter of nutrient solution contained about 5 g nodules for continuous inoculation for 5 days. According to the characteristics of soybean fertilizer demand, the nutrient supply was divided into three periods. The first period was from transplanting the seedlings before the VC stage (the unfolded cotyledon stage), and they were drenched with distilled water once a day, 500 mL per pot. The second period was the unfolded cotyledon to fully expanded stage to the V_4 stage (fourth trifoliate leaf stage), and the basic nutrient solution of 50 mg/L $(NH_4)_2SO_4$ and 31 mg/L KH_2PO_4 was poured once a day, 500 mL each time. The third period was from V4 to harvest, and ^{15}N labeling and different phosphorus supply levels were carried out. This period is divided into three stages, of which the V4 stage to the R1 stage (initial flowering stage) is the first stage, and the nutrient solution is poured once a day, with 500 mL each time. The second stage is from the R1 stage to the R8 stage (maturity stage), and the same nutrition is used as the first stage, but the nutrient solution is poured twice a day, 500 mL each time, that is, 1000 mL of nutrients per pot every day. From the R8 stage to harvest is the third stage, considering the decline in nutrients required by plants, the supply of nutrient solution was reduced to 500 mL once a day (Table S3).

2.5. Sampling and Measurement

Samples were taken 4 times at initial flowering (R1 stage, 42 days), initial pod (R3 stage, 15 days), seed filling (R5 stage, 16 days), and maturity stages (R8 stage, 38 days)., and each treatment was repeated 4 times. The shoots were cut along the boundary between the ground and the ground at 8:00–10:00 AM on a sunny day, and the underground part was washed with distilled water to remove the sand and blotted dry with filter paper, and then the nodule nitrogenase activity was measured [25]. After the measurement, the nodules were removed and recorded. The number of nodules and the dry weight of nodules and roots were determined after drying at 60 °C. The shoots were fixed at 105 °C for 1 h, dried at 65 °C to constant weight, and weighed. After crushing, the ^{15}N abundance and plant nitrogen content of each part were determined. When soybean leaves turned yellow, to reduce the error of dry matter quality caused by falling leaves, the fallen leaves were picked up twice a day, morning and evening, and collected in mesh bags for the calculation of soybean dry matter accumulation in the R8 period.

The nitrogenase activity of nodules was determined by Acetylene reduction activity according to the method of Gremaud and Harper (1986).

Plant nitrogen content was determined by the B324 automatic Kjeldahl nitrogen analyzer produced by Shanghai Shengsheng Automatic Analytical Instrument Co., LTD., China.

^{15}N abundance was determined by a mass spectrometer using a double-channel (DI) measurement (Thermo-Fisher Delta V Advantage IRMS, Waltham, MA, USA).

2.6. Data Calculations

The sand culture was used in the experiment and there was no soil factor interference; therefore, the nitrogen source of the plant was derived from the application of ^{15}N-labeled fertilizer nitrogen and atmospheric nitrogen fixed by nodules.

The nitrogenase activity per unit nodule weight was calculated as follows:

$$SNA = \frac{E_R}{D_N - T_R} \quad (1)$$

where SNA is specific nitrogenase activity (unit is C_2H_4 μmol g^{-1} Nodule Dry Mass h^{-1}), E_R is the amount of reduced ethylene (unit is μmol), D_N is the dry weight of the nodule (unit is g), and T_R is the reaction time (unit is h), the same as below.

The nitrogenase activity per plant was calculated as follows:

$$ARA = \frac{E_R}{T_R} \quad (2)$$

where ARA is acetylene reduction activity (unit is C_2H_4 μmol h^{-1} $Plant^{-1}$).

The percentage of nitrogen fixation originating from nodules in the plant was calculated as follows:

$$RNNF = \frac{f_{fertilizer} - f_{treatment}}{f_{ertilizer} - f_0} \quad (3)$$

The formula source: RNNF × f_0 + (1-RNNF) $f_{fertilizer}$ = $f_{treatment}$, where RNNF is nodule nitrogen fixation rate (unit is %), 1-RNNF is fertilizer utilization rate (unit is %), f_0 is background natural abundance (unit is %), and $f_{treatment}$ is treatment ^{15}N abundance (unit is %).

2.7. Statistical Analyses

All statistical analyses were performed using SPSS 21.0 (SPSS Inc., Chicago, IL, USA). All the data were subjected to a normality test prior to a one-way analysis of variance (ANOVA), and Duncan's multiple range test was used at a significance level of $p < 0.05$.

3. Results

3.1. Effect of Different Phosphorus Supply Levels on Nodulation and Nitrogen Fixation

Table 1 shows the number and dry weight of soybean nodules under different phosphorus supply levels from the initial flowering stage (R1) to the seed-filling stage (R5). From the perspective of the growth process, the dry weight and number of soybean nodules in each treatment showed different trends. When the phosphorus supply level increased from 1 mg/L to 31 mg/L, the number of soybean nodules showed a trend of increasing first and then decreasing with the progress of the growth process, and the peak value appeared in the initial pod stage (R3). When the phosphorus supply level was further increased to 41 mg/L and above, it showed a linear growth trend, and the peak appeared in the seed-filling stage (R5). The dry weight of nodules showed a gradually increasing trend with the advancement of the reproductive process. By comparing the number of nodules and the dry weight of nodules, it was found that the number of nodules decreased under the phosphorus supply level of 1 mg/L–31 mg/L from the initial pod stage to the seed filling stage, but the dry weight of nodules increased. From the perspective of phosphorus supply level, the number of nodules and dry weight of soybean nodules all increased with increasing phosphorus supply level. When the phosphorus supply level increased to 41 mg/L, the number of nodules and dry weight of soybean did not increase significantly when the phosphorus supply level was increased again.

Table 1. Effects of different phosphorus supply levels on soybean nodule number and nodule dry weight.

	Treatments	R1	R3	R5
Nodule Number (Per Plant)	P1	66 ± 8.25 c	70 ± 10.53 c	48 ± 2.65 c
	P11	67 ± 7.17 c	79 ± 9.02 bc	57 ± 0.88 c
	P21	70 ± 0.88 bc	92 ± 13.97 bc	66 ± 10.09 c
	P31	85 ± 4.63 b	124 ± 13.92 bc	120 ± 7.22 b
	P41	120 ± 2.89 a	142 ± 7.22 ab	191 ± 5.49 a
	P51	124 ± 2.03 a	197 ± 17.59 a	207 ± 10.39 a
Nodule Weight (g/Plant)	P1	0.10 ± 0.01 e	0.12 ± 0.07 d	0.13 ± 0.01 c
	P11	0.10 ± 0.01 e	0.18 ± 0.01 d	0.26 ± 0.06 c
	P21	0.15 ± 0.01 d	0.52 ± 0.01 c	0.72 ± 0.02 b
	P31	0.25 ± 0.01 c	0.69 ± 0.15 bc	1.02 ± 0.19 b
	P41	0.26 ± 0.01 b	0.77 ± 0.05 ab	1.41 ± 0.16 a
	P51	0.28 ± 0.01 a	0.88 ± 0.07 a	1.42 ± 0.06 a

The data are represented as the mean values ± standard error and independent measurements with three replicates. Different lowercase letters indicate the differences between treatments at a significance level of 5%.

Table 2 shows soybean SNA and ARA under different phosphorus levels from the initial flowering stage (R1) to the seed-filling stage (R5). From the perspective of the growth process, SNA and ARA showed different trends. SNA showed a linear downward trend with the progress of the growth process at the phosphorus supply level of 1 mg/L-41 mg/L, while at the phosphorus supply level of 51 mg/L, SNA showed a trend of first increasing and then decreasing, and the maximum value appeared in the initial pod stage (R3). ARA showed a linear downward trend with the growth process under the phosphorus supply level of 1 mg/L, first increasing and then decreasing under the phosphorus supply levels of 11 mg/L and 51 mg/L, and the peak appeared at the initial pod stage (R3), while the phosphorus supply levels from 21 mg/L to 41 mg/L showed a linear upward trend. From the perspective of phosphorus supply level, SNA and ARA showed similar trends in the R1 and R5 stages; both showed a unimodal curve with the increase in phosphorus supply level, and the maximum appeared at the phosphorus supply level of 31 mg/L–41 mg/L. SNA and ARA in the R3 stage increased with the increase in phosphorus supply level, and the maximum values appeared at the 51 mg/L phosphorus supply level.

Table 2. Effects of different phosphorus supply levels on soybean SNA and ARA.

	Treatments	R1	R3	R5
SNA(C_2H_4 μmol g^{-1} Nodule Dry Mass h^{-1})	P1	27.62 ± 1.81 d	13.96 ± 0.82 e	7.83 ± 0.23 d
	P11	45.63 ± 6.85 c	31.72 ± 1.04 d	22.25 ± 2.17 c
	P21	78.82 ± 3.97 ab	33.74 ± 0.10 d	33.08 ± 0.22 ab
	P31	92.04 ± 8.50 a	42.01 ± 0.14 c	34.97 ± 0.05 a
	P41	73.28 ± 3.43 b	51.27 ± 6.39 b	32.79 ± 0.40 ab
	P51	71.12 ± 1.51 b	76.36 ± 0.62 a	31.06 ± 1.80 b
ARA(C_2H_4 umol h^{-1} $Plant^{-1}$)	P1	3.00 ± 0.29 d	1.35 ± 0.12 d	0.86 ± 0.02 d
	P11	3.70 ± 0.50 d	5.67 ± 0.39 d	5.44 ± 0.55 d
	P21	11.96 ± 0.74 c	17.60 ± 0.16 cd	23.92 ± 0.77 c
	P31	22.63 ± 2.49 a	29.13 ± 2.07 bc	35.82 ± 0.47 b
	P41	18.79 ± 0.09 b	39.80 ± 7.47 b	46.07 ± 4.92 a
	P51	19.82 ± 0.43 ab	66.89 ± 13.61 a	44.30 ± 4.52 ab

Different lowercase letters indicate the differences between treatments at a significance level of 5%.

3.2. *Effect of Different Phosphorus Supply Levels on Soybean Nitrogen Absorption and Distribution*

3.2.1. The Effect of Different Phosphorus Supply Levels on the Abundance of ^{15}N in Soybean Plants

Table 3 shows that the abundance of ^{15}N in soybean plants showed different changes with the progress of the growth process. The abundance of ^{15}N in shoots and roots showed a decreasing trend from the R1-R5 stage and an increasing trend from the R5–R8 stage. Since ^{15}N is mainly derived from the labeled fertilizer nitrogen, it means that from the initial flowering stage to the seed-filling stage, as the nodules grew and provided nitrogen, the fertilizer nitrogen absorbed by the shoots and roots gradually became unable to meet their own needs, and an increasing amount of nitrogen was fixed through the nodules. To obtain nitrogen, as the nodules continue to decline from the seed-filling stage to the mature stage, the nitrogen absorbed by the shoots and roots comes from fertilizer nitrogen, which in turn causes the ^{15}N abundance to rise in the R8 stage, completing a round of major nitrogen supply. It can also be seen from the data of different treatments in each growth period in the table that the abundance of ^{15}N in each part of soybean increased first and then decreased with the increase in phosphorus supply level, and the phosphorus supply level peaked at 11 mg/L and then increased. The abundance of ^{15}N at the phosphorus supply level was significantly reduced.

Table 3. ^{15}N abundance (%) in plant organs of soybean under different phosphorus supply levels.

Stages	Treatments	Shoot	Root	Nodule
R1	P1	2.27 ± 0.01 b	2.41 ± 0.01 b	1.12 ± 0.03 a
	P11	2.33 ± 0.01 a	2.54 ± 0.01 a	1.15 ± 0.04 a
	P21	2.07 ± 0.01 c	2.31 ± 0.01 c	0.96 ± 0.03 b
	P31	1.97 ± 0.01 d	2.28 ± 0.01 d	0.94 ± 0.09 b
	P41	1.55 ± 0.01 f	1.96 ± 0.01 f	0.78 ± 0.02 c
	P51	1.65 ± 0.01 e	1.97 ± 0.01 e	0.87 ± 0.03 bc
R3	P1	1.89 ± 0.01 c	2.07 ± 0.01 c	0.87 ± 0.01 abc
	P11	1.99 ± 0.01 a	2.17 ± 0.01 a	0.92 ± 0.01 a
	P21	1.91 ± 0.01 b	2.14 ± 0.01 b	0.90 ± 0.01 ab
	P31	1.74 ± 0.01 d	2.04 ± 0.01 d	0.81 ± 0.03 bc
	P41	1.74 ± 0.01 d	2.04 ± 0.01 d	0.83 ± 0.04 abc
	P51	1.50 ± 0.01 e	1.86 ± 0.01 e	0.77 ± 0.06 c
R5	P1	1.38 ± 0.01 c	1.39 ± 0.01 b	0.67 ± 0.02 ab
	P11	1.46 ± 0.01 a	1.45 ± 0.02 a	0.69 ± 0.01 ab
	P21	1.41 ± 0.01 b	1.33 ± 0.01 c	0.68 ± 0.01 a
	P31	1.25 ± 0.01 d	1.26 ± 0.01 d	0.62 ± 0.01 b
	P41	1.19 ± 0.01 e	1.18 ± 0.01 e	0.65 ± 0.05 ab
	P51	1.18 ± 0.01 e	1.15 ± 0.01 e	0.66 ± 0.03 ab
R8	P1	2.41 ± 0.04 a	2.42 ± 0.02 a	-
	P11	2.53 ± 0.04 a	2.49 ± 0.04 a	-
	P21	2.22 ± 0.05 b	2.13 ± 0.01 b	-
	P31	1.95 ± 0.10 c	1.94 ± 0.09 b	-
	P41	2.15 ± 0.05 b	2.01 ± 0.17 b	-
	P51	2.11 ± 0.01 bc	1.96 ± 0.05 b	-

Different lowercase letters indicate the differences between treatments at a significance level of 5%.

3.2.2. Proportion of Nitrogen Sources in Various Organs of Soybean under Different Phosphorus Supply Levels

Table 4 shows the proportion of nitrogen sources in soybean organs under different phosphorus supply levels. It can be seen from the table that the sources of nitrogen in different parts of soybean showed different changes with different growth stages. From the R1 stage to the R5 stage, the proportion of fertilizer nitrogen absorbed by each part of the soybean showed a downward trend and increased from the R5 stage to the R8 stage (except nodules). The proportion of nitrogen fixation from nodules and the proportion of nitrogen absorbed from fertilizers in each part showed an opposite trend with the growth process, and the nitrogen source of each part of the soybean showed different changes with changes in the level of phosphorus supply. In addition, from Table 4, it can be seen that the nitrogen sources of different parts of soybean show different changing rules with the phosphorus supply level. Among them, in the R1 stage, the shoots showed that when the phosphorus supply level increased from 1 mg/L to 31 mg/L, the nitrogen source from fertilizer nitrogen accounted for 53.58–65.52% (>50%), and from a nodule, nitrogen fixation accounted for 34.48–46.42% (<50%). When the phosphorus supply level was 41 mg/L to 51 mg/L, the nitrogen source from fertilizer nitrogen accounted for 39.57–42.67% (<50%), and from nodule nitrogen fixation accounted for 57.33–60.43% (>50%). When the phosphorus supply level increased from 1 mg/L to 21 mg/L in the R3 stage, the nitrogen in the shoots that came from fertilizer nitrogen accounted for 50.08–54.21%, and the nitrogen fixation from nodules accounted for 45.79–49.20%; however when the phosphorus supply level was 31 mg/L–51 mg/L, nitrogen from fertilizers accounted for 37.91–45.85%, and nitrogen from nodules accounted for 54.15–62.09%When the phosphorus supply level increased from 1 mg/L to 51 mg/L in the R5 stage, the shoot nitrogen from fertilizer nitrogen accounted for 27.18–36.50%, and the nitrogen fixation from nodules accounted for 63.50–72.81%. By comparing the changes in nitrogen sources in each treatment in the R1–R5 stages, it can be seen that the main nitrogen supply under low phosphorus treatment will gradually change from fertilizer nitrogen to nitrogen fixation by the nodules as the growth stage progresses. At the R8 stage, the nitrogen in the shoots was 52.68–68.22% from fertilizers and 27.65–47.32% from nodules, The nitrogen source of roots in each growth stage was as follows: when the phosphorus supply level increased from 1 mg/L to 51 mg/L at the R1 stage, the nitrogen source of fertilizer nitrogen accounted for 53.01–72.47%, and the nitrogen fixation from nodules accounted for 27.53–46.99%. During the R3 period, nitrogen from fertilizers accounted for 49.99–60.36%, and nitrogen fixation from nodules accounted for 39.64–50.01%, indicating that nitrogen in the roots with different phosphorus levels from the initial flowering stage to the initial pod's stage was the main nitrogen supply for fertilizers. In the R5 period, when the phosphorus supply level increased from 1 mg/L to 51 mg/L, nitrogen from fertilizers accounted for 26.14–36.04%, and nitrogen fixation from nodules accounted for 63.96–73.86%. In the R8 period, when the phosphorus supply level increased from 1 mg/L to 51 mg/L, the nitrogen of the roots that came from fertilizer nitrogen accounted for 52.41–68.68%, and the nitrogen fixation from nodules accounted for 29.26–47.59%. The nitrogen source from nodules showed that from the R1-R5 stages, nitrogen from self-fixation accounted for 73.95–91.63%, and nitrogen from fertilizer accounted for only 8.37–26.04%.

Table 4. The proportion of nitrogen sources in different parts of soybean under different phosphorus supply levels (%).

Stages	Treatments	Shoot		Root		Nodule	
		Nitrogen Derived from the Fertilizer	Nitrogen Derived from Atmosphere	Nitrogen Derived from the Fertilizer	Nitrogen Derived from Atmosphere	Nitrogen Derived from the Fertilizer	Nitrogen Derived from Atmosphere
R1	P1	63.58 ± 0.14 b	36.42 ± 0.14 e	68.11 ± 0.16 b	31.89 ± 0.16 e	25.21 ± 0.16 a	74.79 ± 0.09 c
	P11	65.52 ± 0.25 a	34.48 ± 0.25 f	72.47 ± 0.06 a	27.53 ± 0.06 f	26.04 ± 1.25 a	73.95 ± 1.25 c
	P21	56.90 ± 0.20 c	43.10 ± 0.20 d	64.88 ± 0.03 c	35.12 ± 0.03 d	19.89 ± 0.88 b	80.11 ± 0.88 b
	P31	53.58 ± 0.02 d	46.42 ± 0.02 c	64.03 ± 0.19 d	35.97 ± 0.19 c	18.94 ± 2.97 b	81.06 ± 2.97 b
	P41	39.57 ± 0.15 f	60.43 ± 0.15 a	53.01 ± 0.13 f	46.99 ± 0.13 a	13.84 ± 0.69 d	86.16 ± 0.69 a
	P51	42.67 ± 0.10 e	57.33 ± 0.10 b	53.50 ± 0.02 e	46.50 ± 0.02 b	16.70 ± 1.09 bc	83.30 ± 1.09 ab
R3	P1	50.80 ± 0.30 c	49.20 ± 0.30c	56.78 ± 0.07 c	43.22 ± 0.07 c	16.75 ± 0.38 abc	83.25 ± 0.38 abc
	P11	54.21 ± 0.38 a	45.79 ± 0.38 e	60.36 ± 0.06 a	39.64 ± 0.06 e	18.43 ± 0.27 a	81.57 ± 0.27 c
	P21	51.51 ± 0.16 b	48.49 ± 0.16 d	59.10 ± 0.16 b	40.90 ± 0.16 d	17.58 ± 0.40 ab	82.42 ± 0.40 bc
	P31	45.72 ± 0.05 d	54.28 ± 0.05 b	55.74 ± 0.27 d	44.26 ± 0.27 b	14.67 ± 0.84 bc	85.33 ± 0.84 ab
	P41	45.85 ± 0.07 d	54.15 ± 0.07 b	55.85 ± 0.35 d	44.15 ± 0.35 b	15.53 ± 1.24 abc	84.47 ± 1.24 abc
	P51	37.91 ± 0.04 e	62.09 ± 0.04 a	49.99 ± 0.08 e	50.01 ± 0.08 a	13.53 ± 1.94 c	86.47 ± 1.94 a
R5	P1	34.03 ± 0.12 c	65.97 ± 0.12 c	34.00 ± 0.12 b	66.00 ± 0.10 d	9.89 ± 0.58 ab	90.11 ± 0.59 ab
	P11	36.50 ± 0.22 a	63.50 ± 0.22 e	36.04 ± 0.69 a	63.96 ± 0.69 e	10.78 ± 0.45 a	89.22 ± 0.45 b
	P21	34.88 ± 0.07 b	65.12 ± 0.07 d	32.15 ± 0.17 c	67.85 ± 0.17 c	10.38 ± 0.26 ab	89.62 ± 0.26 ab
	P31	29.30 ± 0.19 d	70.70 ± 0.19 b	29.91 ± 0.22 d	70.09 ± 0.22 b	8.37 ± 0.28 b	91.63 ± 0.28 a
	P41	27.42 ± 0.15 e	72.58 ± 0.15 a	27.10 ± 0.14 e	72.89 ± 0.80 a	9.33 ± 1.83 ab	90.67 ± 1.83 ab
	P51	27.18 ± 0.08 e	72.81 ± 0.08 a	26.14 ± 0.26 e	73.86 ± 0.26 a	9.57 ± 1.16 ab	90.43 ± 1.16 ab
R8	P1	68.22 ± 1.30 a	31.78 ± 1.30 c	68.68 ± 0.67 a	31.32 ± 0.67 b	-	-
	P11	72.35 ± 1.38 a	27.65 ± 1.38 c	70.74 ± 1.33 a	29.26 ± 1.33 b	-	-
	P21	61.77 ± 1.57 b	38.23 ± 1.57 b	58.73 ± 0.36 b	41.27 ± 0.36 a	-	-
	P31	52.68 ± 3.43 c	47.32 ± 3.43 a	52.41 ± 3.00 b	47.59 ± 3.00 a	-	-
	P41	59.69 ± 1.62 b	40.31 ± 1.62 b	54.95 ± 5.70 b	45.05 ± 5.70 a	-	-
	P51	58.28 ± 0.42 bc	41.71 ± 0.42 ab	53.20 ± 1.71 b	46.80 ± 1.71 a	-	-

Different lowercase letters indicate the differences between treatments at a significance level of 5%.

3.2.3. Nitrogen Accumulation and Nodule Nitrogen Fixation in Different Parts of Soybean under Different Phosphorus Supply Levels

According to the dry matter accumulation in each part of the soybean and the nitrogen content per gram of dry matter, the nitrogen accumulation in each part can be identified, and the nitrogen accumulation in each part multiplied by the nitrogen fixation rate of the nodule (Table 4) can be obtained. Table 5 shows that the nitrogen accumulation in all parts of soybean increased linearly with the growth process, and the nitrogen fixation in nodules increased from the R1 stage to the R5 stage and decreased from the R5 stage to the R8 stage (except nodules). The different phosphorus supply treatments generally showed a significant increase with increasing phosphorus supply levels. Table 5 shows that the nitrogen content of the shoots and roots reached a maximum when the phosphorus supply level increased from 1 mg/L to 31 mg/L at the R1 stage, and then the phosphorus supply level showed a decreasing trend. In the R3 stage, when the phosphorus supply level increased to 21 mg/L, there was no significant upward trend in increasing the phosphorus supply level. The R5 and R8 stages increased with the increase in the phosphorus supply level, but when the phosphorus supply level increased to 31–41 mg/L, there was no significant upward trend in increasing the phosphorus supply level. The initial flowering stage was 31 mg/L, the initial pod stage was 21 mg/L, and the seed-filling stage and maturity stage were both 31 mg/L–41 mg/L. However, the nitrogen accumulation in nodules increased with increasing phosphorus supply levels from the R1 stage to the R5 stage. When the phosphorus supply level increased by 41 mg/L, there was no significant increasing trend when the phosphorus supply level was increased. It can also be concluded from Table 5 that the nitrogen fixation amount of the shoot increases to a maximum of 41 mg/L in the R1 stage, and there is a downward trend when the phosphorus supply level is increased again. From the R3 stage to the R5 stage, it increased with increasing phosphorus supply level, and there was no significant increase between 31 mg/L and 51 mg/L in the R3 stage. There was no significant increase between 41 mg/L and 51 mg/L in the R5 stage. In the R8 stage, when the phosphorus supply level increased to 31 mg/L, the maximum level increased, and the phosphorus supply level showed a decreasing trend. This indicated that it was not the case that the higher the concentration of phosphorus supply was, the better the accumulation of nitrogen fixation in soybean nodules. The phosphorus supply level of 31 mg/L–41 mg/L was the most suitable phosphorus supply level to promote the efficient utilization of nitrogen fixation by nodules in the shoots of soybean. The nitrogen fixation amount by nodules in soybean was the largest in the 31 mg/L treatments in the R1, R3, and R8 stages, while it was 41 mg/L in the R5 stage. The nitrogen fixation of nodules showed that from the R1–R5 stages, it increased with the progression of the growth period and increased with the increase in the phosphorus supply level in the same period, but there was no significant difference when the phosphorus supply level reached 31 mg/L in the R1 stage and no significant increase when the phosphorus supply level reached 41 mg/L in the R3 and R5 stages.

Table 5. Nitrogen accumulation and nodule nitrogen fixation in different parts of soybean under different phosphorus supply levels (mg).

Stages	Treatments	Shoot		Root		Nodule	
		Nitrogen Accumulation	Nodule Nitrogen Fixation	Nitrogen Accumulation	Nodule Nitrogen Fixation	Nitrogen Accumulation	Nodule Nitrogen Fixation
R1	P1	115.89 ± 4.40 c	42.20 ± 1.46 d	47.57 ± 2.39 a	15.18 ± 0.80 b	2.88 ± 0.23 d	2.15 ± 0.17 c
	P11	130.85 ± 9.18 c	45.16 ± 3.46 d	30.62 ± 0.81 c	8.43 ± 0.23 d	3.47 ± 0.30 d	2.57 ± 0.26 c
	P21	160.71 ± 5.04 b	69.24 ± 1.87 c	36.78 ± 1.98 b	12.92 ± 0.71 c	6.29 ± 0.26 c	5.03 ± 0.16 b
	P31	228.20 ± 12.27 a	105.93 ± 5.65 b	48.60 ± 0.31 a	17.48 ± 0.14 a	10.19 ± 0.34 b	8.28 ± 0.55 a
	P41	224.62 ± 6.12 a	135.74 ± 3.86 a	36.55 ± 1.63 b	17.17 ± 0.79 a	10.50 ± 0.66 ab	9.05 ± 0.61 a
	P51	209.76 ± 15.56 a	120.28 ± 9.10 b	31.38 ± 0.69 c	14.59 ± 0.32 bc	11.58 ± 0.54 a	9.65 ± 0.56 a

Table 5. *Cont.*

Stages	Treatments	Shoot		Root		Nodule	
		Nitrogen Accumulation	Nodule Nitrogen Fixation	Nitrogen Accumulation	Nodule Nitrogen Fixation	Nitrogen Accumulation	Nodule Nitrogen Fixation
R3	P1	215.25 ± 5.93 b	109.03 ± 3.14 d	70.37 ± 3.62 b	30.41 ± 1.52 c	4.96 ± 0.33 d	4.13 ± 0.26 d
	P11	342.50 ± 19.47 b	156.84 ± 9.11 cd	70.60 ± 1.10 b	27.98 ± 0.46 c	7.56 ± 0.30 d	6.17 ± 0.23 d
	P21	523.62 ± 51.90 a	253.83 ± 24.76 bc	84.30 ± 3.28 a	34.48 ± 1.42 b	23.75 ± 0.69 c	19.57 ± 0.53 c
	P31	543.99 ± 91.92 a	295.34 ± 50.67 ab	89.23 ± 3.65 a	39.48 ± 1.48 a	30.06 ± 6.79 bc	25.68 ± 5.94 bc
	P41	580.62 ± 31.39 a	314.44 ± 17.18 ab	85.69 ± 2.36 a	37.82 ± 0.76 ab	37.60 ± 2.11 ab	31.72 ± 1.37 ab
	P51	620.71 ± 86.00 a	385.32 ± 53.19 a	71.36 ± 0.66 b	35.68 ± 0.33 b	42.92 ± 2.61 a	37.01 ± 1.41 a
R5	P1	352.59 ± 17.24 d	232.64 ± 11.78 e	90.12 ± 8.83 b	59.48 ± 5.84 c	6.65 ± 0.52 c	5.97 ± 0.43 c
	P11	709.27 ± 28.87 c	450.28 ± 17.11 d	98.52 ± 6.85 ab	63.07 ± 4.83 c	12.45 ± 2.47 c	11.13 ± 2.27 c
	P21	919.26 ± 91.54 b	598.51 ± 59.18 c	105.06 ± 9.12 ab	71.27 ± 6.06 bc	37.87 ± 1.75 b	33.93 ± 1.51 b
	P31	1046.43 ± 9.18 b	739.81 ± 6.68 b	119.72 ± 7.01 a	83.94 ± 5.10 ab	48.97 ± 7.83 b	44.83 ± 7.08 b
	P41	1412.70 ± 68.32 a	1025.37 ± 49.56 a	123.97 ± 6.26 a	90.37 ± 4.23a	69.74 ± 8.22 a	63.16 ± 7.09 a
	P51	1425.14 ± 78.06 a	1037.73 ± 56.37 a	102.40 ± 7.08 ab	75.65 ± 5.34 abc	77.19 ± 5.54 a	69.76 ± 4.70 a
R8	P1	365.04 ± 6.40 e	115.86 ± 2.70 d	80.63 ± 1.64 c	25.27 ± 1.03 c	-	-
	P11	912.96 ± 54.74 d	253.77 ± 26.37 d	94.43 ± 2.10 bc	27.64 ± 1.56 c	-	-
	P21	1388.97 ± 63.41 c	532.34 ± 42.69 c	113.47 ± 7.86 abc	46.77 ± 2.86 b	-	-
	P31	1949.42 ± 160.16 ab	927.12 ± 123.85 a	145.47 ± 7.13 a	69.06 ± 4.48 a	-	-
	P41	2022.57 ± 70.08 a	815.01 ± 41.46 ab	125.81 ± 16.35 ab	55.20 ± 5.60 ab	-	-
	P51	1711.01 ± 73.11 b	713.99 ± 33.22 b	106.95 ± 23.99 ab	49.28 ± 9.24 b	-	-

Different lowercase letters indicate the differences between treatments at a significance level of 5%.

3.3. Comparisons between ARA, RNNF%, and Naccumulation of Nodules

Figure 1 shows the changing trend of the soybean ARA nitrogen fixation rate and nitrogen fixation amount of soybean nodules under different phosphorus supply levels at different growth stages. In terms of the growth process, the ARA, the nitrogen fixation rate of the whole nodule, and the nitrogen fixation amount of the whole nodule of soybean in the R1, R5, and R8 stages increased first and then decreased with the increase in the phosphorus supply level, peaked at 31 mg/L–41 mg/L phosphorus supply level.

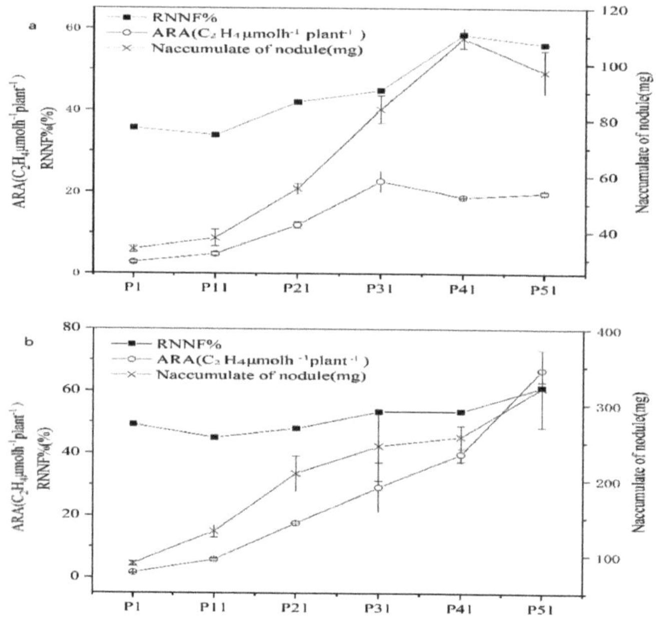

Figure 1. *Cont.*

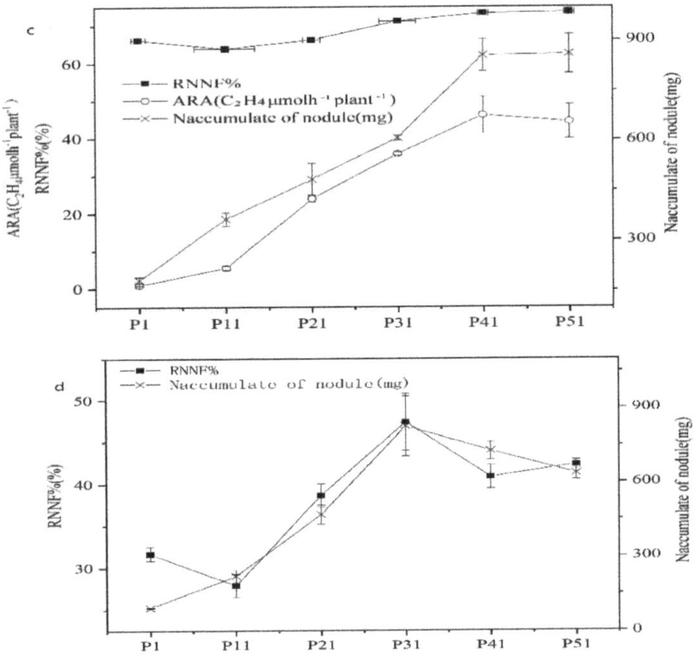

Figure 1. Variation trends of ARA, root nodule nitrogen fixation rate, and root nodule nitrogen fixation in soybean under different phosphorus supply levels. (**a**): initial flowering period; (**b**): initial pod stage; (**c**): filling period; (**d**): mature period.

3.4. Correlations Analysis between Different Phosphorus Supply Levels and Soybean ARA, Nitrogen Fixation Rate of Nodules, and Nitrogen Fixation of Nodules

Table 6 shows the correlation between the phosphorus supply level at different growth stages and ARA, whole nodule nitrogen fixation rate, and whole nodule nitrogen fixation amount. The correlation performance changes in each growth period were consistent, which all showed that the level of phosphorus supply was significantly positively correlated with ARA, the nitrogen fixation rate of the whole nodule, and the nitrogen fixation rate of the whole nodule ($R^2 \geq 0.803$). There was a very significant positive correlation ($R^2 \geq 0.791$) in the nitrogen fixation of the nodules of the whole plant, while the nitrogen fixation rate of the whole plant nodules and the nitrogen fixation of the whole plant nodules had a very significant positive correlation ($R^2 \geq 0.847$) in the R1 to R5 stages and were significantly positively correlated with the nitrogen fixation amount of whole nodules in the R8 stage ($R^2 = 0.725$).

Table 6. Correlation of different phosphorus supply levels, ARA, nitrogen fixation rate of nodules, and nitrogen fixation of nodules.

Stages		Phosphorus Supply Level	ARA	Whole Nodule Nitrogen Fixation Rate
R1	nodule nitrogen fixation amount	0.913 **	0.876 **	0.966 **
	whole nodule nitrogen fixation rate	0.902 **	0.791 **	-
	ARA	0.978 **	-	-
R3	nodule nitrogen fixation amount	0.987 **	0.930 **	0.847 **
	whole nodule nitrogen fixation rate	0.883 **	0.977 **	-
	ARA	0.966 **	-	-
R5	nodule nitrogen fixation amount	0.992 **	0.981 **	0.875 **
	whole nodule nitrogen fixation rate	0.847 **	0.837 **	-
	ARA	0.971 **	-	-
R8	nodule nitrogen fixation amount	0.952 **	-	0.725 *
	whole nodule nitrogen fixation rate	0.803 **	-	-

* denote a significant difference at the 5% level; ** denote a significant difference at the 1% level.

4. Discussion

Most studies have shown that phosphorus can promote nodule growth and nitrogenase activity in leguminous crops, showing that with the increase in phosphorus supply, the number and dry weight of nodules and the activity of nodule nitrogenase in leguminous crops are significantly increased [26–33]. In this experiment, it was found that 41 mg/L was the optimum phosphorus supply level for soybean nodule growth, and low phosphorus concentration (1 mg/L–31 mg/L) promoted the phenomenon of "survival of the fittest" in soybean nodules. This is generally consistent with the results of Miao et al. [34] and Jemo et al. [29], who found that soybean nodule dry weight, nodule number, and nodule nitrogenase all decreased with the decrease in phosphorus supply level, and Yao et al. [23] found that phosphorus levels that were too high or too low inhibited the nitrogen fixation of soybean nodules; this was similar to the conclusion of Tsvetkova et al. [35] and Magadlela et al. [7] that low phosphorus concentration inhibited the nitrogen fixation of nodules by inhibiting the growth and phosphorus absorption of soybean plants, but differed from the conclusion of Almeida et al. [13] that low phosphorus concentration can stimulate the growth of soybean nodules and inhibit nitrogen fixation by nodules. This may be due to the different levels of low-phosphorus stress or the change in the direction of phosphorus transport caused by low-phosphorus stress. This indicates that different phosphorus supply levels will significantly affect the nitrogen fixation of soybean nodules. In addition, this experiment also found that the sources of nitrogen in different parts of soybean showed different trends with the growth period and phosphorus supply concentration. Among them, from the initial flowering stage to the filling stage, the main body of nitrogen supply in the low phosphorus treatment (1 mg/L–31 mg/L) will gradually change from fertilizer nitrogen to nodule nitrogen fixation, while the main body of nitrogen supply in the high phosphorus treatment (41 mg/L–51 mg/L) will always be nodule nitrogen fixation and will be transformed into fertilizer nitrogen at the mature stage. The main nitrogen supply of soybean roots at different levels of phosphorus supply from the initial flowering to the initial pods and maturity stage was fertilizer nitrogen, and the main nitrogen supply at the seed filling stage was nodule nitrogen fixation (as shown in Figure 2). This is different from the findings of Raji et al. [36], Schulze et al. [37], and Cavard et al. [38] that under soil culture conditions, the phosphorus supply level had no significant effect on the nitrogen fixation rate of leguminous nodules and is also different from the research conclusion drawn by Magadlela et al. [28] that 40–50% of the nitrogen nutrition absorbed by plants comes from symbiotic nitrogen fixation. This may be because the soil itself contains a certain amount of phosphorus under soil culture conditions, which affects the phosphorus supply gradient of the experimental setup. Then increased, the level of phosphorus indicators with a decreasing trend, while the indicators in the R3 stage increased with the increase in phosphorus supply level. This shows that the phosphorus supply level required for nitrogen fixation of nodules at different growth stages is different. The phosphorus supply level of 31 mg/L–41 mg/L in the R1 and R5-R8 stages is the optimal phosphorus supply level to promote nitrogen fixation of soybean nodules, while the R3 stage is the period when the nodule grows vigorously. The phosphorus supply required for nitrogen fixation by nodules was significantly higher than that in other growth periods. From the changing trend of each index, the nitrogen fixation of ARA and the nitrogen fixation of the whole plant showed the same change rule in each growth period. This shows that both ARA and ^{15}N tracer methods determined by the acetylene reduction method are effective methods for determining nitrogen fixation in nodules. At the 1 mg/L–11 mg/L phosphorus supply level, ARA and the nitrogen fixation rate of the whole root nodule showed opposite trends. This result indicated that the low phosphorus supply would stimulate nodule nitrogen fixation, resulting in the opposite trend of the nitrogen fixation rate of ARA and whole nodules under the phosphorus supply level of 1 mg/L–11 mg/L. In addition, through the acetylene reduction method in this experiment, it was concluded that the nitrogen fixation of soybean nodules in different growth stages has different requirements for phosphorus supply. The initial flowering stage and seed-filling stage were 31 mg/L–41 mg/L, and the initial pod stage was 51 mg/L. The ^{15}N tracer method showed that the optimal phosphorus supply concentration for promoting nitrogen accumulation and efficient utilization of nitrogen fixation in soybean nodules at each growth stage was 31 mg/L–41 mg/L. The optimum phosphorus concentrations for nitrogen fixation of soybean nodules at different growth stages measured by the two methods are relatively close, which indicates that the acetylene reduction method and the ^{15}N tracer method are both effective methods for determining the nitrogen fixation capacity of soybean nodules.

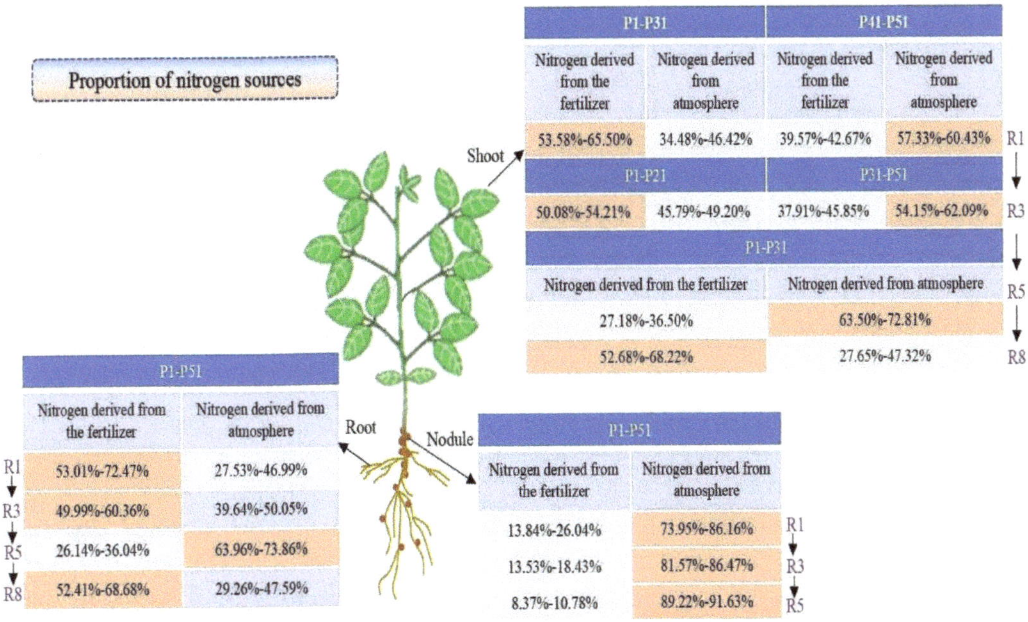

Figure 2. The proportion of nitrogen sources in soybean organs in different growth stages under different phosphorus supply levels.

5. Conclusions

1. The optimum phosphorus supply level for soybean nodule growth is 41 mg/L, and the acetylene reduction method also shows that the nitrogen fixation of soybean nodules at different growth stages has different requirements for phosphorus supply level, which is 31 mg/L–41 mg/L at the initial flowering and seed filling stages and 51 mg/L at the initial pod stage.

2. The nitrogen source in different parts of soybean showed different trends with different growth periods and phosphorus supply concentrations. Among them, from the initial flowering stage to the seed filling stage, the main body of the nitrogen supply of soybean shoots in the low phosphorus treatment (1 mg/L–31 mg/L) will gradually change from fertilizer nitrogen to nodule nitrogen fixation, while the main body of the nitrogen supply of soybean shoots in the high phosphorus treatment (41 mg/L–51 mg/L) will always be nodule nitrogen fixation and will be transformed into fertilizer nitrogen at the mature stage. The main nitrogen supply of soybean roots at different levels of phosphorus supply from the initial flowering to the initial pods and maturity stage was fertilizer nitrogen, and the main nitrogen supply at the seed filling stage was nodule nitrogen fixation. The nitrogen supply to the main body of soybean nodules is always nodule nitrogen fixation.

3. Different phosphorus supply levels can significantly affect the nitrogen fixation of soybean nodules ($R^2 \geq 0.803$), and both the acetylene reduction method and the ^{15}N tracer method can be used to determine the nitrogen fixation ability of soybean nodules.

Supplementary Materials: The following supporting information can be downloaded at: https://www.mdpi.com/article/10.3390/agronomy12112802/s1, Table S1: The elements of P or K in different P level treatments; Table S2: Concentrations of elements in nutrient medium of the sand culture; Table S3: Amount of nutrient solution for soybean at different growth stages.

Author Contributions: Writing review and editing, H.L. and Z.Z., funding acquisition and investigation, L.W., A.Y. and D.L. All authors have read and agreed to the published version of the manuscript.

Funding: We are grateful for the "Young Talents" Project of Northeast Agricultural University, award number: 19QC16, and the Postdoctoral Scientific Research Development Fund of Heilongjiang Province, award number: LBH-Z17033.

Institutional Review Board Statement: The study was conducted in accordance with the Declaration of Helsinki, and approved by the Institutional Review Board (or Ethics Committee) of NAME OF INSTITUTE.

Informed Consent Statement: Informed consent was obtained from all subjects involved in the study.

Data Availability Statement: Not Applicable.

Conflicts of Interest: The authors declare no conflict of interest.

References

1. Elser, J.J.; Bracken, M.E.; Cleland, E.E.; Gruner, D.S.; Harpole, W.S.; Hillebrand, H.; Ngai, J.T.; Seabloom, E.W.; Shurin, J.B.; Smith, J.E. Global analysis of nitrogen and phosphorus limitation of primary producers in freshwater, marine and terrestrial ecosystems. *Ecol. Lett.* **2007**, *10*, 1135–1142. [CrossRef] [PubMed]
2. Xue, Y.; Zhuang, Q.; Zhu, S.; Xiao, B.; Liang, C.; Liao, H.; Tian, J. Genome Wide Transcriptome Analysis Reveals Complex Regulatory Mechanisms Underlying Phosphate Homeostasis in Soybean Nodules. *Int. J. Mol. Sci.* **2018**, *19*, 2924. [CrossRef] [PubMed]
3. Sulieman, S.; Ha, C.V.; Schulze, J.; Tran, L.P. Growth and nodulation of symbiotic Medicago truncatula at different levels of phosphorus availability. *J. Exp. Bot.* **2013**, *64*, 2701–2712. [CrossRef]
4. Vance, C.P.; Uhde-Stone, C.; Allan, D.L. Phosphorus acquisition and use: Critical adaptations by plantsfor securing a nonrenewable resource. *New Phytol.* **2010**, *157*, 423–447. [CrossRef] [PubMed]
5. Tang, C.; Hinsinger, P.; Drevon, J.J.; Jaillard, B. Phosphorus Deficiency Impairs Early Nodule Functioning and Enhances Proton Release in Roots of *Medicago truncatula* L. *Ann. Bot.* **2001**, *88*, 131–138. [CrossRef]
6. Chen, Z.; Cui, Q.Q.; Liang, C.Y.; Sun, L.L.; Tian, J.; Liao, H. Identification of differentially expressed proteins in soybean nodules under phosphorus deficiency through proteomic analysis. *Proteomics* **2011**, *11*, 4648–4659. [CrossRef]

7. Magadlela, A.; Pérez-Fernández, M.A.; Kleinert, A.; Dreyer, L.L.; Valentine, A.J. Source of inorganic N affects the cost of growth in a legume tree species (*Virgilia divaricata*) from the Mediterrean-type Fynbos ecosystem. *J. Plant Ecol.* **2016**, *9*, 752–761. [CrossRef]
8. Taliman, N.A.; Dong, Q.; Echigo, K.; Raboy, V.; Saneoka, H. Effect of Phosphorus Fertilization on the Growth, Photosynthesis, Nitrogen Fixation, Mineral Accumulation, Seed Yield, and Seed Quality of a Soybean Low-Phytate Line. *Plants* **2019**, *8*, 119. [CrossRef] [PubMed]
9. Bulgarelli, R.G.; De Oliveira, V.H.; de Andrade, S.A.L. Arbuscular mycorrhizal symbiosis alters the expression of *PHT1* phosphate transporters in roots and nodules of P-starved soybean plants. *Theor. Exp. Plant Physiol.* **2020**, *32*, 243–253. [CrossRef]
10. Qin, L.; Zhao, J.; Tian, J.; Chen, L.; Sun, Z.; Guo, Y.; Lu, X.; Gu, M.; Xu, G.; Liao, H. The High-Affinity Phosphate Transporter *GmPT5* Regulates Phosphate Transport to Nodules and Nodulation in Soybean. *Plant Physiol.* **2012**, *159*, 1634–1643. [CrossRef]
11. Lu, M.; Cheng, Z.; Zhang, X.M.; Huang, P.; Fan, C.; Yu, G.; Chen, F.; Xu, K.; Chen, Q.; Miao, Y. Spatial Divergence of *PHR-PHT1* Modules Maintains Phosphorus Homeostasis in Soybean Nodules. *Plant Physiol.* **2020**, *184*, 01209–02019. [CrossRef] [PubMed]
12. Esfahani, M.N.; Kusano, M.; Nguyen, K.H.; Watanabe, Y.; Tran, L. Adaptation of the symbiotic Mesorhizobium chickpea relationship to phosphate deficiency relies on reprogramming of whole plant metabolism. *Proc. Natl. Acad. Sci. USA* **2016**, *113*, 4610–4619. [CrossRef] [PubMed]
13. Almeida, J.P.; Hartwig, U.A.; Frehner, M.; Nösberger, J.; Lüscher, A. Evidence that P deficiency induces N feedback regulation of symbiotic N_2 fixation in white clover (*Trifolium repens* L.). *J. Exp. Bot.* **2000**, *51*, 1289–1297. [CrossRef] [PubMed]
14. Sulieman, S.; Schulzec, J.; Trana, L.S. N-feedback regulation is synchronized with nodule carbon alteration in Medicago truncatula under excessive nitrate or low phosphorus conditions. *J. Plant Physiol.* **2014**, *171*, 407–410. [CrossRef] [PubMed]
15. Sa, T.M.; Israel, D.W. Energy status and functioning of phosphorus- deficient soybean nodules. *Plant Physiol.* **1991**, *97*, 928–935. [CrossRef]
16. Miao, S.J.; Qiao, Y.F.; Han, X.Z.; An, M. Nodule formation and development in soybeans *Glycine max* L.) in response to phosphorus supply in solution culture. *Pedosphere* **2007**, *17*, 36–43. [CrossRef]
17. Le Roux, M.R.; Khan, S.; Valentine, A.J. Organic acid accumulation may inhibit N_2 fixation in phosphorus-stressed lupin nodules. *New Phytol.* **2008**, *177*, 956–964. [CrossRef]
18. Mandri, B.; Drevon, J.J.; Bargaz, A.; Oufdou, K.; Faghire, M.; Plassard, C.; Payre, H.; Ghoulam, C. Interactions between common bean genotypes and Rhizobia strains isolated from Morocan soils for growth, phosphatase and phytase activities under phosphorus deficiency conditions. *J. Plant Nutr.* **2012**, *35*, 1477–1490. [CrossRef]
19. Araújo, A.P.; Plassard, C.; Drevon, J.J. Phosphatase and phytase activities in nodules of common bean genotypes at different levels of phosphorus supply. *Plant Soil* **2008**, *312*, 129–138. [CrossRef]
20. Graciano, C.; Goyab, J.F.; Frangi, J.L.; Guiamet, J.J. Fertilization with phosphorus increases soil nitrogen absorption in young plants of Eucalyptus grandis. *For. Ecol. Manag.* **2006**, *236*, 202–210. [CrossRef]
21. Reed, S.C.; Seastedt, T.R.; Mann, C.M.; Suding, K.N.; Townsend, A.R.; Townsend, K.L. Phosphorus fertilization stimulates nitrogen fixation and increases inorganic nitrogen concentrations in a restored prairie. *Appl. Soil Ecol.* **2007**, *36*, 238–242. [CrossRef]
22. Divito, G.A.; Sadras, V.O. How do phosphorus, potassium and sulphur affect plant growth and biological nitrogen fixation in crop and pasture legumes? *A Meta-Anal. Field Crops Res.* **2014**, *156*, 161–171. [CrossRef]
23. Yao, Y.B.; Wu, D.T.; Gong, Z.P.; Zhao, J.K.; Ma, C.M. Variation of nitrogen accumulation and yield in response to phosphorus nutrition of soybean (*Glycine max* L. Merr.). *J. Plant Nutr.* **2018**, *41*, 1138–1147. [CrossRef]
24. Li, H.Y.; Wang, X.X.; Liang, Q.X.; Lyu, X.C.; Li, S.; Gong, Z.P.; Dong, S.K.; Yan, C.; Ma, C.M. Regulation of phosphorus supply on nodulation and nitrogen fixation in soybean plants with dual-root systems. *Agronomy* **2021**, *11*, 2354. [CrossRef]
25. Gremaud, M.F.; Harper, J.E. Selection and initial characterization of partially nitrate tolerant nodulation mutants of soybean. *Plant Physiol.* **1989**, *89*, 169–173. [CrossRef]
26. Le Roux, M.R.; Khan, S.; Valentine, A.J. Nitrogen and carbon costs of soybean and lupin root systems during phosphate starvation. *Symbiosis* **2008**, *48*, 102–109. [CrossRef]
27. Naeem, M.; Khan, M.M.A.; Mohd Idrees, M.; Aftab, T. Phosphorus ameliorates crop productivity, photosynthetic efficiency, nitrogen-fixation, activities of the enzymes and content of nutraceuticals of *Lablab purpureus* L. *Sci. Hortic.* **2010**, *126*, 205–214. [CrossRef]
28. Magadlela, A.; Vardien, W.; Kleinert, A.; Dreyer, L.; Valentine, A.J. The role of phosphorus deficiency in nodule microbial composition, and carbon and nitrogen nutrition of a native legume tree in the Cape fynbos ecosystem. *Aust. J. Bot.* **2015**, *63*, 379–386. [CrossRef]
29. Jemo, M.; Sulieman, S.; Bekkaoui, F.; Olomide, O.A.K.; Hashem, A.; Abd Allah, E.F.; Alqarawi, A.A.; Tran, L.P. Comparative Analysis of the Combined Effects of Different Water and Phosphate Levels on Growth and Biological Nitrogen Fixation of Nine Cowpea Varieties. *Front. Plant Sci.* **2017**, *8*, 2111. [CrossRef]
30. Wang, Q.; Wang, J.; Li, Y.; Chen, D.; Ao, J.; Zhou, W.; Shen, D.; Li, Q.; Huang, Z.; Jiang, Y. Influence of nitrogen and phosphorus additions on N_2-fixation activity, abundance, and composition of diazotrophic communities in a Chinese fir plantation. *Sci. Total Environ.* **2018**, *619*, 1530–1537. [CrossRef]
31. Qiao, Y.F.; Tang, C.X.; Han, X.Z.; Miao, S.J. Phosphorus deficiency delays the onset of nodule function in soybean. *J. Plant Nutr.* **2007**, *30*, 1341–1353. [CrossRef]
32. Pérez-Fernández, M.; Míguez-Montero, Á.; Valentine, A. Phosphorus and Nitrogen Modulate Plant Performance in Shrubby Legumes from the Iberian Peninsula. *Plants* **2019**, *8*, 334. [CrossRef] [PubMed]

33. Lazali, M.; Zaman-Allah, M.; Amencm, L.; Ounane, G.; Abadie, J.; Drevon, J. A phytase gene is overexpressed in root nodules cortex of *Phaseolus vulgaris*–rhizobia symbiosis under phosphorus deficiency. *Planta* **2013**, *238*, 317–324. [CrossRef] [PubMed]
34. Miao, S.J.; Han, X.Z.; Liu, X.B.; Qiao, Y.F. Seedling treatments and phosphorus solution concentrations affect nodulation and nodule functions in soybean (*Glycine max* L.). *Plant Soil Environ.* **2007**, *53*, 65–71. [CrossRef]
35. Tsvetkova, G.E.; Georgiev, G.I. Effect of phosphorus nutrition on the nodulation, nitrogen fixation and nutrient use efficiency of bradyrhizobium japonicum soybean(*Glycine max* L. merr.) symbiosis. *Bulg. J. Plant Physiol.* **2003**, *3*, 331–335.
36. Raji, S.G.; Tzanakakis, V.; Dörsch, P. Bradyrhizobial inoculation and P application effects on haricot and mung beans in the Ethiopian Rift Valley. *Plant Soil* **2019**, *442*, 271–284. [CrossRef]
37. Schulze, J.; Temple, G.; Temple, S.J.; Beschow, H.; Vance, C.P. Nitrogen fixation by white lupin under phosphorus deficiency. *Ann. Bot.* **2006**, *98*, 731–740. [CrossRef]
38. Cavard, X.; Augusto, F.; Saur, E.; Trichet, P. Field effect of P fertilization on N_2 fixation rate of *Ulex europaeus*. *Ann. For. Sci.* **2007**, *64*, 875–881. [CrossRef]

agronomy
MDPI

Article

Tillage Crop Establishment and Irrigation Methods Improve the Productivity of Wheat (*Triticum aestivum*): Water Use Studies, and the Biological Properties and Fertility Status of Soil

Rajendra Kumar [1], Ram Krishan Naresh [2], Rajan Bhatt [3], Mandapelli Sharath Chandra [4], Deepak Kumar [5], Saud Alamri [6], Manzer H. Siddiqui [6,*], Alanoud T. Alfagham [6] and Hazem M. Kalaji [7]

1 M. S. Swaminathan School of Agriculture, Shoolini University of Biotechnology & Management Sciences Solan, Solan 508976, Himachal Pradesh, India; rajdiwakar1990@gmail.com
2 Department of Agronomy, Sardar Vallabhbhai Patel University of Agriculture & Technology, Meerut 250110, Uttar Pradesh, India; r.knaresh@yahoo.com
3 Krishi Vigyan Kendra Amritsar, Punjab Agricultural University, Amritsar 143601, Punjab, India; rajansoils@pau.edu
4 AICRP on Integrated Farming System, Professor Jayashankar Telangana State Agricultural University, Hyderabad 500030, Telangana, India; sharathagrico@gmail.com
5 Department of Agronomy, CSSS PG College, Machhra, Chaudhary Charan Singh University, Meerut 250001, Uttar Pradesh, India; deepak09675@gmail.com
6 Department of Botany and Microbiology, College of Science, King Saud University, Riyadh 11451, Saudi Arabia; saualamri@ksu.edu.sa (S.A.); aalfaghom@ksu.edu.sa (A.T.A.)
7 Department of Plant Physiology, Institute of Biology, Warsaw University of Life Sciences SGGW, 159 Nowoursynowska 159, 02-776 Warsaw, Poland; hazem@kalaji.pl
* Correspondence: mhsiddiqui@ksu.edu.sa

Abstract: The Crop Research Centre of Sardar Vallabhbhai Patel University of Agriculture and Technology in Meerut (U.P.), India, conducted field experiments in a randomised block design, comprising three replicates, one late sown variety (DBW-90), and eight treatments, viz.: T1 was a conventional flood irrigation (CFI); T2, furrow irrigated with gated-pipe raised beds (FIGPRB); T3, all furrow irrigation (AFI); T4, alternate furrow irrigation (Alt. FI); T5, wide bed furrow irrigation (WBFI); T6, skip furrow irrigated (SFI); T7, Sprinkler irrigation (SI); and T8, Zero-till flat-irrigated using gated pipe/controlled-flood irrigation (ZTFIGP). These field experiments were conducted during the Rabi seasons of 2017–2018 and 2018–2019. The purpose of this study was to evaluate the yield, water productivity, and soil health under different tillage crop establishment methods. Test weight, spike length, and productive tillers were all considerably enhanced in treatment T5, with the treatment's statistical significance being similar to that of treatments T8 and T2. Treatment T5 considerably outperformed the other treatments in terms of grain yield, straw yield, biological yield (44.32, 61.88, and 106.19 q ha^{-1}, respectively), as well as harvest index (41.73). Thirty to sixty centimetres of soil were mined for the most water, followed by fifteen to thirty centimetres, zero to fifteen centimetres, and sixty to ninety centimetres. Both water-use efficiency (2.86 q ha^{-1} cm) and water productivity (1.91 kg cm^{-3}) were highest under T7 (Sprinkler irrigation). The maximum total NPK (113.69; 27.45; 127.33 kg ha^{-1}) was found in crops grown with wide bed furrow irrigation. The data also showed that treatment T6 (skip furrow irrigated) had the highest levels of accessible NPK in soil, followed closely by treatment T4 (alternate furrow irrigated). Treatment T8 (zero-till flat-irrigated using gated-pipe/controlled flood irrigation) had the highest bacterial, fungal, and actinomycete populations, followed by T5 (wide bed furrow irrigated) and T2 (furrow irrigated with gated-pipe/elevated bed). Our research showed that there may be more options for maintaining wheat crop water productivity and soil health under different agroecological conditions, including crop productivity, conservation tillage-based establishing methods, and irrigation regimes.

Keywords: wheat; tillage; establishment methods; yield; water productivity; soil health

Citation: Kumar, R.; Naresh, R.K.; Bhatt, R.; Chandra, M.S.; Kumar, D.; Alamri, S.; Siddiqui, M.H.; Alfagham, A.T.; Kalaji, H.M. Tillage Crop Establishment and Irrigation Methods Improve the Productivity of Wheat (*Triticum aestivum*): Water Use Studies, and the Biological Properties and Fertility Status of Soil. *Agronomy* **2023**, *13*, 1839. https://doi.org/10.3390/agronomy13071839

Academic Editors: Gang Li, Dong-Xing Guan and Daniel Menezes-Blackburn

Received: 15 June 2023
Revised: 9 July 2023
Accepted: 9 July 2023
Published: 12 July 2023

1. Introduction

Wheat (*Triticum aestivum* L.) can grow in a wide range of climates and soil types, from sandy loam to heavy black cotton, making it India's principal grain crop. Wheat may be grown at any altitude between sea level and 3658 m in the Himalayas, and from 11 degrees North latitude to 30 degrees North latitude. From the wet soils of the deltaic coastal areas to the dry soils of Rajasthan, it is cultivated in a wide range of environments and soil types [1]. Wheat is a healthy option because it is a rich source of carbohydrates, protein, and fat. Thiamine, niacin, iron, riboflavin, calcium, and fibre are just some of the minerals it has in abundance. India's most extensively cultivated crop, wheat, feeds 35% of the world's population. About 215.5 million hectares are used for wheat production worldwide, leading to a harvest of 764.5 million metric tons and a productivity of 3.39 t per ha^{-1} [2]. India cultivates wheat on 29.65 million hectares of land, with a yearly production of 99.9 million metric tons (at a productivity of 3371 kg per hectare) [2]. About a third of the world's food grain supply comes from wheat. With 1.35 billion inhabitants, India is only slightly less populous than China (1.41 billion). In about seven years, it will have surpassed China's population and is expected to peak at roughly 1.7 billion by 2050. Therefore, wheat will probably continue to play a crucial role in reaching this. The ever-increasing human population makes steady wheat production a need.

It is crucial that agricultural water management be factored into irrigation schedules, as more than a third of the world's population will face absolute water scarcity by 2025 [3]. Global wheat output has to grow by 1.6% to 2.6% per year to meet demand, and this growth may be accomplished mostly through enhanced input utilisation efficiency. The Indo-Gangetic Plains are experiencing a rate of groundwater loss that ranges from 13 to 17 km^3 on an annual basis [4]. Two primary crop management practices can be implemented to increase the efficiency with which wheat inputs are used. Today, 70% of the world's freshwater is utilised in agriculture, although only 40% of the world's food is grown in irrigated soils. Aquifer water is used for about 10% of irrigation, which is an excessive and unsustainable amount [5]. It is generally agreed that adjusting irrigation systems for maximum efficiency saves water and improves crop yields and quality. Sprinkler irrigation should be implemented to minimise water loss, enhance water-use efficiency, and increase agricultural water yield. This method allows for the accurately application water during the presowing and subsequent watering stages. New hydro-physical features of the soil can be strongly influenced using soil tillage. As of late, there has been a lot of buzz about conservation tillage (sometimes known as "zero tillage") and the assertions that it invariably improves soil properties for plant development and water retention. Management strategies that prioritise resource conservation are gaining favour in the rice–wheat cropping system. This approach improves soil organic matter, moisture availability, aggregation, and water transmission capacity [3].

To improve the soil's microbial population, conservation agriculture practises such as zero tillage are preferable under a rice–wheat cropping system from both an economic and ecological standpoint. Zero-till farming has gained popularity among farmers in recent years [6,7] due to its ability to increase crop yields while reducing soil disturbance and protecting soil carbon. Conventional tillage methods are important to India's farming tradition and have made substantial contributions to India's food security. Before planting crops, CT entails a number of procedures, such as clearing the land of residue (either by removing it or by burning it) plough tillage (PT), harrowing, and levelling the ground. CT's effect on the soil's physicochemical and biological properties can have an effect on soil production and longevity [8]. Long-term mechanical disturbances of soil, such as those induced by inversion tillage or severe CT throughout the entire crop growing season, can lead to soil erosion and mycelium network damage [9]. Conventional tillage practices, such as improper straw management, can reduce soil organic carbon (SOC) storage and endanger sustainable crop production [10]. Higher C stocks in agricultural soils can be produced by returning crop residue to the soil [11], which is a key indicator of the soil's environmental quality, and agronomic sustainability [12]. In comparison to the traditional tillage system, the rice–wheat cropping system with zero tillage had a much greater population of bacteria, fungus, actinomycetes, microbial-C, microbial-N, and SOC sequestration. Avoiding tillage can have an impact on crop productivity and, by extension, food security [13]. Therefore, methods of soil and crop management that increase organic carbon and microbial biomass carbon (while keeping yields constant) are of importance. Soil microbial biomass variations as a result of changes in soil management and environmental stresses are indicative of changes in the chemical and physical properties of agricultural ecosystems. Since the soil microbiome is responsible for supplying nutrients to plants, it regulates the availability and production of nutrients in agroecosystems. The capacity of an ecosystem to sequester the carbon fixed during photosynthesis in soil organic matter is related to its net primary output [14]. Since appropriate irrigation practises improve soil health and maximise water consumption without reducing output, they are vital for wheat production, and must be implemented by farmers. Due to the diminishing ground water level caused by improper usage and over-extraction from the ground, this study on tillage crop establishment and irrigation methods in wheat was undertaken, as we know water is the most vital natural resource for humans, animals, and the production of food-producing crops. In light of these facts, wheat (*Triticum aestivum*) productivity, water use efficiency, nutrient uptake, and soil health in sandy loam soils in western Uttar Pradesh, India were investigated as a result of this interest by researchers at the SVPUA&T in Meerut, U.P.

2. Materials and Methods

2.1. Selected Site

The study's research was conducted at the Crop Research Centre of the Sardar Vallabhbhai Patel University of Agriculture and Technology in Meerut, Uttar Pradesh (U.P.). Based in the centre of western Uttar Pradesh, Meerut has a subtropical climate and is 237 m above mean sea level. Its coordinates are 29°08′ N latitude and 77°41′ E longitude. The trial field was completely level, and it had well-developed irrigation and drainage systems already in place. The experimental soil was sandy loam in texture, low in available nitrogen and organic carbon, with a medium level of available phosphorus and potassium, while being alkaline in reaction.

2.2. Climate and Weather Condition

This area experiences a semiarid, subtropical climate, with extremely hot summers and freezing winters. In the first year of the study, 2017–2018, the minimum mean temperature was recorded in the month of January 2018 at 4.80 °C (Figure 1A), and in the second year, 2018–2019, it was recorded in the month of December 2018 at 2.90 °C (Figure 1B). During both research years, the month of April saw the mean maximum temperature. During the crop seasons of 2018 and 2019, the mean relative humidity was found to be at its highest points in the months of January (95.7 and 96.7%, respectively) and its lowest points in the months of December (30.3 and 38.9%, respectively). In the first year of the experiment, the minimum evapotranspiration throughout the crop period was 1.7 mm in the month of December, and 1.3 mm in the month of January. In total, the crop years 2017–2018 and 2018–2019 saw rainfall totals of 20.5 and 127.5 mm, respectively (Figure 1A,B).

2.3. Treatments Description

In total, there were nine different treatments, eight of which were different combinations of irrigation and tillage practises, and the ninth was a late-planted variety that was given the name DBW-90. The following treatments were utilised: T1, conventional flood irrigation (CFI); T2, furrow irrigated with gated-pipe raised beds (FIGPRB); T3, all furrow irrigation (AFI); T4, alternate furrow irrigation (Alt. FI); T5, wide bed furrow irrigation (WBFI); T6, skip furrow; T7, Sprinkler irrigation (SI); and T8, Zero-till flat-irrigated using gated pipe/controlled flood irrigation (ZTFIGP). A randomised block design (RBD) with three independent replications was used to conduct controlled flood irrigation (ZTFIGP) over the 2017–2018 and 2018–2019 growing seasons.

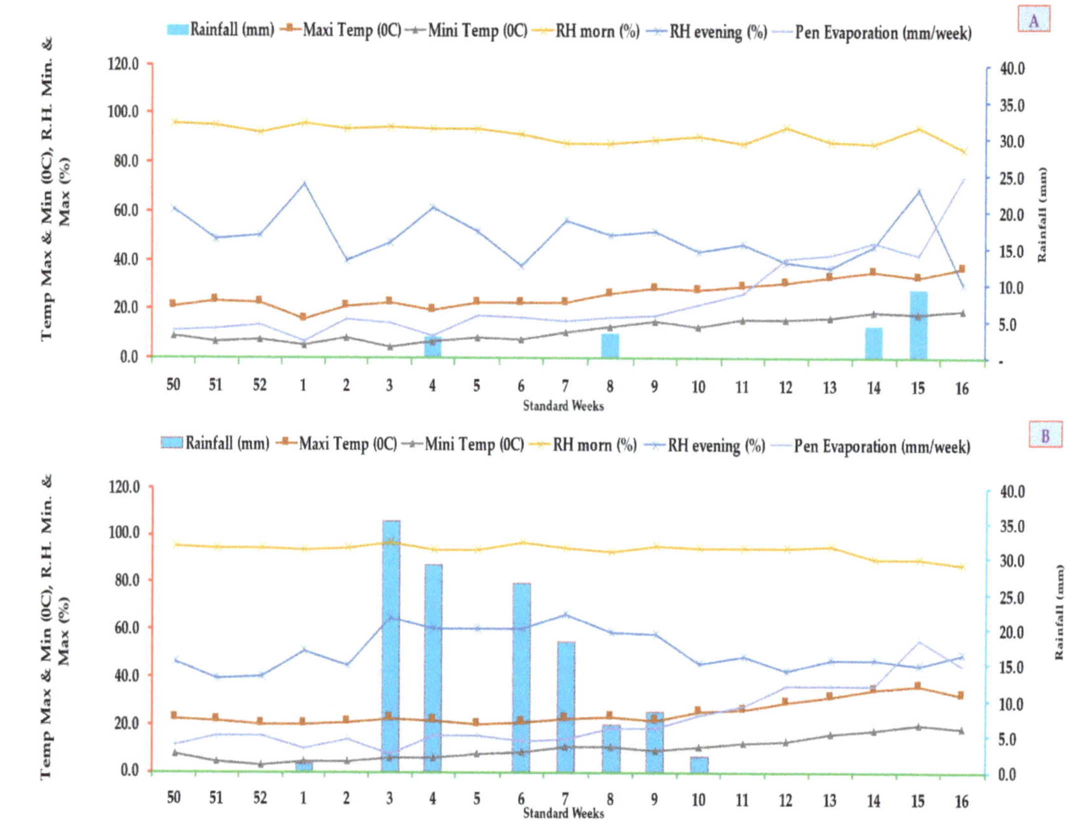

Figure 1. Mean weekly Agro-meteorological data during the crop growing *rabi* season for (**A**) 2017–2018 and (**B**) 2018–2019.

2.4. Cultural Practices

Wheat was planted using a seed drill equipped with a dry fertiliser attachment in rows 20 cm apart, after presowing irrigation and a CT method comprising two harrowings, three ploughings (using a cultivator), and then planking. Raised beds were constructed utilising a tractor-drawn multicrop raised bed planter outfitted with inclined plate seed metering systems while using furrow-irrigated raised-bed tillage (FIRB). Narrow beds were 40 cm wide, while broad beds were 100 cm wide. Irrigated furrows were 12 cm high and 30 cm wide at the top, and a 70 cm spacing was kept between the centres of neighbouring furrows. Wheat was planted in a staggered pattern of three rows per raised bed. The crops were planted utilising the ZT method, which involves minimal tilling of the ground, and the zero-till seed drill was utilised to achieve this. With this implement, farmers could sow seeds directly into narrow slots in the ground that were only a few millimetres wide, and four to seven cm deep.

2.5. Management of Fertilisers and Crops

All of the fields were fertilised with the proper ratio of nitrogen to phosphorus to potassium (150:60:60 kg ha^{-1}), in order to achieve optimal crop yield. At the time of planting, a full dose of phosphorus and potassium (as well as half the recommended dose of nitrogen) were applied using a seed-cum-fertiliser drill. Urea, DAP, and MOP were used together as a source of N, P, and K. The rest of the nitrogen was applied along with the urea at 25 and then at 55 days following planting. The herbicide Sulfosulfuron (postemergence) at 33.3 g a.i. per ha^{-1} was applied to the standing crop at 30 days, and then one round of hand weeding was performed at 45 days to control the weed population. The soil used in the experiment was a sandy loam with medium levels of accessible phosphorus and potassium and a mildly alkaline pH.

2.6. Yield Attributing Characters and Yield (q ha^{-1})

For each net plot, we calculated the number of effective tillers per metre of row length in a random sample of marked rows, and then translated the results to m^2. The average length of each spike was determined by measuring ten spikes at random from each plot. Ten separate spikes were counted to provide an average for the number of grains per spike. By isolating the grains from the spikelets, the density of grains per unit of spikelet area could be calculated. The number of grains in a composite sample obtained from the harvest of each plot was counted using an automatic seed counter, and the weight of 1000 grains was reported in grammes.

The biological harvest from the whole net area of each plot was collected using miniature plot threshers. After harvesting the net plot area, wheat bundles were sun-dried for four days, and their final weights were converted to kilogrammes per hectare (ha) in order to calculate the biological yield (q ha^{-1}). Straw yield (in q ha^{-1}) was determined by subtracting the biological yield per net plot area from the grain yield. Grain yields were originally recorded in kilogrammes per hectare for the net plot area, but were converted to kilogrammes per hectare after being normalised to 14% moisture. The harvest index was determined by dividing the economic yield by the biological yield and then expressing the result as a percentage.

$$\text{Harvest index}(\%) = \frac{\text{Economic yield}}{\text{Biological yield}} \times 100$$

2.7. Water Use Studies

2.7.1. Consumptive Use of Water

The consumptive use was worked out from the loss in soil moisture, effective rainfall, and potential evapotranspiration for 2 days following irrigation. The seasonal consumptive use was calculated using the formula given below.

$$U = \sum_{i}^{n} (Eo \times 0.8) + (M_1 - M_2) + ER$$

2.7.2. Water-Use Efficiency (WUE)

The economic yield (kg ha^{-1}) was divided by the total amount of water used (cm) from the relevant plots to calculate the water-use efficiency (WUE) of various treatments.

$$WUE = \frac{\text{Economic yield}\left(\text{kg ha}^{-1}\right)}{\text{Total consumptive use of water (cm)}} \text{kg ha}^{-1}\text{cm}^{-1}$$

2.7.3. Water Productivity (WP)

By dividing the economic yield (kg $ha^{-1)}$ by the depth of irrigation water applied (cm) from separate plots, the WP of various treatments was calculated. It is expressed in kg m^{-3}.

$$WP = \frac{\text{Economic yield}\left(\text{kg ha}^{-1}\right)}{\text{Amount of water applied (cm)}} \text{kg m}^{-3}$$

2.8. Plant Analysis

Nutrient concentrations in harvested grains and straw were examined and computed separately from estimates from selected plants in each plot. To evaluate dry matter production (i.e., grains and straw), representative samples of plants were dried in a hot air oven at 60 °C after harvest.

Dried samples were pulverised in a Wiley mill and kept in polythene bags for further examination.

2.8.1. Nutrient Uptake (kg ha^{-1})

In order to calculate nutrient uptake, we multiplied the grain yield and straw yield by the percentage of each nutrient they contained.

$$\text{Nutrient uptake (kg ha}^{-1}\text{)} = \text{Content (\%) in grains/straw} \times \text{grains/straw yield}$$

$$\text{Total uptake (kg ha}^{-1}\text{)} = \text{Uptake from grains + nutrient uptake from straw}$$

2.8.2. Nutrient Harvest Index (NHI)

Nutrient harvest index (NHI) is the ratio of nutrient uptake in economic part of the crop plants to the total nutrient uptake in biological part of the crop plants. NHI of nitrogen, phosphorus and potassium were computed by using the formula given below:

$$\text{NHI} = \frac{\text{Nutrients uptake in grains}\left(\text{kg ha}^{-1}\right)}{\text{Total nutrients uptake(grains + straw)}\left(\text{kg ha}^{-1}\right)}$$

2.9. Economic Nutrients Use Efficiency (ENUE)

Economic Nutrients Use Efficiency is defined as the amount of INR (₹) invested on production of per kg grain yield. ENUE was calculated by using the formula given below:

$$\text{NHI} = \frac{\text{Grain Yield}\left(\text{kg ha}^{-1}\right)}{\text{₹ Invested on nutrient}}$$

2.10. Biological Properties

The population of bacteria, fungi, and actinomycetes were counted using a serial soil dilution method. In the beginning and at the end, soil samples were taken from the field while receiving the designated treatments, and they were then screened using a 2 mm sieve. To create a representative sample, the samples were properly combined and blended. For the purpose of identifying and isolating live bacteria, fungus, and actinomycetes count, the serial dilution approach was used, which is outlined as follows: Set up the media to support the required microbiota. Fill sterile petri plates with the cooled (45 °C) and autoclaved medium. Permit the medium to set. Then, 9 mL of sterile water blank and 1 g of sieved (2 mm) soil should be shaken for 15 to 20 min. Prepare dilutions 10^{-2}, 10^{-3}, 10^{-4}, 10^{-5}, 10^{-6}, 10^{-7}, and 10^{-8} in serial order. Add 1 mL aliquots of various dilutions to the medium in Petri plates after it has cooled and solidified. To ensure that spores are distributed equally, rotate the plates for 3–4 days at 28 °C. Check the plate for colonies growing on the medium's surface. According to [15], the population counts of bacteria, fungi, and actinomycetes were determined using the dilution plate technique, with Martin's rose agar, Bengal agar, and Ken Knight's agar media, respectively.

2.11. Soil Fertility Status

The fertility status of soil was estimated using the Walkley–Black wet oxidation Method [16] for organic carbon, the alkaline potassium permanganate method [17] for the available nitrogen, Olsen's method for the available phosphorus, and the 1 N NH4OAC extraction method [16] for the available potassium in soil after wheat harvesting.

2.12. Statistical Analysis

OPSTAT was utilised for the investigation's comprehensive analysis of variance (ANOVA). The statistical significance level used to compare the treatment means was p 0.05.

3. Results

3.1. Yield Attributing Characters

The number of productive tillers m^{-2} (i.e., tillers with fertile spike) is an important yield attribute, accounting for major variation in grain yield of wheat (Figure 2). Among the tillage crop establishment methods, the greatest number of productive tillers (283.50 m^{-2}) was recorded under T_5 compared to all other treatments, with the exception of T_2 and T_8 during experimentation. However, treatments T_1 and T_7 were recorded as being superior to the remainder of the treatments, and on par with each other. Treatments T_3, T_4, and T_6 were recorded to be on par with each other. Treatment T_6 produced the lowest number of productive tillers (252.50 m^{-2}) during our investigation. Tillage crop establishment methods exhibited a significant effect on spike length during this study. The spike length of wheat varied from 10.40 to 14.00 cm during experimentation. The maximum spike length (14.00 cm) of the wheat was recorded in T_5, which was higher than all other treatments except T_8 during study. However, T_2 and T_7 were superior to the rest of the treatments for this parameter. Treatments T_1, T_3 and T_6 were on par with each other during experimentation, while treatment T6 was found to exhibit the minimum spike length (10.40 cm).

Number of grains per spike^{-1} is an important yield attribute, which directly affects the grain yield. The number of grains per spike^{-1} registered significant variation during this study. Among the tillage crop establishment methods, a significantly greater number of grains (53 grains per spike^{-1}) was produced under treatment T_5 than under other treatments during experimentation, with the exception of T_2 and T_8. T_1 and T_7 were both equally effective, though well beyond the other treatments. The experimental results showed that Treatments T_3, T_4, and T_6 were all similar, with Treatment T_6 yielding the fewest grains (42 grains per spike^{-1}). An essential aspect of yield, test weight is the weight of 1000 grains divided by the weight of a single grain, so as to determine how effective the grain filling process was.

The difference in test weight of wheat varied significantly in relation to the tillage crop establishment method during both the years of study. The maximum test weight was recorded in treatment T_5 than all other treatments except T_1, T_2 and T_8. However, treatments T_3, T_4, T_6 and T_7 were on par with each other. Meanwhile, treatment T_4 recorded the lowest test weight during investigation.

3.2. Yield (q ha^{-1})

The harvest index and yield (grain, straw, and biological, measured in q ha^{-1}) are shown for a variety of tillage methods in Figure 2. All treatments (except T_5, which produced a yield of 44.32 q ha^{-1}) reported significantly lower yields. Compared to the other treatments, both T_1 and T_7 were shown to be comparably effective. After T_5, the next highest grain yield was in Treatment 3 (30.10 q ha^{-1}), followed by Treatment 4 (29.01 q ha^{-1}), and Treatment 6 (28.29 q ha^{-1}). The production of wheat grains differed by 20.99% between T_1 and T_5, 17.79% between T_8 and T_3, and 16.21% between T_3 and T_5.

Wheat straw yield varied from 46.23 to 61.88 q ha^{-1}. While T_2 and T_8 had similar straw yields, T_5 had the highest yield (61.88 q ha^{-1}). There was a clear preference for T_1 and T_7 over the other therapies. While treatments T_4 and T_6 produced the most straw, treatment T_3 produced the least. However, T_3, T_4, and T_6 all functioned to about the same extent.

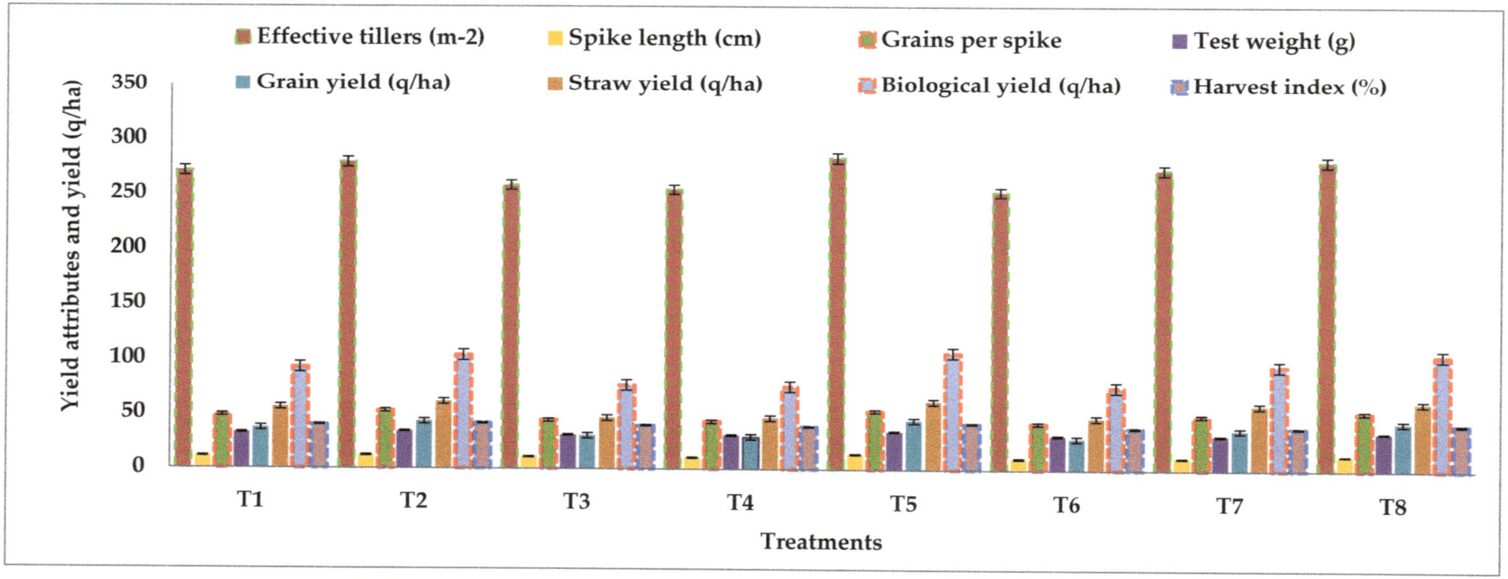

Figure 2. Impact of tillage and crop establishment practices on yield attributing characters and yield of wheat (2017–2018 and 2018–2019 pooled) (Bars represent standard error).

Tillage crop establishment with Treatment T5 was shown to be the most effective, followed by T_8 and T_2. T_3, T_4, and T_6 were all equivalent in every measurable way. The T6 treatment had the lowest observed biological yield, namely, 74.98 q ha^{-1}. Increases in grain, straw, and biological yield have all been linked to refinements in tillage methods. Increased growth and dry matter accumulation occurred because FIRB and Zero-till rapidly meet the crop's water needs. Grain output, as indicated with metrics like effective tiller count, grain count per spike, and test weight, grew at a higher pace for plants that grow at a faster rate. The treatment with the greatest harvest index was T_5, at 41.73 percent; however, treatments T_8 and T_2 were quite close behind. The harvest index ranged from 48.22% in Treatment $_7$, to 37.27% in Treatment 6. However, T_1, T_3, T_4, and T_7 all functioned similarly.

3.3. Water Input Studies

Soil moisture extraction pattern within layers was analysed, and it was found that the greatest amount of water was extracted (absorbed) from the 30–60 cm layer during the experiment, followed by the 15–30 cm, 0–15 cm, and the 60–90 cm layers (Table 1). Land arrangement under furrow-irrigated raised-beds practises enhanced the moisture extraction from the surface layer (0–15 cm) in both study years. Similarly, a modest drop in moisture extraction was seen with increasing profile depth, with the largest decrease (1.93) happening in the 60–90 cm soil layer under spray irrigation practises, due to a moisture deficit at shallower depths. The data also showed that in both years of analysis, the FIRB and zero-till plots drank more heavily from the deeper profile layer than the typical flood irrigation practise plots did.

Table 1. Effect of tillage and crop establishment practices on soil moisture depletion, consumptive use, and water-use efficiency of wheat (2017–2018 and 2018–2019, pooled).

Treatments	Soil Moisture Depletion Depth of Soil (cm)				Total Soil Moisture Depletion (cm)	Consumptive Use (cm)	Water-Use Efficiency (q ha^{-1} cm)	Water Productivity (kg cm^{-3})
	0–15	15–30	30–60	60–90				
T_1	3.66	3.36	2.83	2.39	12.23	24.83	1.47	0.94
T_2	2.27	3.18	3.57	2.78	11.79	21.10	2.02	1.46
T_3	2.25	3.75	3.21	2.40	11.60	19.00	1.58	1.01
T_4	2.23	3.48	2.68	2.30	10.68	16.88	1.72	1.14
T_5	3.04	3.86	4.42	3.46	14.77	18.25	2.43	1.72
T_6	2.13	3.31	2.37	2.22	10.02	16.05	1.76	1.23
T_7	4.30	3.34	2.61	1.93	12.17	12.55	2.86	1.91
T_8	3.25	4.11	2.57	2.93	12.86	18.95	2.28	1.40
Mean	*2.89*	*3.55*	*3.03*	*2.55*	*12.00*	*18.45*	*2.01*	*1.31*

Under plots where higher tillage procedures were used, crop water-use rose. Treatment T_1 used significantly more water than treatments T_2 and T_5, while treatment T5 used significantly less water. Treatment T_7 showed the highest water use efficiency, followed by T_5, T_8, and T_2. Researchers found that as production grew, water productivity grew as well. In both years, T1 had significantly lower water productivity compared to T_7, T_5, T_2, and T8. In terms of water output, the rankings were as follows: $T_7 > T_5 > T_2 > T_8 > T_6$.

3.4. Nutrient Uptake (kg ha^{-1})

3.4.1. Nitrogen

Wheat's nitrogen level ranged from 1.52% in the grains to 1.74% in the straw, and from 0.59 to 0.59 percentage points. Among the tillage crop establishment treatments, significant maximum nitrogen percentage levels in grains (1.74%) and straw (0.59%) was recorded in treatment T_5, which was recorded as being statistically at par with treatment T_8 (Figure 3). However, treatment T_2 was recorded to be statistically superior, on a par with T_7 over rest of the treatments and across all parameters, with the exception of grain yield. However, the minimum nitrogen percentage contents in the grains (1.52%) and straw (0.48%) of the wheat were recorded in treatment T_6, followed by T_4, T_3 and T_1. Among the tillage crop establishment treatments, significant maximum nitrogen uptake in grains (76.9 kg ha^{-1}) and straw (36.8 kg ha^{-1}) of wheat was recorded in treatment T_5, followed by T_8 and T_2 (Figure 4). However, statistically speaking, T_8 and T_2 were on level with one another. Statistically speaking, T_7 and T_1 were equally the best treatments overall; however, for straw, T_3 was the worst. Furthermore, with treatment T6, nitrogen uptake was measured at 45.2 kg ha^{-1} (grains) and 24.2 kg ha^{-1} (straw), significantly lower than the uptake under treatment T4 (52.4 kg ha^{-1} in grains, and 25.6 kg ha^{-1} in straw). However, uptake of nitrogen into straw under treatments T_6 and T_4 were recorded as being statistically on par with each other during investigation.

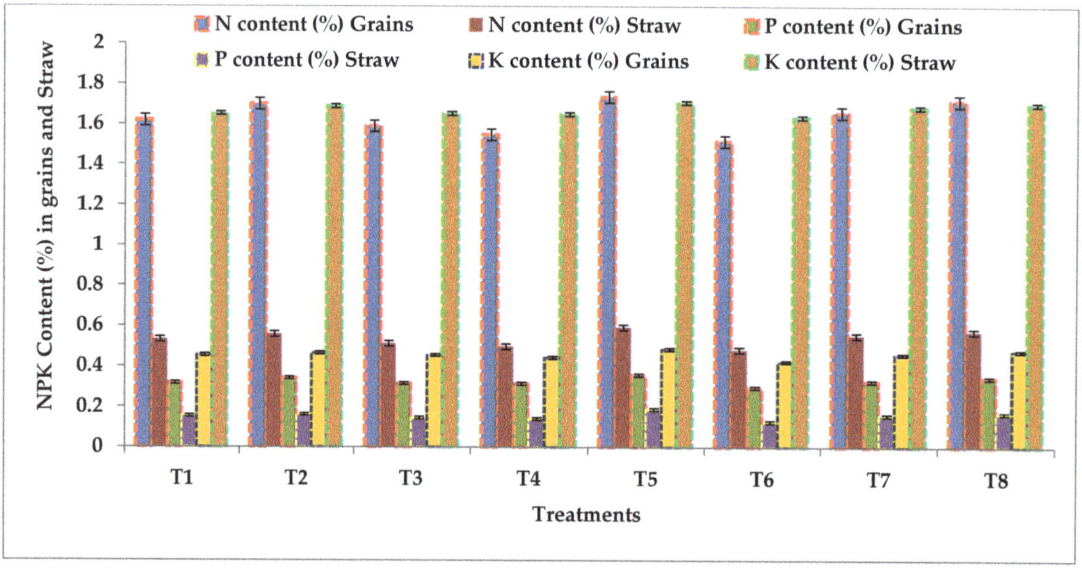

Figure 3. Nutrient (N, P, and K) content per cent in grains and straw of wheat affected by tillage crop establishment methods (Pooled data: 2017–2018 and 2018–2019) (Bars represent standard error bars).

The total nitrogen uptake varied from 69.4 to 113.7 kg ha^{-1}. Treatments T_5, T_8 and T_2 were recorded as having 27.97, 22.73, and 19.89% more total Nitrogen uptake as compared to treatment T_1 (Figure 5). However, treatment T6 was recorded as having the lowest total Nitrogen uptake (69.4 kg ha^{-1}), followed by T_4, T_3 and T_7 at values of 78.0, 83.4, and 92.3 kg ha^{-1}, respectively. The nitrogen harvest index of wheat was affected by tillage crop establishment practices. The maximum nitrogen harvest index was recorded in treatments T_2 and T_8, with the value of 0.7, which was recorded as being statistically on par with treatments T_1, T_4, T_3, and T_6 (Figure 6). However, the minimum nitrogen harvest index (with the value of 0.6) during the course of our investigation was recorded under treatment T_7, followed by T_6.

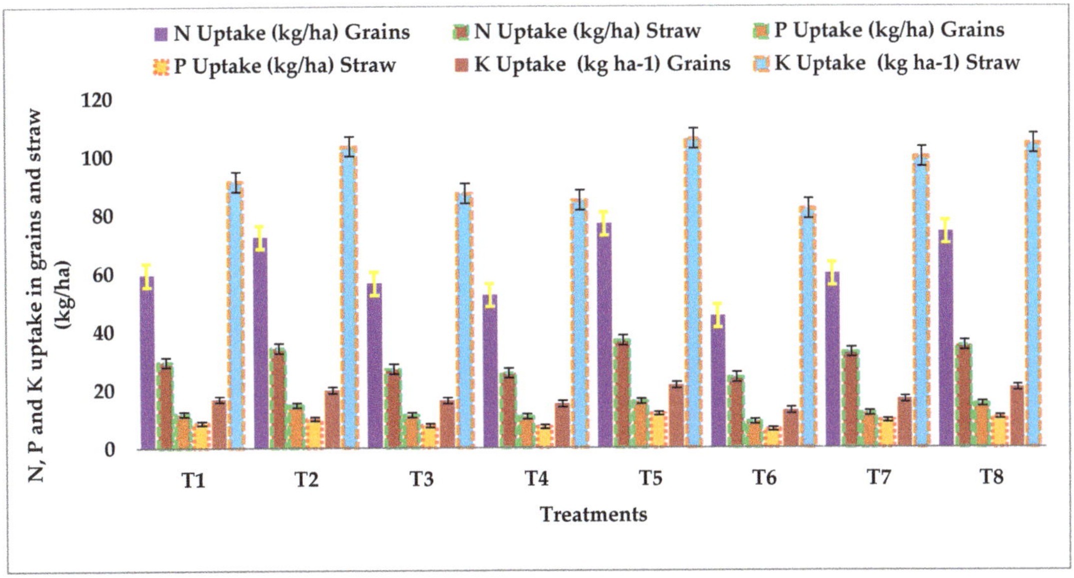

Figure 4. Nutrients (N, P, and K) uptake in grains and straw of wheat affected by tillage crop establishment methods (Pooled data: 2017–2018 and 2018–2019 (Bars represent standard error bars).

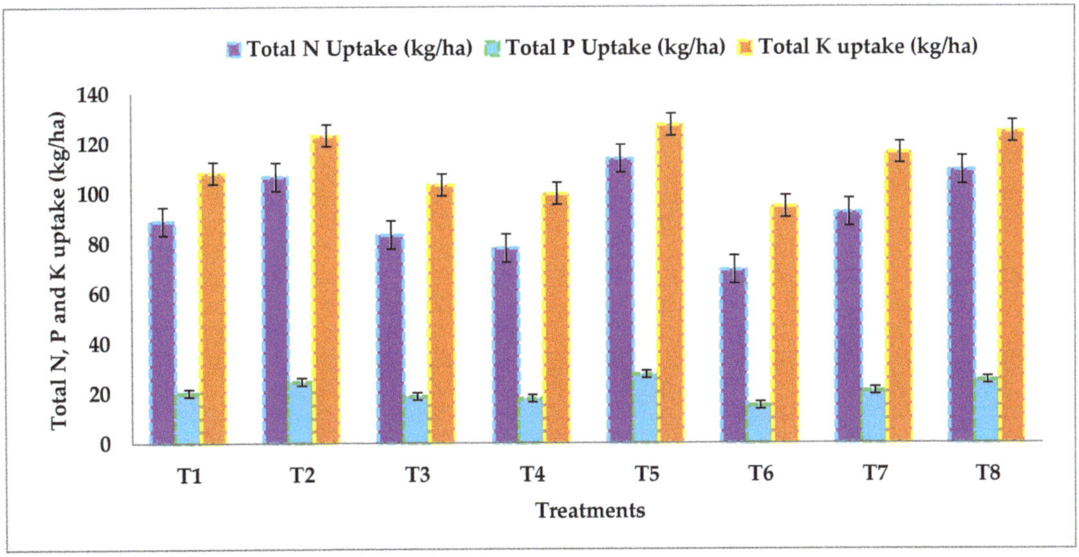

Figure 5. Effect of tillage crop establishment methods on total N, P, and K uptake in wheat (Pooled data: 2017–2018 and 2018–2019) (Bars represent standard error bars).

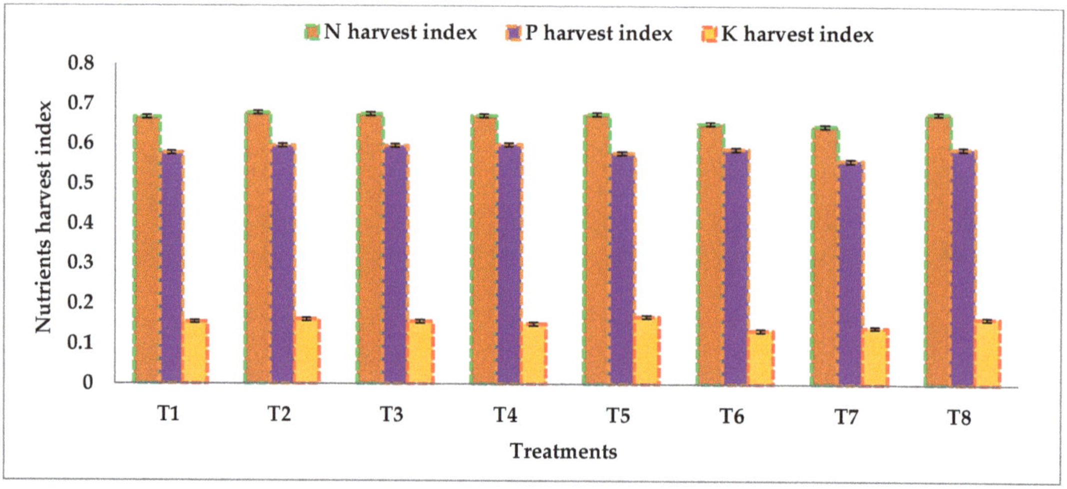

Figure 6. Nutrients (N, P, and K) harvest index of wheat influenced by tillage crop establishment methods (2017–2018 and 2018–2019, pooled) (Bars represent standard error bars).

3.4.2. Phosphorus

The percentage content, uptake, and harvest index of phosphorous were significantly affected by various treatments. Among the treatments, significant maximum Phosphorus percentage content in grains (0.36%) and straw (0.19%) of wheat were recorded in treatment T_5 (Figure 3). However, treatments T_2, T_7, and T_8 were found to be statistically superior over the rest of the treatments in case of wheat straw, being about on par with each other. However, significantly low percentages of phosphorus content in grains (0.30%) and straw (0.12%) were recorded under treatment T_6; this was followed by T_4, T_3, and T_1 with the value of 0.32, 0.14, and 0.32 per cent for grains, and 0.14, 0.32, and 0.15 per cent in straw, respectively. The uptake of phosphorus in the grains and straw of the wheat was affected by various treatments (Figure 4). Among the treatments, significant maximum phosphorus uptakes in the grains and straw of wheat were recorded in treatment T_5 compared to all other treatments, with value of 15.9 and 11.6 kg ha^{-1}. This was followed by T_8 and T_2, which were recorded as statistically superior over rest of the treatments and on par with each other. Likewise, treatments T_7 and T_1 were recorded on par with each other, and superior to T_3, T_4, and T_6. However, treatment T_6 recorded significantly low phosphorus uptakes, with the values of 8.87 and 6.22 kg ha^{-1} in grains and straw, respectively.

Total phosphorus uptake varied from 15.1 to 27.5 kg ha^{-1}. Among the tillage crop establishment practices, significant maximum total phosphorus uptakes were recorded in treatment T_5 (27.5 kg ha^{-1}) compared to all other treatments, followed by T_2 and T_8 (Figure 5). However, the treatments T_2 (24.6 kg ha^{-1}) and T_8 (25.2 kg ha^{-1}) were recorded statistically superior over the rest of the treatments and on par with each other, followed by T_1 and T_7 which were also recorded on par with each other. However, treatment T_6 was recorded significant minimum total phosphorus uptakes, followed by T_3 and T_4. The phosphorus harvest index differed significantly due to treatments' effects. The phosphorus harvest index was highest in treatment T_4, then T_2, then T_3, then T_8, then T_6, then T_5, then Treatment$_1$ (Figure 6), and was ultimately lowest in treatment T_7.

3.4.3. Potassium

Among the treatments, significant maximum percentages of potassium content in the grains and straw of wheat were recorded in treatment T_5, with the values of 0.48 and 1.71 per cent; this was recorded as being statistically on par with treatments T_2 and T_8 (Figure 3). However, treatments T_7 and T_1 were recorded as being similarly statistically superior, being on par with each other, followed by T_3 and T_4. Meanwhile, the minimum per cent potassium content in the grains and straw of wheat were recorded in treatment T_6. As is evident from the data, potassium uptake in the grains and straw of wheat differed significantly due to treatments' effects (Figure 4). Significant maximum potassium uptakes in both grains (21.4 kg ha^{-1}) and straw (105.9 kg ha^{-1}) of wheat was recorded in treatment T_5 compared to all other treatments except T_2 and T_8, which were recorded as being statistically on par, with the exception of T_2 in the case of grains. However, treatments T_7, T_3 and T_1 were recorded as statistically superior (on par with each other) over rest of the treatments except wheat straw. Meanwhile, the minimum potassium uptake in the grains (12.7 kg ha^{-1}) and straw (81.9 kg ha^{-1}) were recorded in treatment T_6, followed by T_4. Among the treatments, significant maximum total potassium uptakes were recorded into treatment T_5 (127.3 kg ha^{-1}), which was recorded as being statistically on par with treatment T_8 (Figure 5). However, the treatment T_2 was recorded as statistically superior to the rest of the treatments, followed by T_7, T_1, and T_3. However, significantly low total potassium uptake was recorded into treatment T_6 (94.5 kg ha^{-1}), followed by T_4. Treatments T_5 and T_8 exhibited an increased total uptake of potassium compared to T_1 and T_6, with difference of 17.8, 34.7, and 15.4, 31.9 per cent, respectively, during course of investigation. Among the treatments, the maximum potassium harvest index was noted in treatment T_5, which was recorded as being statistically on par with treatments T_2 and T_8, all three being greater than the rest of the treatments (Figure 6). However, the treatments T_1, T_3, and T_4 were also recorded as possessing statistical superiority, being on par with each other over the remainder of the treatments. However, the minimum potassium harvest index of wheat was recorded in treatment T_6, followed by T_7.

3.5. Economic Nutrients Use Efficiency

The economic nutrient-use efficiency of wheat was significantly affected by tillage crop establishment methods during course of investigation (Figure 7). All the treatments received equal amount of nutrient, which is why the investment (INR/ha) for nutrients across all treatments was the same. An economic nutrient-use efficiency is dependent upon obtaining the ratio of productivity to the amount of INR invested on the nutrients applied. Among the treatments, maximum economic nitrogen-use efficiency (ENUE) was observed in treatment T_5 (2.68), which was recorded as being statistically on par with treatments T_2 and T_8. After this, the treatments T_1 and T_7 were recorded as the next most statistically superior compared to the remainder of the treatments, being on par with each other. However, minimum ENUE was recorded under treatment T_6 (1.71), followed by T_4 and T_3. A significant maximum Economic Phosphorus Use Efficiency was observed in treatment T_5 (1.42) compared to all other treatments, with the exception of T_2 and T_8, which were recorded as being statistically on par. However, treatments T_1 and T_7 were recorded as being similarly statistically superior, on par with each other over rest of the treatments. The minimum EPUE was recorded in treatment T_6, followed by T_4 and T_3. Economic Potassium Use Efficiency (EPUE) varied from 1.49 to 2.33. Among the treatments, the maximum EPUE was recorded in treatment T_5 (2.33) over all other treatments, with the exception of T_2 and T8, which were recorded as being statistically on par with it. After those, the treatments T_1 and T_7 were recorded as the next most statistically superior, *on par* with each other over rest of the other treatments. Meanwhile, the minimum EPUE was recorded in treatment T_6, followed by T_4 and T_3.

3.6. Biological Properties of Soil

Biological properties of soil were significantly influenced by tillage crop establishment methods during experimentation (Table 2). Among the treatments, the highest bacterial population was recorded under T_8 compared to other treatments. However, treatments T_2 and T_5 were recorded as superior to the remaining the treatments. Treatments T_3, T_4, T_6, and T_1 were recorded as being on par with each other, but treatment T_7 had the lowest recorded bacterial population during both years of study. Tillage crop establishment methods resulted in significant differences in the population of fungi after wheat harvesting. Treatment T_8 recorded the highest fungi population, followed by T_5. However, treatments T_2, T_3, and T_4 were recorded as similarly superior, being about on par with each other over rest of the treatments. The lowest fungi population was recorded in treatment T_7, followed by T_6 and T_1. The highest population of actinomycetes following the wheat harvest was noted in treatment T_8 compared to other treatments. However, treatments T_2 and T_5 were recorded as superior over remaining the treatments. Treatments T_1, T_4, and T_3 were recorded as being on par with each other. The lowest population of actinomycetes was recorded during T_7, followed by T_6. Tillage crop establishment methods recorded significant differences in microbial carbon during the investigation. Treatment T_8 recorded the highest microbial biomass carbon (160.21 μg g^{-1}) compared to other treatments. However, treatments T_2 and T_5 were recorded as being similarly superior, being on par with each other. Treatments T_3 and T_4 were likewise recorded as being on par with each other. The lowest microbial-C during experimentation was recorded in treatment T_7, followed by T_6 and T_1. The microbial biomass nitrogen was significantly affected by tillage crop establishment methods. Maximum microbial nitrogen (23.40 μg g^{-1}) was recorded in T_8 compared to other treatments, followed by T_5 and T_2. In addition, treatments T_3 and T_4 were recorded as being on par with each other. The lowest microbial-N during experimentation was recorded in treatment T_7, followed by T_6 and T_1.

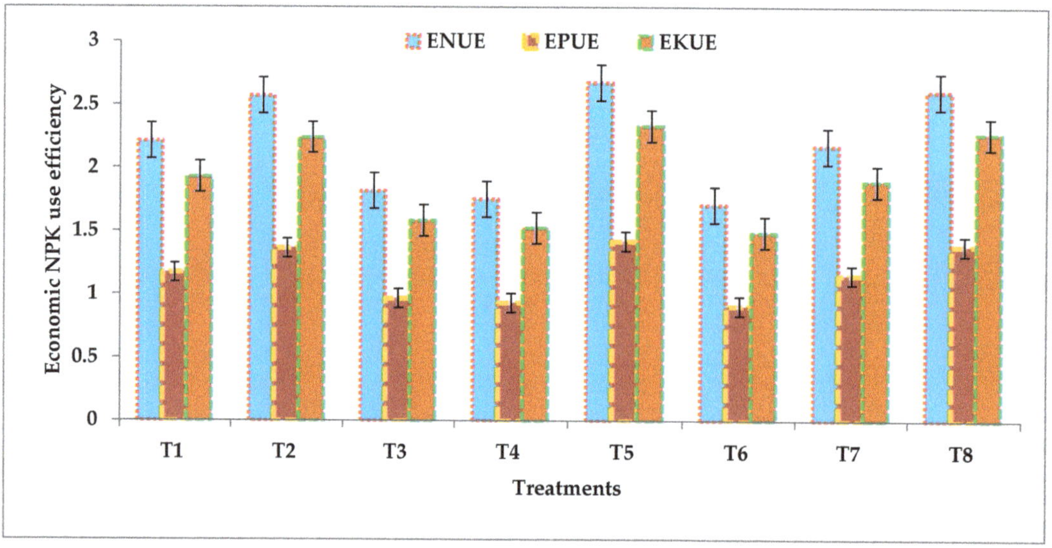

Figure 7. Economic Nitrogen, Phosphorus and Potassium Use Efficiency of wheat, as influenced by tillage crop establishment methods (pooled data: 2017–18 and 2018–19) (Bars represent standard error bars). ENUE: Economic Nitrogen Use Efficiency; EPUE: Economic Phosphorus Use Efficiency; and EKUE: Economic Potassium Use Efficiency. Nitrogen through Urea: 276 kg ha^{-1} at INR 6 kg^{-1}, Phosphorus through DAP: 130 kg ha^{-1} at INR 24 kg^{-1} and Potassium through MOP: 100 kg ha^{-1} at INR 19 kg^{-1}.

3.7. Fertility Status of Soil

Data showed that the status of soil organic carbon percentage differed only insignificantly (Table 3). Treatment T_8 registered the highest organic carbon percentage compared to all other treatments. However, treatment T_4 recorded the lowest organic carbon content percentage during experimentation, followed by T_6, T_7, T_3, T_2, T_5, and T_1. The tillage crop establishment method registered a significant difference in the fertility status of soil during the investigation (Table 3). Among the treatments, the maximum available nitrogen (228.95 kg ha^{-1}) was recorded in T_6, compared to all other treatments. However, treatments T_3 and T_4 were recorded as superior to the remainder of the treatments. Treatments T_1, T_2, and T_7 were on par with each other over the rest of the treatments. Treatment T_5 recorded the minimum available nitrogen during the investigation, followed by T_8. The maximum available phosphorus (15.70 kg ha^{-1}) was recorded under T_6 compared to other treatments (except T_4). However, treatments T_1 and T_3 were similarly superior, being on par with each other. Treatments T_2, T_5, T_7, and T_8 were on par with each other, while treatment T_5 recorded the minimum available phosphorus. Treatment T_6 (skip furrow irrigated) recorded the maximum (209.62 kg ha^{-1}) available potassium compared to other treatments. However, treatment T_5 recorded the minimum available potassium than rest of the treatments during experimentation, followed by T_8, T_2, T_7, T_3, T_4, and T_1.

Table 2. Biological properties of soil as influenced by tillage crop establishment methods (2017–2018 and 2018–2019, pooled).

Treatments	Biological Properties of Soil				
	Bacteria (10^5 CFU g^{-1})	Fungi (10^4 CFU g^{-1})	Actinomycetes (10^6 CFU g^{-1})	Microbial-C (μg/g^{-1} Soil)	Microbial-N (μg/g^{-1} Soil)
T_1	0.73	0.56	0.53	151.63	18.71
T_2	0.77	0.62	0.57	156.03	20.88
T_3	0.76	0.60	0.56	155.60	20.28
T_4	0.75	0.57	0.54	154.57	19.62
T_5	0.80	0.65	0.59	157.31	21.91
T_6	0.72	0.52	0.50	147.80	17.38
T_7	0.69	0.50	0.48	144.94	16.14
T_8	0.84	0.66	0.62	160.21	23.40
SEm±	0.01	0.01	0.01	0.62	0.34
CD ($p = 0.05$)	0.02	0.02	0.02	1.80	0.98

Table 3. Effect of tillage crop establishment methods on fertility status of soil (2017–2018 and 2018–2019, pooled).

Treatments	Organic Carbon (%)	Available Nutrients (kg ha^{-1})		
		Nitrogen	Phosphorus	Potassium
T_1	0.49	215.17	12.94	206.37
T_2	0.48	213.64	12.30	204.30
T_3	0.49	219.10	14.15	205.72
T_4	0.47	221.35	14.45	206.07
T_5	0.49	204.52	10.99	200.84
T_6	0.48	228.95	15.70	209.62
T_7	0.49	213.68	12.64	205.03
T_8	0.50	210.64	11.82	203.48
SEm±	0.001	1.33	0.44	0.94
CD ($p = 0.05$)	NS	3.88	1.27	2.74

4. Discussion

The rice plant height, number of tillers, dry matter accumulation, CGR, RGR, AGR, LAI, and NAR were recorded as being most significantly high in the wide-bed furrow-irrigated (T_5) treatment, being on par with both zero-till flat irrigated using gated pipes (T_8) and furrow irrigated with gated-pipe raised bed (T_2). It has been hypothesised that increased rates of dry matter production, translocation, and photosynthesis conversion were responsible. Better light penetration from bed seeding wheat resulted in stronger plants and more efficient photosynthesis, both of which boosted yields [18,19]. There were more spikes produced per square metre when there were more tillers present. Larger spikes and heavier grains resulted from a bigger proportion of biomass being assigned to spikes, which in turn resulted from a greater leaf area index, allowing the crop to absorb more solar energy for dry matter formation via photosynthesis. These results are discussed further in [20–25].

Of the studied yield attributes, productive tillers, grains spike^{-1}, spike length, and test weight were significantly increased in T_5, which was statistically on par with T_8 and T_2. The grains, straw, and biological yield, as well as the harvest index, were significantly higher under treatment T_5. The slow supply of moisture for longer times in order to enhance root and shoot growth of the crop is directly reflected in the source-to-sink transformation and boosting the metabolic activity of the crop plants. The yield per hectare rose as a result of the favourable effect of the increase in available moisture on per-plant productivity. Grain yield per plant increased as a result of an increase in moisture supply in three ways: the number of effective tillers, the number of grains per spike, and the test weight. Treatment T_5 (the wide-bed furrow-irrigated treatment) was recorded as producing 20.96, 10.99 and 14.94 per cent higher grain, straw, and biological yields, respectively, over T_1 (conventional flood irrigation). The same trend has been confirmed by several other investigations [26–30].

The significant impact that water had on the vegetative growth of the crop plant suggested that straw crop output may have increased as a result. Improved water distribution led to more vegetative growth, which in turn increased straw production. Straw production has reportedly followed a similar trend in other studies [22,31,32].

Due to the lack of surface moisture, conventional irrigation may be the leading cause of water loss. Many writers [33–35] have reported the same results.

The presence of soil moisture in the root zone of the crop increased root growth, which in turn enhanced nutrient uptake and boosted crop growth. Wide-bed furrow-irrigated treatment had the maximum recorded NPK content percentage and uptake in both grain and straw, as well as having the highest nutrient harvest index. Total NPK uptake was increased by approximately 27.97, 35.35, and 17.82 percent, respectively, in T_5 wide-bed furrow-irrigated wheat compared to T_1 conventional flood irrigation, which also boosted grain and biomass yield. The results are corroborated by the published research [35–40].

An increased porosity, increased availability of nutrients (especially P), and greater water availability in the soil profile available to be used by plants may have improved the soil's microbiological properties. These results are also affirmed in [41,42].

The maximum population of bacteria, fungi, and actinomycetes, microbial-C, and microbial-N were recorded in treatment T_8 (zero-till flat, irrigated using gated pipe/controlled-flood irrigation treatment) followed by T_5 (wide-bed furrow-irrigated treatment) and T_2 (furrow-irrigated with gated-pipe raised-bed treatment). The retention of previous crop residue in wheat under the rice–wheat cropping system on the surface of the soil (i.e., Conservation Agriculture, or Zero-till farming) maximised organic carbon level and microbial population, and minimised soil moisture loss, which helped the multiplication of microbes in the soil. Microbial development and, by extension, microbial circulation, are both influenced by the availability of organic matter in the soil [43]. Because of the decreased soil disturbance, the soil microbial biomass and activity were increased with full CA-based management using conservation-agriculture practices. Conservation agriculture may lead to higher microbial biomass nitrogen levels than conventional farming [44] due to the increased associated C inputs, residue retention, and decreased tillage.

The leached-down N, P, and K during treatment T_5 was the outcome of increased growth, yield, nutrient uptake, and soil moisture availability over a longer period of time. The findings of [35,37,38] were consistent with this.

5. Conclusions

This study recommends sowing wheat in a wide-bed furrow-irrigated system, or a zero-till flat-irrigated using gated-pipe/controlled-flood irrigation system for the best outcomes when growing rice and wheat together under irrigated conditions. Wide-bed furrow irrigation increased the total nutrient NPK uptake in wheat, while sprinkler irrigation maximised water-use efficiency, water productivity, and consumption. The zero-till and gated-pipe/controlled-flood irrigation treatment also had larger bacterial, fungal, and actinomycete populations, as well as higher microbial-C and microbial-N levels than traditional flood irrigation. Our research suggests that farmers should use conservation tillage-based establishing tactics and irrigation strategies to increase crop productivity, water- and nutrient-use efficiency, nutrient uptake, and soil fertility in wheat-growing regions, in order to maintain crop water productivity and soil health under different agroecological conditions.

Author Contributions: Conceptualisation, R.K. and R.K.N.; methodology, R.K. and D.K.; software, R.K.; validation, R.K., R.K.N., M.S.C., R.B. and H.M.K. formal analysis, R.K. and R.B.; investigation, R.K., R.K.N. and R.B.; resources, S.A., M.H.S. and A.T.A.; data curation, R.K.; writing—original draft preparation., R.K., R.B., S.A. and M.H.S.; writing—review and editing, H.M.K., M.H.S., D.K., S.A. and A.T.A.; visualisation, R.K. and R.B.; supervision, M.S.C. and D.K.; funding acquisition, S.A. and A.T.A. All authors have read and agreed to the published version of the manuscript.

Funding: This study was supported by the Researchers Supporting Project number (RSP2023R194), King Saud University, Riyadh, Saudi Arabia.

Data Availability Statement: The data presented in this study are available in the article.

Acknowledgments: The authors would like to extend their sincere appreciation to the Researchers Supporting Project number (RSP2023R194), King Saud University, Riyadh, Saudi Arabia.

Conflicts of Interest: The authors declare no conflict of interest.

References

1. Singh, M.; Supriya, K. Growth Rate and Trend Analysis of Wheat Crop in Uttar Pradesh, India. *Int. J. Curr. Microbiol. Appl. Sci.* **2017**, *6*, 2295–2301. [CrossRef]
2. *USDA Report (2019–2020)*; USDA: New York, NY, USA, 2019; pp. 11–12. Available online: https://www.usda.gov/media/agency-reports (accessed on 10 June 2023).
3. Alla, M.A.; Nadine, N.; Khudhair, A.J.; Radhi, K. Effect of irrigation methods and tillage system, seed level on water use efficiency and wheat (*Triticum aestivum* L.) growth. *Period. Eng. Nat. Sci.* **2020**, *8*, 1701–1715.
4. Rodell, M.; Velicogna, I.; Famiglietti, J.S. Satellite-based estimates of groundwater depletion in India. *Nature* **2009**, *460*, 999–1002. [CrossRef] [PubMed]
5. Somerville, C.; Briscoe, J. Genetic engineering and water. *Science* **2001**, *292*, 2217. [CrossRef] [PubMed]
6. Lal, R.; Follett, R.F.; Kimble, J.M. Achieving soil carbon sequestration in the United States: A challenge to thepolicy makers. *Soil Sci.* **2003**, *1680*, 827–845. [CrossRef]
7. Husnjak, S.; Filipovie, D.; Kosutiae, S. Influence of different tillage systems on soil physical properties and crop yield. *Rostl. Vyrob.-UZPI* **2002**, *48*, 249–254. [CrossRef]
8. Palm, C.; Blanco-Canqui, H.; DeClerck, F.; Gatere, L.; Grace, P. Conservation agriculture and ecosystem services: An overview. *Agric. Ecosyst. Environ.* **2014**, *187*, 87–105. [CrossRef]
9. Borie, F.; Rubio, R.; Rouanet, J.L.; Morales, A.; Borie, G.; Rojas, C. Effects of tillage systems on soil characteristics, glomalin and mycorrhizal propagules in a Chilean Ultisol. *Soil Tillage Res.* **2006**, *88*, 253–261. [CrossRef]
10. Mathew, R.P.; Feng, Y.; Githinji, L.; Ankumah, R.; Balkcom, K.S. Impact of No-tillage and conventional tillage systems on soil microbial communities. *Appl. Environ. Soil Sci.* **2012**, *2012*, 548620 . [CrossRef]
11. Hobbs, P.R.; Sayre, K.D.; Gupta, R.K. The role of conservation agriculture in sustainable agriculture. *Philos. Trans. R. Soc. B* **2008**, *363*, 543–555. [CrossRef]
12. Sharma, K.L.; Grace, J.K.; Mandal, U.K.; Gajbhiye, P.N.; Srinivas, K.; Korwar, G.R.; Bindu, V.H.; Ramesh, V.; Ramachandran, K.; Yadav, S.K. Evaluation of long-term soil management practices using key indicators and soil quality indices in a semi-arid tropical. *Alfisol. Soil Res.* **2008**, *46*, 368–377. [CrossRef]
13. Huang, Z.Q.; Xu, Z.H.; Chen, C.R. Effect of mulching on labile soil organic matter pools, microbial community functional diversity and nitrogen transformations in two hardwood plantations of subtropical Australia. *Appl. Soil Ecol.* **2008**, *40*, 229–239. [CrossRef]

14. Bolinder, M.A.; Janzen, H.H.; Gregorich, E.G.; Angers, D.A.; Vanden and Bygaart, A.J. An approach for estimating net primary productivity and annual carbon inputs to soil for common agricultural crops in Canada. *Agric. Ecosyst. Environ.* **2007**, *118*, 29–42. [CrossRef]
15. Rangaswami, G. *Agricultural Microbiology*; Asia Publishing House: London, UK, 1966; p. 413.
16. Jackson, M.L. *Soil Chemical Analysis*; Prentice Hall of India Pvt. Ltd.: New Delhi, India, 1973.
17. Subbiah, B.V.; Asija, G.L. A rapid procedure for the estimation of available nitrogen in soil. *Curr. Sci.* **1956**, *25*, 259–260.
18. Dhillon, S.S.; Prashar, A.; Thaman, S. Studies on bed planted wheat (*Triticum aestivum* L.) under different nitrogen levels and tillage methods. *Curr. Sci.* **2004**, *5*, 253–256.
19. Tanveer, S.K.; Hussain, I.; Sohail, M.; Kissana, N.S.; Abbas, S.G. Effect of different planting methods on yield and yield component of wheat. *Asian J. Plant Sci.* **2003**, *2*, 811–813. [CrossRef]
20. Ahmad, R.N.; Mahmood, N. Impact of Raised Bed Technology on Water Productivity and Lodging of Wheat. *Pak. J. Water Resour.* **2005**, *9*, 7–15.
21. Ali, M.; Ali, L.; Waqar, M.Q.; Ali, M.A. Bed planting a new establishment method for wheat (*Triticum aestivum* L.) in cotton-wheat cropping system of southern Punjab. *Int. J. Agric. Appl. Sci.* **2016**, *4*, 1–7.
22. Atikullah, M.N.; Sikder, R.K.; Asif, M.I.; Mehraj, H.; Jamaluddin, A.F.M. Effect of irrigation levels on growth, yield attributes and yield of wheat. *J. Biosci. Agric. Res.* **2014**, *2*, 83–89. [CrossRef]
23. Sepat, R.N.; Rai, R.K.; Dhar, S. Planting systems and integrated nutrient management for enhanced wheat (*Triticum aestivum*) productivity. *Indian J. Agron.* **2010**, *55*, 114–118.
24. Idnani, L.K.; Kumar, A. Relative efficiency of different irrigation schedules for conventional, ridge and raised bed seeding of wheat (*Triticum aestivum* L.). *Indian J. Agron.* **2012**, *57*, 148–151.
25. Ram, H.; Dadhwal, V.; Vashist, K.; Kaur, H. Grain yield and water use efficiency of wheat (*Triticum aestivum* L.) in relation to irrigation levels and rice straw mulching in North West India. *Agric. Water Manag.* **2013**, *128*, 92–101. [CrossRef]
26. Choudhary, R.L.; Behera, U.K. Conservation agricultural and nitrogen management in maize-wheat cropping system: Effect on growth, productivity and economics of wheat. *Int. J. Chem. Stud.* **2020**, *8*, 2432–2438. [CrossRef]
27. Singh, K.; Dwivedi, B.S.; Shukla, A.K.; Mishra, R.P. Permanent raised bed planting of the pigeon pea–wheat system on a typic-ustochrept, Effects on soil fertility, yield and water and nutrient use efficiencies. *Field Crops Res.* **2010**, *116*, 127–139. [CrossRef]
28. Singh, V.; Naresh, R.K.; Kumar, R.; Singh, A.; Shahi, U.P.; Kumar, V. Enhancing yield and water productivity of wheat (*Triticum aestivum*) through sowing methods and irrigation schedules under light textured soil of western Uttar Pradesh, India. *Int. J. Curr. Microbiol. Appl. Sci.* **2017**, *4*, 1400–1411.
29. Mollah, M.I.U.; Bhuiya, M.S.U.; Hossain, M.S.; Hossain, S.M.A. Growth of wheat (*Triticum aestivum* L.) under raised bed planting method in rice-wheat cropping system. *Bangladesh Rice J.* **2015**, *19*, 47–56. [CrossRef]
30. Naresh, R.K.; Singh, B.; Singh, S.P.; Singh, P.K.; Kumar, A.; Kumar, A. Furrow irrigated raised bed (FIRB) planting technique for diversification of rice-wheat system for western IGP region. *Int. J. Life Sci. Biotechnol. Pharma Res.* **2012**, *1*, 134–141.
31. Kumar, V.; Kumar, P.; Singh, R. Growth and yield of rice-wheat cropping sequence in raised bed planting system. *Indian J. Agric. Res.* **2013**, *47*, 157–162.
32. Kumar, R.; Pandey, D.S.; Singh, V.P. Wheat (*Triticum aestivum*) productivity under different tillage practices and legume options in rice (*Oryza sativa*) and wheat cropping sequence. *Indian J. Agric. Sci.* **2014**, *84*, 101–106.
33. Parihar, S.S.; Tiwari, R.B. Effect of irrigation and nitrogen level on yield, nutrient uptake and water use of late-sown wheat (*Triticum aestivum*). *Indian J. Agron.* **2003**, *48*, 103–107.
34. Naresh, R.K.; Singh, S.P.; Kumar, V. Crop establishment, tillage and water management technologies on crop and water productivity in rice-wheat cropping system of North West India. *Int. J. Sci. Life Sci. Biotechnol. Pharma. Res.* **2013**, *2*, 1–12.
35. Idnani, L.K.; Kumar, A. Performance of wheat (*Triticum aestivum* L.) under different irrigation schedules and sowing methods. *Indian J. Agric. Sci.* **2013**, *83*, 37–40.
36. Talukder, A.S.; Mominul Haque, M.; Meisner, C.; Kabir, M.J.; Hossain, A.B.S.; Rashid, M.H. Productivity of multi-crops sown on permanent raised beds in the tropics. In *New Directions for a Diverse Planet: Proceedings of the 4th International Crop Science Congress, Brisbane, Australia, 26 September–1 October 2004*; Crop Science Society of America: Madison, WI, USA, 2004.
37. Naresh, R.K.; Rathore, R.S.; Kumar, P.; Singh, S.P.; Singh, A.; Shahi, U.P. Effect of precision land leveling and permanent raised bed planting on soil properties, input use efficiency, productivity and profitability under maize (*Zea mays*)—Wheat (*Triticum aestivum* L.) cropping system. *Indian J. Agric. Sci.* **2014**, *84*, 725–732.
38. Jat, M.L.; Gupta, R.; Saharawat, Y.S.; Khosla, R. Layering precision land leveling and furrow irrigated raised bed planting: Productivity and input use efficiency of irrigated bread wheat in Indo-Gangetic Plains. *Am. J. Plant Sci.* **2011**, *2*, 578–588. [CrossRef]
39. Hossain, M.I.; Islam, K.; Sufian, A.; Abu, M.; Meisner, C.A.; Islam, S.M. Effect of planting method and nitrogen levels on the yield and yield attributes of wheat. *J. Bio Sci.* **2006**, *14*, 127–130. [CrossRef]
40. Rajanna, G.A.; Dhindwal, A.S.; Narender, N.; Patil, M.D.; Shiva kumar, L. Alleviating moisture stress under irrigation scheduling and crop establishment techniques on productivity and profitability of wheat (*Triticum aestivum*) under semi-arid conditions of western India. *Indian J. Agric. Sci.* **2018**, *88*, 32–38.

41. Sharma, A.R.; Singh, R.; Dhyani, S.K. Conservation tillage and mulching for optimizing productivity in maize-wheat cropping system in the outer western Himalayan region—A review. *Indian J. Soil Conserv.* **2005**, *33*, 35–43.
42. Samal, S.K.; Rao, K.K.; Poonia, S.P.; Kumar, R.; Mishra, J.S.; Prakash, V.; Mondal, S.; Dwivedi, S.K.; Bhatt, B.P.; Sushanta, N.K.; et al. Evaluation of long-term conservation agriculture and crop intensification in rice-wheat rotation of Indo-Gangetic Plains of South Asia: Carbon dynamics and productivity. *Eur. J. Agron.* **2017**, *90*, 198–208. [CrossRef]
43. Wang, Z.; Chen, Q.; Liu, L.; Wen, X.; Liao, Y. Responses of soil fungi to 5-year conservation tillage treatments in the drylands of northern China. *Appl. Soil Ecol. J.* **2016**, *101*, 132–140. [CrossRef]
44. Wang, J.J.; Li, X.Y.; Zhu, A.N.; Zhang, X.K.; Zhang, H.W.; Liang, W.J. Effect of tillage and residue management on soil microbial communities in North China. *Plant Soil Environ.* **2012**, *58*, 28–33. [CrossRef]

agronomy

Article

Mineral Fertilization and Maize Cultivation as Factors Which Determine the Content of Trace Elements in Soil

Marzena S. Brodowska [1,*], Mirosław Wyszkowski [2,*] and Barbara Bujanowicz-Haraś [3]

[1] Department of Agricultural and Environmental Chemistry, Faculty of Agrobioengineering, University of Life Sciences in Lublin, Akademicka 15 Str., 20-950 Lublin, Poland
[2] Department of Agricultural and Environmental Chemistry, Faculty of Agriculture and Forestry, University of Warmia and Mazury in Olsztyn, Łódzki 4 Sq., 10-727 Olsztyn, Poland
[3] Department of Management and Marketing, Faculty of Agrobioengineering, University of Life Sciences in Lublin, B. Dobrzańskiego 37 Str., 20-262 Lublin, Poland; barbara.bujanowicz-haras@up.lublin.pl
* Correspondence: marzena.brodowska@up.lublin.pl (M.S.B.); miroslaw.wyszkowski@uwm.edu.pl (M.W.)

Abstract: This study has been carried out in order to determine the effect of increasingly intensive fertilization with potassium, applied in combination with nitrogen, on the content of trace elements in soil after the harvest of maize (*Zea mays* L.). The soil content of trace elements depended on the fertilization with potassium and nitrogen. Potassium fertilization had a stronger effect on the content of trace elements in the pots fertilized with the lower nitrogen dose (130 mg N kg^{-1} of soil). The increasing doses of potassium led to a higher soil content of zinc (Zn), and especially of nickel (Ni). The impact of potassium fertilization on the content of the remaining trace elements in the soil was less unambiguous, and depended on the dose of potassium and nitrogen fertilization. Nitrogen fertilization resulted in a higher soil content of manganese (Mn), chromium (Cr), nickel (Ni) and cadmium (Cd), as well as a decreased soil content of lead (Pb). It needs to be underlined that changes in the soil content of Ni, Cd, and Pb, effected by nitrogen fertilization, were larger than in the cases of the other trace elements. The influence of potassium and nitrogen fertilization did not result in exceeding the current threshold amounts of trace elements set for agriculturally used soil. An increase in the contents of some trace elements in soil is beneficial from an agricultural point of view. Some of these elements are necessary for the correct growth and development of arable plants.

Keywords: mineral fertilization; maize; trace elements in soil

Citation: Brodowska, M.S.; Wyszkowski, M.; Bujanowicz-Haraś, B. Mineral Fertilization and Maize Cultivation as Factors Which Determine the Content of Trace Elements in Soil. *Agronomy* **2022**, *12*, 286. https://doi.org/10.3390/agronomy12020286

Academic Editors: Gang Li, Dong-Xing Guan, Daniel Menezes-Blackburn and Fuyong Wu

Received: 7 October 2021
Accepted: 20 January 2022
Published: 23 January 2022

1. Introduction

The constantly growing human population requires an increased production of food and feeds. To secure food for people and to achieve high yields in agriculture, it is necessary to apply mineral fertilizers in plant production, as they will meet the nutritional requirements of crops while acting more rapidly than natural or organic fertilisers. Fertilisation with basic elements is a sufficient measure to achieve high crop yields, but they may not be of high quality. It is therefore recommended to supply cultivated plants with a larger array of macro- and micronutrients [1]. In recent years, multi-nutrient fertilizers, specifically designed for individual plant species by accounting for their nutritional requirements, have been gaining increasing interest. Meanwhile, eating habits have changed and a demand for high quality food products has been increasing, which is closely connected with the improved life comfort and wealth of the human population [2].

While fertilization with basic macronutrients affects mainly the volume of yields, the application of trace elements decides about their quality. Small quantities of some trace elements (Cr, Cu, Ni, Zn, Mn, iron—Fe or cobalt—Co) are essential for the proper growth and development of plants and other living organisms [3,4]. On the other hand, little is known about the positive impact of other trace elements (such as Cd, Pb, Hg or As) on living organisms, and it is believed that these elements do not play a beneficial physiological role [5]. The influence of trace elements, however, depends on quantities in which they enter the plants' environment, mainly the soil [6]. In an uncontaminated soil environment, the main source of trace elements, which may become a potential threat to plants and other living organisms, are fertilizers, especially mineral ones [7]. Even mineral fertilizers which are not enriched with trace minerals added during their manufacturing process import some amounts of these elements, usually small ones, into the soil [8]. Under specific conditions, e.g., in areas with relatively severe air pollution, this additional dose of trace elements incorporated to soil might be the factor that will result in exceeding the permissible threshold amounts of these contaminants in soil [9]. A risk then arises that trace elements will migrate to subsequent links in the trophic chain [7]. They can become hazardous to the development of living organisms [10,11]. The literature data most often point to phosphate fertilizers as a source of trace elements in soil [12]. However, it should be emphasized that mineral fertilizers most polluted with trace elements can be placed as follows: phosphoric > calcium > potassium > nitrogen. The biggest impact on the degree of pollution of mineral fertilizers with trace elements has a raw material and technological process of their production [13]. Phosphorus fertilizers have a particularly large impact on the increase in the content of trace elements in soil [14]. The applied phosphorus fertilizers can be a significant source of soil contamination with heavy metals, mainly Cd, Cu, Pb, Ni and Zn. However, there is a significant variation in the content of trace elements depending on the form of the fertilizer [13]. According to Bracher et al. [15], the content of some trace elements in soil, mainly Cd, increases with an increasing dose of phosphorus fertilizers. Then, not only its total content in soil increases, but also the content of available forms for plants increases [16]. There are far fewer reports implicating potassium or nitrogen fertilizers in this regard [17]. Long-term mineral fertilization can raise the soil content of some trace elements, including Cd, Pb, Cu, Zn or Mn [18]. In their study, Zao et al. [19] found that a persistent application of fertilizers increased the content of trace elements in soil, with organic fertilizers having a stronger impact than mineral ones. It is worth drawing attention to several factors that determine the availability of trace elements for plants [20], such as soil acidity [21,22], or soil sorptivity, associated with the presence of organic matter, clay minerals [21], or hydrated iron and aluminium oxides [23,24], as well as soil microorganisms [25], which participate in the cycling of elements in nature [26].

In view of the above, this study has been carried out in order to determine the effect of increasingly intensive fertilization with potassium, applied in combination with nitrogen, on the content of trace elements in soil.

2. Materials and Methods

2.1. Methodological Design

An experiment was carried out in a greenhouse, with plants grown in polyethylene pots filled with soil from the humic horizon of proper brown soil (Eutric Cambisol) which, in terms of texture, was classified as loamy sand (sand > 0.05 mm—75.47%, silt 0.002–0.05 mm—21.30% and clay < 0.002 mm—3.23%) according to the taxonomy by the United States Department of Agriculture [27]. The basic soil properties were as follows: pH in 1 M KCl dm^{-3}—5.43; hydrolytic acidity—30.00 mM(+) kg^{-1}; total exchangeable bases—56.0 mM(+) kg^{-1}; cation exchange capacity—86.5 mM(+) kg^{-1}; base saturation—64.7%; content of total organic carbon (TOC)—5.386 g kg^{-1}; total nitrogen (total-N)—1.225 g kg^{-1}; available forms of phosphorus (P)—26.58 mg kg^{-1}; potassium (K)—125.36 mg kg^{-1}; magnesium (Mg)—27.97 mg kg^{-1}; sulphur (S) 12.86 mg $S\text{-}SO_4$ kg^{-1}; Cd—0.312 mg kg^{-1}; Pb—28.49 mg kg^{-1}; Cr—42.30 mg kg^{-1}; Co—5.741 mg kg^{-1}; Ni—16.05 mg kg^{-1}; Zn—36.18 mg kg^{-1}; Cu—6.658 mg kg^{-1}; Mn—318.9 mg kg^{-1}; Fe—11,906 mg kg^{-1} d.m. of soil. Increasing doses of potassium, 0, 140, 190 and 240 mg K_2O kg^{-1} of soil, were tested in combination with a lower and higher dose of nitrogen: 130 and 170 mg N kg^{-1} of soil. Potassium was applied as potassium sulphate (500 g K_2O kg^{-1} and 450 g S kg^{-1}), while nitrogen was added to soil as urea and ammonium nitrate solution (UAN) (280 g N kg^{-1}), with half the dose applied before sowing and the other half during the growth of plants. Same amounts of phosphorus, 85 mg P_2O_5 kg^{-1} of soil, and amounts of the micronutrients, 2.9 mg Zn [$ZnCl_2 \cdot 7H_2O$], 3.4 mg Cu [$CuSO_4 \cdot 5H_2O$], 1 mg B [H_3BO_3], 2.7 mg Mn [$MnCl_2 \cdot H_2O$], 0.02 mg Mo kg^{-1} of soil [$(NH_4)_6Mo_7O_{24} \cdot 4H_2O$], were added in soil in each pot. The influence of potassium and nitrogen fertilizers was tested on maize (*Zea mays* L.). A 9-kg batch of soil was carefully mixed with the mineral fertilizers and placed in a pot. Afterwards, maize was sown to grow 8 plants per pot. During the growth of the maize plants, at the 4–6-leaf stage, the second dose of potassium and nitrogen fertilizers was added to the soil. During the entire experiment, the soil moisture was maintained at a constant level of 60% of the water capillary capacity. Maize was harvested at the stage of the middle of tassel emergence (BBCH 55), which was also when soil samples for laboratory analyses were collected.

2.2. Methods of Laboratory and Statistical Analyses

Soil was prepared for laboratory analyses by drying and sifting through a sieve with the mesh opening size of 1 mm. Then, each soil sample was wet-digested in a mixture of concentrated hydrochloric acid (HCl AR—1.18 g cm^{-3}) and nitric (HNO_3 AR—1.40 g cm^{-3}) in a MARS 6 microwave digestion system (CEM Corporation, Matthews, NC, USA), in Xpress Teflon vessels (CEM Corporation, Matthews, NC, USA), according to the method US-EPA3051 [28]. The digested soil samples were analysed to determine the total content of Cd, Pb, Cr, Co, Ni, Zn, Mn and Fe using flame atomic absorption spectrometry (FAAS) with an air–acetylene flame [29]. Correctness of the laboratory analyses was verified against reference solutions by Fluka denoted as: Cd 51994, Pb 16595, Cr 02733, Co 119785.0100, Ni 42242, Zn 188227, Cu 38996, Mn 63534 and Fe 16596, and Certified Analytical Soil Reference Material from the AGH University of Science and Technology in Kraków, Poland.

Before starting the experiment, soil underwent the following determinations: the textural composition by the aerometric method [30] and laser diffraction, pH in 1 M KCl with the potentiometric method [31], TOC on a total organic analyser coupled with a solids analyser Shimadzu TOC-L (Shimadzu Corporation, Kyoto, Japan) [32], total-N by the Kjeldahl method [33], available forms of P and K by the Egner–Riehm method [34,35], Mg by the Schachtschabel method [36], and S by nephelometry according to the procedure by Bardsley and Lancaster [37]. The results were submitted to statistical verification, using a two-factorial analysis of variance ANOVA, principal component analysis PCA, Pearson's simple correlation coefficients and the $\eta 2$ coefficient, calculating the observed variance percentages according to the ANOVA method in Statistica (StatSoft, Inc., Tulsa, OK, USA) [38].

3. Results and Discussion

The content of trace elements in soil is essential for the proper growth and development of plants [3]. In uncontaminated soils, the most common source of trace elements are fertilizers, including minerals, especially multi-nutrient ones [1]. Single-nutrient fertilizers are also a source of micronutrients found in ballast, but their amounts are much smaller than in multi-nutrient fertilizers [8]. The uptake of fertilizers by plants is affected by soil properties, as well as the type of a crop, and even its cultivar [20]. The uptake of trace elements, and other macronutrients, by plants results in the depletion of soils, which must be enriched each year through the application of fertilizers. This is a prerequisite for high and good-quality yields of crops [39].

Consistent application of fertilizers tends to satisfy the demand of crops for micronutrients [40]. However, it is worth emphasising that the use of fertilizers containing basic macronutrients may have an antagonistic effect on the availability of some trace elements to plants. According to Symanowicz [41], excessively high doses of potassium fertilizers can trigger ionic antagonism between potassium and some trace elements.

In our experiment, the content of trace elements in soil proved to be statistically significantly dependent on the potassium and nitrogen fertilization (Tables 1 and 2). The effect of potassium fertilization on the content of trace elements in soil was stronger in the series in the lower (130 mg N kg^{-1} of soil) than in the higher dose of nitrogen (170 mg N kg^{-1} of soil).

In the series with the lower nitrogen dose (130 mg N kg^{-1} of soil), higher doses of potassium fertilizer resulted in an increase in the content of Zn, Co, Ni and Cu in soil, as well as a small but significant decrease in the accumulation of Fe in soil, compared with the soil not fertilized with this element (Tables 1 and 2). Differences in the content of these trace elements in the control versus the soil fertilized with the highest potassium dose (240 mg K_2O kg^{-1} of soil) reached 11% for Zn (r = 0.779), 29% for Co (r = 0.806), 38% for Ni (r = 0.857), 39% for Cu (r = 0.906) and 7% for Fe (r = −0.900). The lowest potassium dose (140 mg K_2O kg^{-1} of soil) caused a small increase in the content of Cd (by 5%, r = −0.743), while the medium dose (190 mg K_2O kg^{-1} of soil) also raised the amount of Cr in soil by 5% (r = −0.060). In the latter case, the increase was not significant statistically. Any further rise in supplied doses of potassium had a negative effect on the soil content of these two elements, especially of Cd. Changes in the soil content of Pb were small and irregular, while those in the amounts of accumulated Mn were insignificant.

Table 1. Trace elements (Cd, Pb, Cr, Co and Ni) content in soil (mg kg^{-1} D.M.).

Nitrogen Dose (mg kg^{-1} of Soil)	Potassium Dose (mg kg^{-1} of Soil)				Average	r
	0	140	190	240		
			Cd			
130	0.292	0.308	0.229	0.104	0.233	−0.743 **
170	0.300	0.325	0.271	0.254	0.288	−0.608 *
Average	0.296	0.317	0.250	0.179	0.260	−0.722 **
LSD for:	N dose—0.039 *, K dose—0.055 **, interaction—n.s.					
			Pb			
130	25.48	25.61	23.64	26.51	25.31	0.034
170	21.41	21.81	21.62	24.62	22.37	0.686 **
Average	23.45	23.71	22.63	25.57	23.84	0.435
LSD for:	N dose—0.69 **, K dose—0.97 **, interaction—1.37 *					
			Cr			
130	40.60	42.21	42.50	39.44	41.19	−0.060
170	43.81	44.53	40.31	45.35	43.50	−0.026
Average	42.21	43.37	41.41	42.40	42.34	−0.089
LSD for:	N dose—1.46 **, K dose—n.s., interaction—2.91 **					
			Co			
130	3.516	3.545	3.929	4.544	3.884	0.806 **
170	3.525	3.910	4.102	3.564	3.775	0.351
Average	3.521	3.728	4.016	4.054	3.829	0.959 **
LSD for:	N dose—n.s., K dose—0.107 **, interaction—n.s.					
			Ni			
130	14.67	15.00	19.04	20.20	17.23	0.857 **
170	17.53	17.99	19.03	19.63	18.55	0.925 **
Average	16.10	16.50	19.04	19.92	17.89	0.877 **
LSD for:	N dose—1.31 *, K dose—1.86 **, interaction—2.63 *					

LSD (least squares deviation). Significant for: ** $p \leq 0.01$, * $p \leq 0.05$, n.s. non-significant; r—correlation coefficient.

Table 2. Trace elements (Zn, Cu, Mn and Fe) content in soil (mg kg^{-1} D.M.).

Nitrogen Dose (mg kg^{-1} of Soil)	Potassium Dose (mg kg^{-1} of Soil)				Average	r
	0	140	190	240		
			Zn			
130	35.11	35.60	36.09	39.14	36.49	0.779 **
170	33.59	36.59	37.69	39.03	36.73	0.999 **
Average	34.35	36.10	36.89	39.09	36.61	0.948 **
LSD for:	N dose—n.s., K dose—1.09 **, interaction—1.54 *					
			Cu			
130	6.611	7.098	8.105	9.186	7.750	0.906 **
170	8.228	8.179	7.565	6.398	7.593	−0.794 **
Average	7.420	7.639	7.835	7.792	7.671	0.953 **
LSD for:	N dose—n.s., K dose—0.271 **, interaction—0.383 **					
			Mn			
130	321.8	332.8	321.2	320.0	324.0	−0.125
170	332.4	338.8	339.1	344.5	338.7	0.965 **
Average	327.1	335.8	330.2	332.3	331.3	0.551
LSD for:	N dose—8.3 *, K dose—n.s., interaction—n.s.					
			Fe			
130	12,224	11,384	11,186	11,384	11,545	−0.900 **
170	11,494	11,287	11,264	11,912	11,489	0.326
Average	11,859	11,336	11,225	11,648	11,517	−0.549
LSD for:	N dose—n.s., K dose—392 *, interaction—554 *					

LSD (least squares deviation). Significant for: ** $p \leq 0.01$, * $p \leq 0.05$, n.s. non-significant; r—correlation coefficient.

The application of the higher nitrogen dose to soil (170 mg N kg^{-1} of soil) distorted the direction of changes in the content of some trace elements in soil fertilized with potassium (Tables 1 and 2). It was only for Cd, Ni and Zn that the tendencies were the same as in the first experimental series. Under the influence of the increasing potassium doses, the content of Mn rose by 4% (r = 0.965), Ni by 12% (r = 0.925), Pb by15% (r = 0.686) and Zn by 16% (r = 0.999), while the content of Cu in soil declined by 22% (r = −0.794). The first dose of potassium (140 mg K_2O kg^{-1} of soil) contributed to a small increase in Cd, by 8% (r = −0.608), and the second (190 mg K_2O kg^{-1} of soil) to a 16% (r = 0.351) increase in the Co content in soil. Changes in the content of Mn and Cr in soil were insignificant, those in the Fe content were small, not exceeding 4%, and irregular.

Nitrogen fertilization caused the greatest changes in the content of Ni, Cd and Pb, contributing to an average increase by up to 8% (Ni) and 24% (Cd), as well as a decrease by 12% (Pb) in their accumulation in soil fertilized with the higher dose of this element—170 mg N kg^{-1} of soil, in comparison to the lower dose—130 mg N kg^{-1} of soil (Tables 1 and 2). A relatively small increase (5–6%) in the soil content of Mn and Cr in response to nitrogen fertilization was recorded.

Mineral fertilization, especially with nitrogen, decreased the soil pH and thereby enhanced the mobility of many trace elements, such as Cu, Zn, Mn or Fe [42]. In a study by Rutkowska et al. [43], the highest content of Zn, Cu, Mn and Fe was observed in soil fertilized with NPK that had the lowest pH. The analogous influence of the NPK fertilization on the mobility of the aforementioned elements was determined by Li et al. [42]. Singh et al. [44] found that NPK fertilization caused elevated mobility of Cu and Mn in soil.

According to Gudžić et al. [45], long-term (33-year-long) mineral fertilization, apart from increasing the level of soil acidity as well as the soil content of P and K, and decreasing the content of TOC and total-N, raised the mobility of Mn and Fe in the top horizon of soil, but did not have a significant effect on the changes in the chemical composition of a deeper soil horizon (20–40 cm). NK fertilization had a significant effect on the content of Zn, while the fertilization with NP and NPK significantly affected the soil content of Cu and Mn.

Jaskulska et al. [46] maintained that organic fertilization had a stronger effect on soil properties, including the content of macronutrients and trace elements, than mineral fertilization. However, mineral fertilization has been observed to cause positive effects as well. Mazur and Mazur [47] obtained similar results. In their experiment, mineral fertilization led to higher soil concentrations of Cu, Zn and Pb. Organic fertilization had a stronger impact than mineral fertilization, and resulted in an increase in the content of most of the analysed elements.

In a study by Park et al. [14], the long-term (40-years) application of mineral fertilizers increased the content of trace elements in soil. Phosphate fertilizers had the strongest effect. The application of phosphorus fertilizers caused an over 2-fold increase in the Cd content in soil. In the experiment of Zahoor et al. [48], a significantly greater increase in the content of Cu, Zn, Mn and Fe in soil was observed compared with the authors' own research on the effect of NPK fertilization. NPK fertilization had a greater effect than NP fertilization on the content of these trace elements in soil. According to Sungur et al. [49] mineral fertilizers mainly increase the content of Cu and Cd in soil. The trace elements are uptaken by plants and incorporated into the food chain [50].

The content of trace elements in soil under the influence of both nitrogen and potassium fertilization may significantly increase [51]. This was confirmed by our own research. Nitrogen fertilization modifies the soil's sorptive capacity and bioavailability of trace elements for plants, especially Cd. Nitrogen fertilization, regardless of its form, increases the uptake of Cd and other trace elements, and their translocation and accumulation in plants. However, nitrate fertilizers have the strongest effect [52]. The relatively small effect of nitrogen fertilization in the form of UAN on the content of trace elements in soil may also result from the chemical composition of this fertilizer. UAN contains nitrogen in the form of NH_4^+-N and NO_3^--N, which regulates the pH in soil and the availability of trace elements for plants [53]. Similar results were achieved in our own studies. Higher doses of nitrogen caused a relatively small increase in the contents of some trace elements in soil. Of the nine studied elements, only in the case of four elements (Ni, Cd, Cr and Mn) was a small several-percent increase in content in the soil recorded. The increase in the content of trace elements in soil after mineral-fertilizer application to soil was also confirmed in many experiments of other authors [41,42,44–46,51,52]. Potassium sulphate, as a product obtained from severe potassium salts by way of a physical mining process, contains greater amounts of tracing elements than a urea and ammonium nitrate solution (UAN), which is obtained by chemical synthesis [54]. This has been confirmed in our own studies, in which potassium fertilizers caused greater changes in the content of trace elements in soil than nitrogen fertilizers. Bąk et al. [55] found an increase in the content of Zn in maize fertilized with potassium.

In our experiment, the PCA results (Figures 1 and 2) and Pearson's correlation coefficients (Table 3) revealed significant relationships between the content of some trace elements in the soil. The cumulative impact of fertilization with potassium and nitrogen on the content of trace elements in soil was illustrated in Figure 1, in the form of PCA vector variables. The total correlation of the set of data on Co, Cd, Cu and Zn was 44.32%, and on Pb, Fe, Mn, Cr and Ni it equalled 25.07%. The longest vectors corresponded to Co and Cd, which signifies their greatest importance, while the shortest vector was plotted for Zn, which illustrates its least contribution to variance. Vectors of the other trace elements had an approximately similar length. The distribution of vectors points to quite strong positive correlations between Cr versus Mn, Cu and Zn versus Co and Ni, Ni versus Co, and negative correlations between Cd versus Co, Ni versus Pb, Pb versus Cr and Mn, and the weakest relationship between Ni and Fe.

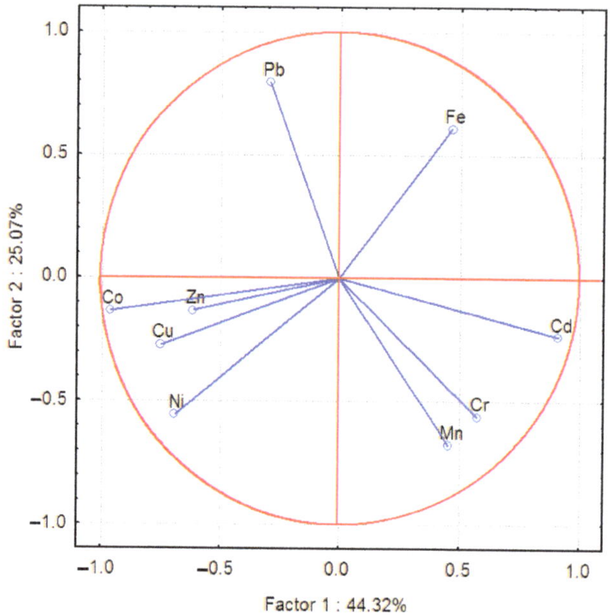

Figure 1. Trace element content in soil calculated with the PCA method. Vectors represent trace elements (content of Cd, Pb, Cr, Co, Ni, Zn, Cu, Mn and Fe).

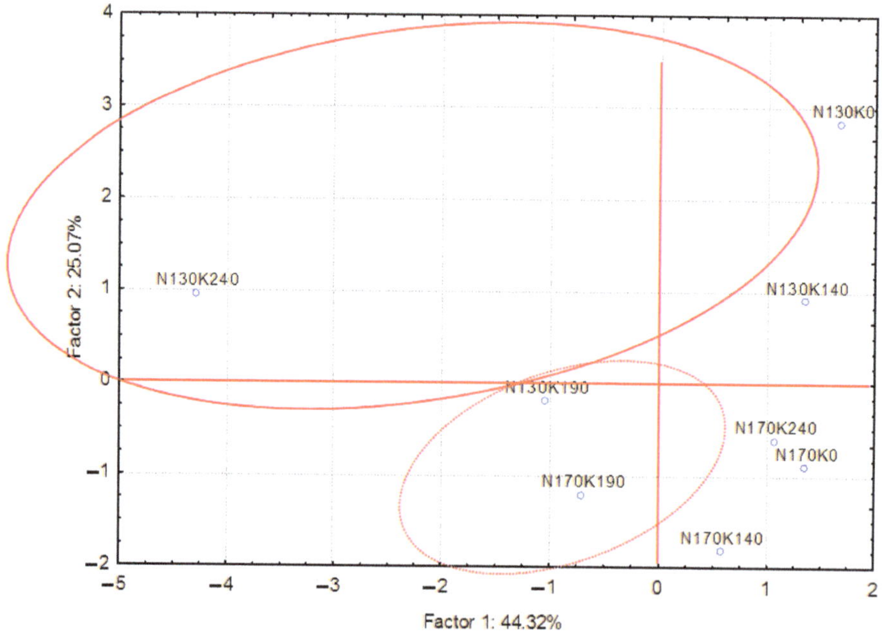

Figure 2. Effect of mineral (N and K) fertilization on trace elements content in soil calculated with the PCA method. Points show all trace elements content in soil (N 130—130 mg N kg^{-1}, N 170—170 mg N kg^{-1}; K 0—0 mg K kg^{-1}, K 140—140 mg K kg^{-1}, K 190—190 mg K kg^{-1}, K 240—240 mg kg^{-1} of soil).

Table 3. Correlation coefficients between content of trace elements in soil.

Element	Cd	Pb	Cr	Co	Ni	Zn	Cu	Mn
Pb	−0.440 *							
Cr	0.392	−0.427 *						
Co	−0.484 *	0.102	−0.366					
Ni	−0.532 **	−0.219	0.157	0.555 **				
Zn	−0.243	0.152	0.110	0.582 **	0.569 **			
Cu	−0.352	−0.116	−0.259	0.706 **	0.422 *	0.184		
Mn	0.474 *	−0.442 *	0.606 **	0.018	0.206	0.380	−0.117	
Fe	0.034	0.314	0.108	−0.040	−0.285	0.078	−0.361	0.099

Significant at ** $p \leq 0.01$ * $p \leq 0.05$; r—correlation coefficient.

The distribution of the research results displayed in Figure 2 justifies the conclusion that the second (190 K_2O kg^{-1} of soil) and especially the highest dose of potassium (240 mg K_2O kg^{-1} of soil) had the strongest effect on the soil content of trace elements. It was much stronger in the series fertilized with the lower nitrogen dose (130 mg N kg^{-1} of soil) than in the one treated with its higher dose (170 mg N kg^{-1} of soil).

The percentage of observed variance, calculated with the help of η2 coefficient from the ANOVA approach, indicates a stronger effect of potassium fertilization on the content of Co, Fe, Zn, Cd and Ni in soil (Figure 3). The percentage contribution of potassium to the variance of these elements was 25.5%, 28.7%, 35.3%, 49.4% and 60.3%, respectively. Quite a high value of this indicator was also achieved for Pb, 30.8%. The highest contribution of nitrogen fertilization was determined for Mn, 40.6%, and Pb, 57.8%, while a moderate one was for Cr, 28.1%.

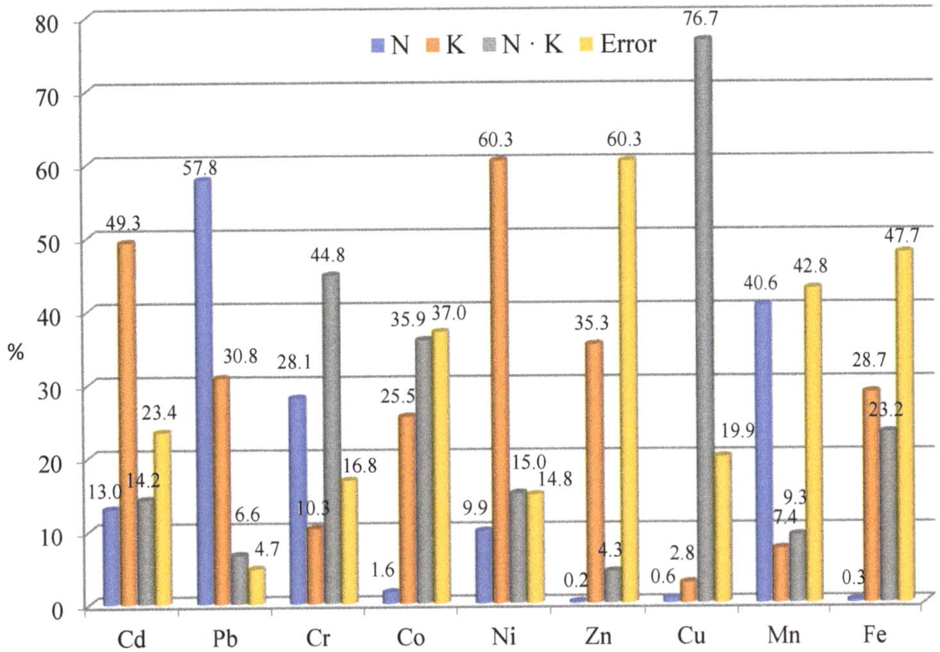

Figure 3. Percent contribution of mineral fertilization according to the content of trace elements in soil: N—nitrogen dose, K—potassium dose, interaction (N · K).

Concentrations of micronutrients in soil tend to be sufficient to satisfy the nutritional requirements of crops during their growth and development [39]. The content of trace elements determined in this study was sufficient to nourish the maize grown in this soil; this fact was confirmed by the amounts of these trace elements determined after the maize harvest and reported in this paper.

Strong relationships emerged in this experiment between concentrations of some trace elements in soil, confirmed statistically (PCA, Pearson's correlation coefficients). Mazur and Mazur [47] implicate strong relationships between Mn and Cd or Ni, between Cd and Ni in lighter soil, between Mn and Cd, and between Cd and Zn versus Pb in heavier soil.

In this experiment, fertilization with potassium and nitrogen caused an increase in soil content of some trace elements, but never above the thresholds set for agriculturally used soil, stipulated by law [9].

Mineral fertilizers usually cause a small increase in the contents of trace elements in soil. Of course, the uptake of trace elements by plants has an effect on the contents of trace elements in soil. This may result in reducing the effect of fertilizers on the content of trace elements in soil after harvesting plants. Therefore, there is a correlation between the content of trace elements in soil and the yield of plants. It is also an explanation of a small increase in the content of trace elements in soil, resulting from the influence of mineral fertilizers [52]. Reducing the content of trace elements in soil under the influence of higher doses of mineral fertilizers results from their favorable effect on the yield of plants. The plants uptake larger quantities of trace elements and decrease their contents in soil.

4. Conclusions

The soil content of trace elements depended on the fertilization with potassium and nitrogen. Potassium fertilization had a stronger effect on the content of trace elements in plots fertilized with the lower nitrogen dose (130 mg N kg^{-1} of soil).

The increasing doses of potassium led to a higher soil content of Ni and Zn. The biggest changes were observed for Ni. The impact of potassium fertilization on the content of the remaining trace elements in soil was less unambiguous, and depended on the dose of potassium and nitrogen fertilization. Nitrogen fertilization resulted in a higher soil content of Mn, Cr, Ni and Cd, as well as a decreased soil content of Pb. It needs to be underlined that changes in the soil content of Ni, Cd and Pb, effected by nitrogen fertilization, were bigger than in the case of the other trace elements.

The presence of significant relationships was noticed between the content of some trace elements in soil, which was confirmed statistically after performing the PCA and calculating Pearson's simple correlation coefficients.

The influence of potassium and nitrogen fertilization did not lead to exceeding the current threshold amounts of trace elements set for agriculturally used soil.

Application potassium and nitrogen fertilizers increased the contents of some trace elements in the soil. This is beneficial from an agricultural point of view, as some of these elements are necessary for the correct growth and development of arable plants.

Author Contributions: Conceptualization, M.S.B. and M.W.; methodology, M.S.B. and M.W.; software, M.W.; analysis, M.S.B. and M.W.; references collect, M.W. and B.B.-H.; writing—review and editing, M.S.B. and M.W.; supervision, M.W. and M.S.B.; M.S.B., corresponding author. All authors contributed significantly to the discussion of the results and the preparation of the manuscript. All authors have read and agreed to the published version of the manuscript.

Funding: The results presented in this paper were obtained as part of a comprehensive study financed by the University of Life Sciences in Lublin, Faculty of Agrobioengineering, Department of Agricultural and Environmental Chemistry (grant No. RKC/S/59/2021) and by the University of Warmia and Mazury in Olsztyn, Faculty of Agriculture and Forestry, Department of Agricultural and Environmental Chemistry (grant No. 30.610.006-110).

Institutional Review Board Statement: Not applicable.

Informed Consent Statement: Not applicable.

Data Availability Statement: Data are available by contacting the authors.

Conflicts of Interest: The authors declare no conflict of interest. The funders had no role in the design of the study; in the collection, analyses, or interpretation of data; in the writing of the manuscript, or in the decision to publish the results.

References

1. Lemaire, G.; Tang, L.; Bélanger, G.; Zhu, Y.; Jeuffroy, M.-H. Forward new paradigms for crop mineral nutrition and fertilization towards sustainable agriculture. *Eur. J. Agron.* **2021**, *125*, 126248. [CrossRef]
2. Buchner, B.; Fischler, C.; Gustafson, E.; Reilly, J.; Riccardi, G.; Ricordi, C.; Veronesi, U. Eating in 2030: Trends and Perspectives. Barilla Center for Food and Nutrition, 2012; p. 52. Available online: https://www.barillacfn.com/m/publications/eating-in-2030-trends-and-perspectives.pdf (accessed on 10 September 2021).
3. Nagajyoti, P.C.; Lee, K.D.; Sreekanth, T.V. Heavy metals, occurrence and toxicity for plants: A review. *Environ. Chem. Lett.* **2010**, *8*, 199–216. [CrossRef]
4. Kabata-Pendias, A. *Trace Elements in Soils and Plants*, 4th ed.; CRC Press: Boca Raton, FL, USA, 2011; p. 403. Available online: http://base.dnsgb.com.ua/files/book/Agriculture/Soil/Trace-Elements-in-Soils-and-Plants.pdf (accessed on 10 September 2021).
5. Latifi, Z.; Jalali, M. Trace element contaminants in mineral fertilizers used in Iran. *Environ. Sci. Pollut. Res.* **2018**, *25*, 31917–31928. [CrossRef] [PubMed]
6. Chibuike, G.U.; Obiora, S.C. Heavy metal polluted soils: Effect on plants and bioremediation methods. *Appl. Environ. Soil Sci.* **2014**, *2014*, 1–12. [CrossRef]
7. Belon, E.; Boisson, M.; Deportes, I.Z.; Eglin, T.K.; Feix, I.; Bispo, A.O.; Galsomies, L.; Leblond, S.; Guellier, C.R. An inventory of trace elements inputs to French agricultural soils. *Sci. Total Environ.* **2012**, *439*, 87–95. [CrossRef] [PubMed]
8. Alloway, B.J. Sources of heavy metals and metalloids in soils. In *Heavy Metals in Soils: Trace Metals and Metalloids in Soils and Their Bioavailability*; Alloway, B.J., Ed.; Environmental Pollution, 22; Springer: Dordrecht, The Netherlands, 2013; pp. 11–50. [CrossRef]
9. Regulation of Minister of the Environment of 1 September 2016 on the Procedures for the Assessment of Land Surface Contamination. In *Journal of Laws*; 2016; Poz. 1395. Available online: http://prawo.sejm.gov.pl/isap.nsf/download.xsp/WDU20160001395/O/D20161395.pdf (accessed on 10 September 2021).
10. Wuana, R.A.; Okieimen, F.E. Heavy metals in contaminated soils: A review of sources, chemistry, risks and best available strategies for remediation. *Int. Sch. Res. Not. Ecol.* **2011**, *2011*, 402647. [CrossRef]
11. Ali, H.; Khan, E.; Ilahi, I. Environmental chemistry and ecotoxicology of hazardous heavy metals: Environmental persistence, toxicity, and bioaccumulation. *J. Chem.* **2019**, *2019*, 6730305. [CrossRef]
12. Jiao, W.; Chen, W.; Chang, A.C.; Page, A.L. Environmental risks of trace elements associated with long-term phosphate fertilizers applications: A review. *Environ. Pollut.* **2012**, *168*, 44–53. [CrossRef]
13. Gorlach, E.; Gambuś, F. Phosphorus and compound fertilizers as a source of soil contamination with heavy metals. *Probl. J. Adv. Agric. Sci.* **1997**, *448a*, 139–146.
14. Park, H.J.; Kim, S.U.; Jung, K.Y.; Lee, S.; Choi, Y.D.; Owens, V.N.; Kumar, S.; Yun, S.W.; Hong, C.O. Cadmium phytoavailability from 1976 through 2016: Changes in soil amended with phosphate fertilizer and compost. *Sci. Total Environ.* **2021**, *762*, 143132. [CrossRef] [PubMed]
15. Bracher, C.; Frossard, E.; Bigalke, M.; Imseng, M.; Mayer, J.; Wiggenhauser, M. Tracing the fate of phosphorus fertilizer derived cadmium in soil-fertilizer-wheat systems using enriched stable isotope labeling. *Environ. Pollut.* **2021**, *287*, 117314. [CrossRef] [PubMed]
16. Gray, C.W.; McLaren, R.G.; Roberts, A.H.C. An assessment of the effect of contact time on cadmium phytoavailability in a pasture soil. *Environ. Sci. Pollut. Res.* **2016**, *23*, 22212–22217. [CrossRef] [PubMed]
17. Wyszkowski, M.; Brodowska, M.S. Potassium and nitrogen fertilization vs. trace element content of maize (*Zea mays* L.). *Agriculture* **2021**, *11*, 96. [CrossRef]
18. Czarnecki, S.; Düring, R.-A. Influence of long-term mineral fertilization on metal contents and properties of soil samples taken from different locations in Hesse, Germany. *Soil* **2015**, *1*, 23–33. [CrossRef]
19. Zhao, S.; Qiu, S.; He, P. Changes of heavy metals in soil and wheat grain under long-term environmental impact and fertilization practices in North China. *J. Plant Nutr.* **2018**, *41*, 1970–1979. [CrossRef]
20. Rolka, E.; Wyszkowski, M. Availability of trace elements in soil with simulated cadmium, lead and zinc pollution. *Minerals* **2021**, *11*, 879. [CrossRef]
21. Nunes, J.R.; Ramos-Miras, J.; Lopez-Piñeiro, A.; Loures, L.; Gil, C.; Coelho, J.; Loures, A. Concentrations of available heavy metals in Mediterranean agricultural soils and their relation with some soil selected properties: A case study in typical Mediterranean soils. *Sustainability* **2014**, *6*, 9124–9138. [CrossRef]
22. Ok, Y.S.; Usman, A.R.A.; Lee, S.S.; El-Azeem, S.A.M.A.; Choi, B.; Hashimoto, Y.; Yang, J.E. Effects of rapeseed residue on lead and cadmium availability and uptake by rice plants in heavy metal contaminated paddy soil. *Chemosphere* **2011**, *85*, 677–682. [CrossRef]
23. Kumpiene, J.; Lagerkvist, A.; Maurice, C. Stabilization of As, Cr, Cu, Pb and Zn in soil using amendments—A review. *Waste Manag.* **2008**, *28*, 215–225. [CrossRef] [PubMed]
24. Różański, S. Fractionation of selected heavy metals in agricultural soils. *Ecol. Chem. Eng. S* **2013**, *20*, 117–125.

25. Wyszkowska, J.; Borowik, A.; Kucharski, M.; Kucharski, J. Effect of cadmium, copper and zinc on plants, soil microorganisms and soil enzymes. *J. Elem.* **2013**, *18*, 769–796. [CrossRef]
26. Wang, L.; Chen, H.; Wu, J.; Huang, L.; Brookes, P.C.; Rodrigues, M.J.L.; Xu, J.; Liu, X. Effects of magnetic biochar-microbe composite on Cd remediation and microbial responses in paddy soil. *J. Hazard. Mater.* **2021**, *414*, 125494. [CrossRef] [PubMed]
27. IUSS Working Group WRB. World Reference base for soil resources. In *International Soil Classification System for Naming Soils and Creating Legends for Soil Maps*; WRB: London, UK, 2014; Volume 106, p. 182.
28. *US-EPA Method 3051A*; Microwave Assisted Acid Digestion of Sediment, Sludges, Soils and Oils. Office of Solid Waste and Emergency Response, United States Government Printing Office: Washington, DC, USA, 2007; 1–30. Available online: https://www.epa.gov/sites/production/files/2015-12/documents/3051a.pdf(accessed on 23 June 2021).
29. Ostrowska, A.; Gawliński, S.; Szczubiałka, Z. *Methods for Analysis and Evaluation of Soil and Plant Properties*; IOŚ: Warszawa, Poland, 1991; pp. 1–334. (In Polish)
30. *PN-R-04032*; Soil and Mineral Materials—Sampling and Determination of Particle Size Distribution. Polish Committee for Standardization: Warszawa, Poland, 1998.
31. *ISO 10390*; Soil Quality—Determination of Ph. International Organization for Standardization: Geneva, Switzerland, 2005.
32. Shimadzu, Co. *Analytical and Measuring Instruments*; Shimadzu Corporation: Kyoto, Japan, 2019. Available online: https://solutions.shimadzu.co.jp/an/n/en/etc/jpz19014.pdf?_ga=2.50821161.1231336941.1597769507-1298426863.1597769507 (accessed on 11 September 2021).
33. *ISO 11261*; Soil Quality—Determination of Total Nitrogen—Modified Kjeldahl Method. International Organization for Standardization: Geneva, Switzerland, 1995.
34. *PN-R-04023*; Chemical and Agricultural Analysis—Determination of the Content of Available Phosphorus in Mineral Soils. Polish Standards Committee: Warszawa, Poland, 1996.
35. *PN-R-04022*; Chemical and Agricultural Analysis—Determination of the Content Available Potassium in Mineral Soils. Polish Standards Committee: Warszawa, Poland, 1996.
36. *PN-R-04020*; Chemical and Agricultural Analysis—Determination of the Content Available Magnesium. Polish Standards Committee: Warszawa, Poland, 1994.
37. Boratyński, K.; Grom, A.; Ziętecka, M. Research on the content of sulfur in soil. Part I. *Rocz. Gleboz.* **1975**, *3*, 121–139.
38. Dell Inc. Dell Statistica (Data Analysis Software System), Version 13. 2016. Available online: http://software.dell.com (accessed on 29 July 2021).
39. Alloway, B.J. Micronutrients and crop production: An introduction. In *Micronutrient Deficiencies in Global Crop Production*; Alloway, B.J., Ed.; Springer Science + Business Media, B.V.: Dordrecht, The Netherlands, 2008; p. 370.
40. Symanowicz, B.; Kalembasa, S.; Becher, M.; Toczko, M.; Skwarek, K. Effect of varied levels of fertilization with potassium on field pea yield and content and uptake of nitrogen. *Acta Sci. Pol. Agric.* **2017**, *16*, 163–173. [CrossRef]
41. Symanowicz, B. Antagonistic changes in the content of molybdenum and boron in field pea and in soil under of the influence potassium fertilisation. *J. Elem.* **2020**, *25*, 193–203. [CrossRef]
42. Li, B.Y.; Zhou, D.M.; Cang, L.; Zhang, H.L.; Fan, X.H.; Qin, S.W. Soil micronutrient availability to crops as affected by long-term inorganic and organic fertilizer applications. *Soil Tillage Res.* **2007**, *96*, 166–173. [CrossRef]
43. Rutkowska, B.; Szulc, W.; Sosulski, T.; Stępień, W. Soil micronutrient availability to crops affected by long-term inorganic and organic fertilizer applications. *Plant Soil Environ.* **2014**, *60*, 198–203. [CrossRef]
44. Singh, A.; Agrawal, M.; Marshall, F.M. The role of organic vs. inorganic fertilizers in reducing phytoavailability of heavy metals in a wastewater-irrigated area. *Ecol. Eng.* **2010**, *36*, 1733–1740. [CrossRef]
45. Gudžić, N.; Šekularac, G.; Djikić, A.; Djekić, V.; Aksić, M.; Gudžić, S. The impact of the long-term fertilisation with mineral fertilizers on the chemical properties of Vertisol (central Serbia). *Appl. Ecol. Environ. Res.* **2019**, *17*, 12385–12396. [CrossRef]
46. Jaskulska, I.; Lemanowicz, J.; Breza-Boruta, B.; Siwik-Ziomek, A.; Radziemska, M.; Dariusz, J.; Białek, M. Chemical and biological properties of sandy loam soil in response to long-term organic–mineral fertilisation in a warm-summer humid continental climate. *Agronomy* **2020**, *10*, 1610. [CrossRef]
47. Mazur, Z.; Mazur, T. The influence of long-term fertilization with slurry, manure and NPK on the soil content of trace elements. *J. Elem.* **2016**, *21*, 131–139. [CrossRef]
48. Zahoor, M.; Afzal, M.; Ali, M.; Mohammad, W.; Khan, N.; Adnan, M.; Ali, A.; Saeed, M. Effect of organic waste and NPK fertilizer on potato yield and soil fertility. *Pure Appl. Biol.* **2016**, *5*, 439–445. [CrossRef]
49. Sungur, A.; Kavdir, Y.; Özcan, H.; İlay, R.; Soylak, M. Geochemical fractions of trace metals in surface and core sections of aggregates in agricultural soils. *Catena* **2021**, *197*, 104995. [CrossRef]
50. McLaughlin, M.J.; Smolders, E.; Zhao, F.J.; Grant, C.; Montalvo, D. Managing cadmium in agricultural systems. *Adv. Agron.* **2020**, *166*, 1–129. [CrossRef]
51. Wyszkowski, M.; Brodowska, M.S. Content of trace elements in soil fertilized with potassium and nitrogen. *Agriculture* **2020**, *10*, 398. [CrossRef]
52. Yang, Y.; Xiong, J.; Tao, L.; Cao, Z.; Tang, W.; Zhang, J.; Yu, X.; Fu, G.; Zhang, X.; Lu, Y. Regulatory mechanisms of nitrogen (N) on cadmium (Cd) uptake and accumulation in plants: A review. *Sci. Total Environ.* **2020**, *708*, 135186. [CrossRef]
53. Hawkesford, M.; Horst, W.; Kichry, T.; Lambers, H.; Schjoerring, J.; Møller, I.S.; White, P. Functions of macronutrient. In *Mineral Nutrition of Higher Plants*; Marschner, P., Ed.; Academic Press: London, UK, 2011; pp. 135–189. [CrossRef]

54. Parkinson, R.; Willson, R. Soils and plant nutrition. In *The Agricultural Notebook*, 21st ed.; Soffe, R.J., Lobley, M., Eds.; John Wiley & Sons Ltd., Wiley-Blackwell: Hoboken, NJ, USA, 2021; pp. 1–50.
55. Bąk, K.; Gaj, R.; Budka, A. Distribution of zinc in maize fertilized with different doses of phosphorus and potassium. *J. Elem.* **2016**, *21*, 989–999. [CrossRef]

agronomy

MDPI

Review

Phosphorus Mobilization in Plant–Soil Environments and Inspired Strategies for Managing Phosphorus: A Review

Muhammad Ibrahim [1,2], Muhammad Iqbal [3], Yu-Ting Tang [4], Sardar Khan [5], Dong-Xing Guan [6] and Gang Li [1,7,8,]*

1 CAS Engineering Laboratory for Recycling Technology of Municipal Solid Waste, CAS Key Lab of Urban Environment and Health, Institute of Urban Environment, Chinese Academy of Sciences, Xiamen 361021, China
2 Department of Environmental Sciences, Faculty of Life Sciences, University of Okara, Okara 56130, Pakistan
3 Department of Botany, Faculty of Life Sciences, University of Okara, Okara 56130, Pakistan
4 School of Geographical Sciences, Faculty of Science and Engineering, University of Nottingham, Ningbo 315100, China
5 Department of Environmental Sciences, University of Peshawar, Peshawar 25120, Pakistan
6 Zhejiang Provincial Key Laboratory of Agricultural Resources and Environment, Institute of Soil and Water Resources and Environmental Science, College of Environmental and Resource Sciences, Zhejiang University, Hangzhou 310058, China
7 Zhejiang Key Lab of Urban Environmental Processes and Pollution Control, CAS Haixi Industrial Technology Innovation Center in Beilun, Ningbo 315830, China
8 Jiaxing Key Lab of Soil Health, Yangtze Delta Region Healthy Agriculture Institute, Jiaxing 314503, China
* Correspondence: gli@iue.ac.cn

Citation: Ibrahim, M.; Iqbal, M.; Tang, Y.-T.; Khan, S.; Guan, D.-X.; Li, G. Phosphorus Mobilization in Plant–Soil Environments and Inspired Strategies for Managing Phosphorus: A Review. *Agronomy* **2022**, *12*, 2539. https://doi.org/10.3390/agronomy12102539

Academic Editor: Wei Zhang

Received: 13 August 2022
Accepted: 10 October 2022
Published: 17 October 2022

Abstract: Crop productivity and yield are adversely affected by the deficiency of P in agricultural soil. Phosphate fertilizers are used at a large scale to improve crop yields globally. With the rapid increase in human population, food demands are also increasing. To see that crop yields meet demands, farmers have continuously added phosphate fertilizers to their arable fields. As the primary source of inorganic phosphorous, rock phosphate is finite and the risk of its being jeopardized in the foreseeable future is high. Therefore, there is a dire need to improve plant-available P in soil, using feasible, environmentally friendly technologies developed on the basis of further understanding of P dynamics between soil and plants. This study systemically reviews the mechanism of P uptake and P-use efficiency by plants under starvation conditions. The recent advances in various strategies, especially imaging techniques, over the period 2012–2021 for the measurement of plant-available P are identified. The study then examines how plants fulfill P requirements from tissue-stored P during P starvation. Following this understanding, various strategies for increasing plant-available P in agricultural soil are evaluated. Finally, an update on novel carriers used to improve the P content of agricultural soil is provided.

Keywords: phosphorus; soil mobilization; speciation; plant uptake; strategies

1. Introduction

Phosphorus (P) is a fundamental macronutrient required for optimum plant growth and development in the agricultural sector [1]. Phosphorus was discovered in 1669 by Hennig Brandt, a German merchant. It is one of the 12 most abundant elements in the Earth's crust. The only isotope that occurs naturally is phosphorus-31 (^{31}P). Phosphorus non-availability is a serious issue and is considered the main limiting factor for decreased crop yields in the modern agricultural ecosystem [2]. In acidic soil, soil P is mostly immobilized in two forms, i.e., inorganic phosphorus (Pi) and organic phosphorus (Po). It may be trapped with mineral compounds of iron (Fe) or aluminum (Al) hydroxides, or it can be incorporated into rocks rich in mineral oxides, such as hematite, goethite, and gibbsite [3]. In alkaline soil, P is trapped in less soluble mineral compounds (variscite, strengite, and apatite) of magnesium (Mg) and calcium (Ca) [4]. Po accounts for 30% to 50% of the total phosphorus (Pt) present in soil, primarily in the form of inositol phosphates, phospholipids, and sugar phosphates [5]. Soil mineralization reactions are generally activated by soil microbes coupled with enzyme phosphatase in the plant rhizosphere to release Pi from fixed P reservoirs. Environmental factors, such as soil pH, temperature, moisture content, and physicochemical properties (texture, organic matter, and electric conductivity), affect the mineralization process. Plants naturally uptake Pi from soil solutions in the form of $H_2PO_4^-$, and HPO_4^{2-}, which are available in a narrow pH range (6.0–6.5). Pi accounts for 35% to 70% of the Pt present in soil solutions [6]. However, if Pi is immobilized in soil solid or transformed to Po, it will largely be unavailable to plants.

In modern agricultural practice, farmers apply chemical phosphate fertilizers to increase crop growth and yield. Intensive use of P fertilizer causes excessive P accumulation in soil [7]. Only 10–20% of the total P applied to soil is taken up by plants as Pi [8]. Part of the accumulated P in fertilizers applied to soil leaches into surrounding water bodies, thus affecting water quality. This is the reason artificially applied P fertilizers are not recommended for soil fertility, due to the risks of underground water contamination. Chemical fertilizers boost soil P levels temporarily. Furthermore, rock phosphate reserves (non-renewable P sources) are continuously declining. In the coming 50–100 years, it is estimated that natural sources of P reservoirs will be exhausted [9]. Therefore, it is essential to understand P dynamics in the soil environment, specifically the way plants uptake P, so as to identify possible alternatives to enhance P uptake from soil. This review first updated knowledge about P immobilization in the soil environment and the mechanism of plant P uptake and then discussed the inspired strategies applied for the improvement of available P in low-P soil in the agricultural sector. The recommendation of a suitable approach (strategy) was then summarized for sustainable soil P management.

2. Phosphorus Mobilization in the Plant Rhizosphere

The rhizosphere is the central and critical zone in the plant–soil environment, where maximum adsorption of nutrients, especially P, takes place from soil to plants [10]. Various biochemical and physiological processes, particularly the excretion of hormones, organic acids, and phosphatases, change the plant rhizosphere zone [11]. These processes are the main drivers of many changes occurring in the plant rhizosphere. Nutrient-use efficiency and crop yields are controlled not only by the physiological and biological processes that take place in the plant rhizosphere but also by microbial dynamics and their abundance in the rhizosphere zone [12]. Phosphorus is gradually being depleted in the plant-root rhizospheric zone due to slow mobility and solubility coupled with root uptake, resulting in P dynamics away from the root rhizosphere to upper plant parts. The decreased mobility of available P results in that plant requirements cannot be met. To maintain the constant requirements of plants, rhizospheric soluble P should be replaced 20 to 50 times per day by P supplements from bulk soil [13]. Soil is a medium for the storage and transformation of nutrients, including P. Many soil biochemical processes occur at various spatial and temporal scales, influencing the bioavailability and cycling of P. Diffusive gradients in thin films (DGT) is an in situ technique that has been used to study P bioavailability in soils and P dynamics in the plant rhizosphere at two-dimensional high resolution [14]. The DGT method employs a passive sampler composed mainly of a binding layer and a diffusive layer. For P imaging using DGT, available P firstly flows through the diffusive layer and is immediately-bound onto the binding layer. Afterwards, post-deployment assessment of P in the binding layer by 1D/2D slicing-colorimetry, computer/colorimetric imaging densitometry (CID), or laser ablation inductively coupled plasma mass spectrometry (LA–ICP–MS) acquires available P distributions at sub-mm spatial resolution. Table 1 represents DGT imaging technique data for different plant rhizospheres. The introduction of DGT as an imaging tool for measuring in-situ high-resolution P distribution in the rhizosphere is a large step toward a comprehensive understanding of P dynamics in soil-plant system.

Table 1. Publications on the diffusive gradients in thin films technique (DGT) over the period 2012–2021.

Soil	Plant Species	References
Low-P soil	*Brassica napus* L.	[15]
Soils and sediments	*Zea mays* L., *Zostera marina* L.	[16]
High- and low-P soils	*Orayza sativa* L.	[17]
Flooded paddy soils	*Orayza sativa* L.	[18]
Soils and sediments	*Zea mays* L.	[19]
Calcareous and non-calcareous soil	*Triticum aestivum* L., *Lupinus albus* L., *Fagopyrum esculentum* L.	[10]
High- and low-P soil	*Brassica napus* L.	[13]
Paddy soil	*Zea mays* L.	[20]
Halosols and Cambosols	*Orayza sativa* L. (grain)	[21]

3. Mechanism of P Uptake and Consumption by Plants

Soil contains a high amount of total phosphorus (Pt), composed of inorganic phosphorus (Pi) and organic phosphorus (Po), but only a small amount of P is available for plant uptake [22]. From soil solutions, plants obtain Pi in the form of orthophosphate anions ($H_2PO_4^-$ and HPO_4^{2-}) (Figure 1). A study [23] reported that, in most agricultural soil, the concentration of orthophosphate is very low. In P-deficient conditions, P might be supplied from other P pools to meet plant nutritional requirements. A large P concentration difference occurs between plant roots and bulk soil when P is rapidly depleted [24]. The rate of P diffusion in plant roots in most soils is not high enough to overcome the localized P difference (created due to P uptake in soil); therefore, P deficiency occurs in most cases. Previous studies have shown that, for plant growth and developmental processes, P uptake capacity is insufficient and has become a limiting factor in many regions globally [25]. Therefore, for sufficient P uptake, the importance of root architecture cannot be ignored. In many cases, roots show active responses towards P deficiency in soil [17]. The root characteristics for high P uptake include sharp, long, and deep root systems, as well as high root-to-shoot biomass ratios [26]. The common characteristics of plants include extensive root branching [27], the occurrence of a large number of root hairs, and long specific roots developed to reach P-nutrient-rich regions (sub-layers) in the soil [28]. *Brassica napus* L., *Zea mays* L., and *Triticum aestivum* L. have fibrous roots and exhibit high P uptake as compared to the plant species *Vicia faba* L., *Glycine max* L., and *Lupinus albus* L. [29].

Nutrient P is the main component of membrane lipids and nucleic acids (DNA) in plant cells and is essential for many biochemical and physiological processes. In the absence of sufficient P, plants sharply develop adaptive mechanisms not only to gain sufficient P by facilitating P uptake but also to utilize internal tissue-stored P by maintaining the P life cycle in the plant–soil environment, thus decreasing P utilization and readjusting old tissue-stored P for better performance of plant functions (Figure 1). In the cellular vacuole, about 85% to 95% of Pi is present. In P-deficient conditions, the vacuolar Pi flux is insufficient to fill the rapid gap due to cytosolic Pi reduction [15]. Therefore, in the Golgi membrane, a phosphate transporter, PHT4;6, is present, which might transport Pi from the Golgi complex [30] to fulfill the shortage of P. In plant-cell chloroplasts, another phosphate transporter, PHT2; 1, is also available, which might maintain concentrations of Pi in plant cells in P-deficient circumstances. A previous study revealed that, under P-deficient conditions, dual-targeted purple acid phosphatase isozyme (At PAP26) was necessary for Pi accumulation and that phosphatase enzymes were needed for the secretion of Pi from phosphate monoesters in *Arabidopsis thaliana* [31]. For the mobilization of P from RNA, two genes (At RNS1 and RNS2) and ribonucleases are essential during P deficiency. Leaf senescence also stimulates these two ribonuclease genes and phosphatase enzymes during P-remobilization processes, further supporting this phenomenon [32]. Under P-scarcity conditions, membrane lipid composition can be changed through increased non-P lipid and decreased phospholipid formation [33]. At the cellular level, the decomposition of diacylglycerol by phospholipases C and D and degradation of phospholipids into Pi is mediated under P deficiency [34]. In addition, by the action of two enzymes, SQD1 and SQD2, cellular diacylglycerol is further converted into sulfolipids and galactolipids [35].

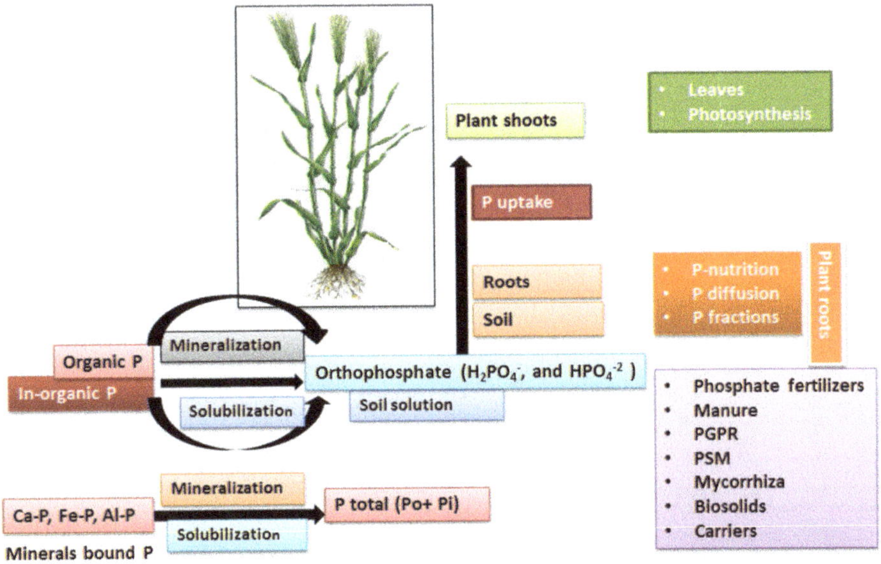

Figure 1. Mechanism of P uptake from soil to plant and associated biochemical process in the plant–soil environment.

Furthermore, plant cells in P-scarcity and starvation conditions could also utilize respiratory cellular routes, bypassing Pi reactions and adenylate [36]. Shortly, for efficient uptake and utilization of available P, plants effectively adopt several biochemical, physiological, and morphological changes [37]. Efficient P utilization and nutrition for the complex network in plants are necessary. Phosphate signaling pathways and their molecular mechanisms respecting P have been reported previously [38]. Transcriptional factors (PHR1), ubiquitin E2 conjugate (PHO_2), and microRNA (miRNA) are important players in this network for the regulation of PHO1, PHT, RNS, and ATPase genes at various levels during P deficiency. The hormonal network and sugar-signaling pathways might also be involved during P deficiency [39].

4. Inspired Strategies for Managing Phosphorus in Agricultural Soil

Several inspired strategies have been used for the improvement of plant-available P and its fractions, as shown in Figure 2. The progress of these strategies is a focus of this subsection.

4.1. Phosphate Fertilizer Application

Chemical P fertilizer application in arable fields is the most widely and long-used strategy to increase as well as precipitate the fixed P in agricultural soil for efficient utilization by crop plants [40]. This strategy is helpful for the initial depletion of fixed P and enhances crop yields to some extent. However, due to the massive applicaton of inorganic phosphate fertilizers by the agricultural sector, the P use efficiency (PUE) in crop farming continue dropping in the developing countries. This makes the P concentration in soil cannot catch up with that required to reach the optimum level for crop growth and development. Anthropogenic activities, such as fertilization, have significantly affected the terrestrial P cycle [41]. In soil, the level of P is much lower (10 μM) than levels of plant tissue-stored P (5–10 mM) [42]. Applications of chemical P fertilizers are required because of low P availability and fixed P in the soil to enhance crop growth and yield [43]. In addition, phosphate fertilizers are applied to low-P soil in the form of monocalcium phosphate (MCP) and monopotassium phosphate (MPP). Monophosphate fertilizer application prominently affects soil biochemical and physiological processes and, through the wetting process, creates huge amounts of protons and phosphate and finally establishes a P-rich zone in the soil to which it is applied [44]. Further processes create precipitation reactions, direct reactions, and adsorption reaction zones.

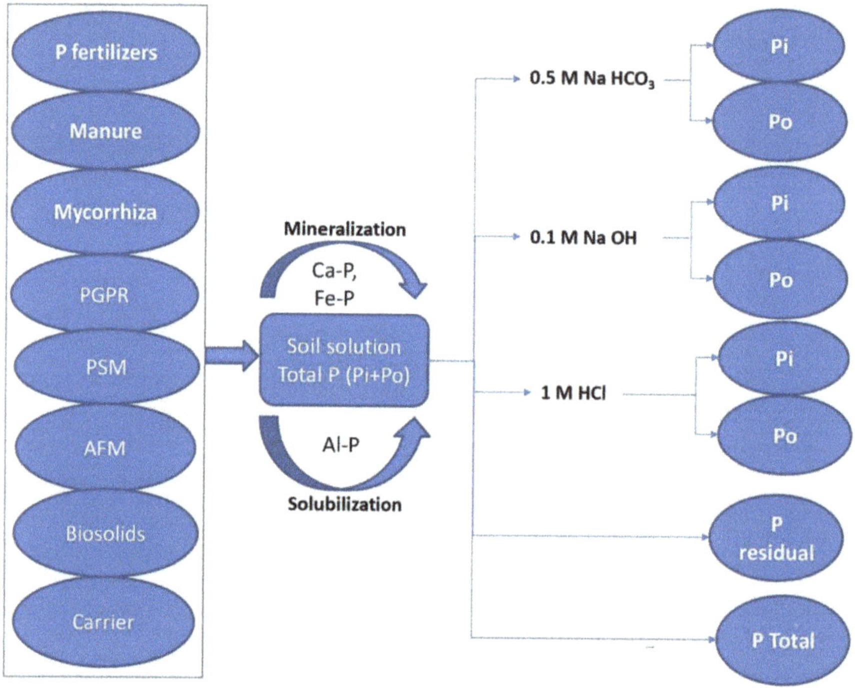

Figure 2. Inspired strategies for managing phosphorus and its fractionation. Abbreviations: PGPR: plant growth-promoting bacteria, PSM: phosphate-solubilizing microorganism, AFM: arbuscular mycorrhizal fungi, Pi: inorganic phosphorus, Po: organic phosphorus, P: phosphorus, Ca-P: calcium phosphate, Fe-P: iron phosphate, Al-P, aluminum phosphate, $NaHCO_3$: sodium bicarbonate, NaOH: sodium hydroxide, HCl: hydrochloric acid.

The P-saturated region (direct reaction zone) is acidic, ranging from pH 1.0 to 1.6 [45]. Due to its acidic nature, rapid movement of metal ions occurs. These released metal ions react with Pi in the direct reaction zone, thus enhancing further precipitation of Pi. Metal ions and P chemically interact and form complex molecules of Fe-P, Al-P, and Mg-P compounds [46]. Phosphorus is tightly bound in such compounds and is scarcely accessible to various plant species. Thus, new aggregates of monocalcium and dicalcium phosphate are produced in calcareous soil gradually. With the passage of time, these aggregates are converted into apatite (a stable form of calcium phosphate). The addition of balanced fertilizer matching the treated soil's physical and chemical characteristics might be an efficient and suitable approach for enhancing plant growth and development in P-deficient agricultural soil.

4.2. Manure Application

Manure application is also a promising strategy to enhance P-use efficiency in agricultural soil. To increase soil fertility and P concentration, manure is frequently used as an amendment in agricultural soil on a routine basis. In manure, the concentration of total P is variable [47]. Manure contains approximately 70% of total P in the labile form, while Pi constitutes 50% to 90% [48]. In addition, organic phosphorus (Po) is also present in the form of nucleic acids and phospholipids [49]. The mineralization process increases the soil Po content in manure-amended soil. Furthermore, mineralization reactions dissolve molecules of organic acids into calcium phosphate [50]. With soil amendment of manure, P adsorption on soil particles greatly decreased. Manure soil application also changes the P availability and pH level. Large molecules of hydroxyls and carboxyls of humic acid, as well as negatively charged sites in manure, strongly compete with Pi in amended soil for adsorption [51]. However, the transformation mechanism of P between Pi and Po pools needs further research in manure-amended soil.

4.3. Mycorrhizal Application

Mycorrhizal fungal application to P-deficient soil is an alternative strategy to the use of phosphate fertilizers for enhancing plant growth and available P. Mycorrhiza, particularly arbuscular mycorrhizal fungi (AMF), establish symbiotic associations (70%) with the roots of higher plants called angiosperms and contribute to P nutrition significantly [52]. Mycorrhiza associated with complex root systems have the same P reservoirs in soil as available to other plant species [53]. A previous study [54] investigated alternative reservoirs of P that are used both by ectomycorrhiza and AMF. In such symbiotic associations, AMF receive carbon (C) sources from host plants while constantly supplying P nutrition through an extensive hyphal system to their host plants. In stress conditions, for sustainable agricultural productivity, the nutritional effect of AMF is very beneficial in the formation of soil aggregates and the establishment of plants. Through a symbiotic relationship, the chief advantage of AMF is increased P uptake. Mycorrhizal plants showed better growth than non-mycorrhizal plants, having AMF pathways as well as increased P uptake, in a previous report [55]. However, in some reported cases, even mycorrhizal plants with AMF pathways did not show sufficient P uptake, and plant growth was stopped. A previous study [56] revealed that, via the solubilization process, AMF hyphal exudates alone solubilized more P from soil reservoirs than plant-root exudates, suggesting that, through the solubilization process, mycorrhiza can enhance P uptaek of the plants. The secretion of phosphatase enzymes from the fungal hyphal mycelia of AMF fungi has been investigated and it was found that it could significantly increase the utilization and mineralization of organic P [57]. Mycorrhizal plant P acquisition increases through extraradical mycelia that penetrate deep into P reservoir sources in the soil and transform it into roots and subsequently into other plant parts. Large surface areas, as well as high mycorrhizal hyphal densities, could also contribute to enhanced orthophosphate uptake by plants. In mycorrhizal-associated plants, elevated uptake can usually be explained in terms of enhanced P utilization through the hyphal system from indigenous soil P reservoirs [58]. Mycorrhizal plants use two pathways (direct and mycelial) for sufficient P uptake. The first one occurs at the soil–plant interface via root hairs called direct pathways. The second pathway occurs through fungal mycelia and is called the mycorrhizal pathway [59].

4.4. Plant Growth-Promoting Bacteria Application

The inoculation of plant growth-promoting bacteria (PGPB), applied as an alternative to chemical fertilizers, is based on P solubilization, N_2 fixation, phytohormone secretion, vitamin synthesis, and organic P mineralization [60]. Plant growth-promoting bacteria application to soil is a natural method for increasing soil P concentrations by the P-solubilization mechanism [61]. Numerous PGPB have been isolated from different soil environments, and their effects on plant growth and development have been reported [62]. Most of the studies have concentrated on PGPR, which perform functions associated with P solubilization to increase P availability to the host plant [63]. A diverse array of PGPR, such as species of *Bacillus*, *Enterobacter*, *Pseudomonas*, and *Serratia*, have been found to be involved in P mobilization, e.g., dissolving Ca-P by lowering pH, unlocking P bound to Al and Fe oxides through the exudation of organic acids, as well as mineralization of organic P through enzyme phosphatase. The mentioned PGPR have been considered as potentially effective inoculants for improved crop growth and yield. However, some researchers have reported that the direct inoculation of PGPR into soil did not enhance plant biomass and P uptake capacity, due to variations in P-related soil properties as well as enzyme activities that were limited by certain environmental factors [64]. Previous findings [65] revealed a low survival rate for PGPR due to low nutrient concentrations as well as competition with indigenous soil microbial, fungal, protozoan, and nematode populations.

Phosphate-solubilizing microorganisms (PSMs) solubilize orthophosphate from different inorganic and organic sources [66]. Phosphate-solubilizing microorganisms include *Rhizobium*, *Bacillus*, and *Pseudomonas* species, as well as different species of fungi, such as *Penicillium* and *Aspergillus* species [67]. In laboratory experiments, rock phosphates (Ca_5 $(PO_4)_3OH$, $Ca_{10}(PO_4)_6F_2$) and calcium compounds, such as tricalcium phosphate ($Ca_3(PO_4)_2$), are mainly used for the identification of PSMs. The quantity of released soluble P is particularly dependent on the P source; therefore, various microorganisms solubilize different sources of P [68]. Fungal species were previously extensively studied with respect to their abilities to effectively solubilize Al and Fe phosphate sources [69]. Variation in the pH of culture media plays an important role in the solubilization of calcium phosphate. The secretion of organic anions, such as lactate, oxalate, citrate, and gluconate, by PSMs is linked to the acidification of solution media [70]. PSMs improve the solubilization and, ultimately, mobilization of poorly soluble P-binding Ca-P and Al-P compounds by chelating organic anions (Table 2). Many studies have reported increased P uptake and solubilization by various crops inoculated with PSMs in controlled conditions [71,72].

Table 2. Different amendment/inoculation methods used and concentrations of available and total P (mg kg^{-1}) in agricultural soil.

Amendment	Soil pH	Available P	Total P	Amendment/ Inoculation	Plant Species	References
Pig manure	6.54	28.07	810	15,000 kg ha^{-1}	*Oryza sativa* L.	[28]
Poultry manure	5.20	45.03	975	42 Mg ha^{-1}	*Lolium perenne* L.	[73]
Compost	7.79	12.20	98	20 t ha^{-1}	*Phaseolus vulgaris* L.	[74]
Biosolids	7.40	25.21	7600	100 kg ha^{-1}	*Saccharum officinarum*	[75]
Willow and pine biochar	6.20	33.31	1981	10 t ha^{-1}	*Lotus pedunculatus* L.	[76]
Cow-dung biochar	5.61	10.61	265	10 t ha^{-1}	*Cyperus esculentus* L.	[77]
PSM	7.20	17.41	145	3×10^{-4} per seed	*Triticum aestivum* L.	[78]
PSM	7.80	39.15	9500	4×10^{-5} per seed	*Lolium perenne* L.	[79]
AMF and PSM	8.50	8.00	183	20 g kg^{-1}	*Solanum lycopersicum* L.	[80]
Nano-rock phosphate and PSB	8.39	ND	296.23	250 kg ha^{-1}, 250 mL seeds kg^{-1}	*Zea mays* L.	[81]
PSB	ND	7.00	ND	1×10^7 C.F.U.mL^{-1}	*Zea mays* L.	[82]
PSB and P_2 O_5	4.70	8.11	ND	180 kg ha,$^{-1}$ 2×10^{-8} C.F.U.mL^{-1}	*Saccharum officinarum*	[83]

Abbreviations: AMF: Arbuscular mycorrhizal fungi, PSM: Phosphate-solubilizing microorganism, PSB: Phosphate-solubilizing bacteria, ND: Not detected.

In field experiments, the functional mechanism of PSM is more complicated and has proved very difficult to explain clearly in the presence of many indigenous soil microbes and varying environmental conditions [84]. This may be due to a lack of suitable and precise methods for the introduction of PSMs into soil environments, gaps in knowledge about the interactions of PSMs with indigenous organisms, and poor understanding of P dynamics in soil [85]. *Penicillium* significantly solubilized P and enhanced wheat growth in laboratory experiments [86]. Microbes play a key role in unlocking soil-bound organic P through a process called mineralization. PSM inoculation enhanced organic P availability as well as plant biomass in greenhouse experiments. In addition, microbial biomass is essential for maintaining an optimum level of Po and Pi in soil solutions. Upon the death and decay of microbial biomass, P is available to plants in soil [87].

The PSM inoculation technique has proved effective, helpful, and efficient respecting plant rhizospheres, where there are large quantities of microbial biomass that are quickly metabolized into carbon [88]. However, P mobilization in the plant rhizosphere associated with PSMs requires additional confirmation and investigation. The mineralization of organic and inorganic P and microbial-activated phosphate solubilization are essential processes through which microbes obtain P from soil. However, in some cases, PSMs are unable to mobilize enough P above their demands to meet plant requirements. The P cycle in soil and microbial biomass indeed demonstrate an essential pathway for P mobilization in different, available, including soil, P pools. Different physiological processes disturb organic P pools and thus render them unavailable to the plant–soil ecosystem [89]. The importance of this mechanism in plant rhizospheres needs further research in detail in complex environments. The inoculation of crops with PSMs is a method frequently used, and some PSMs are already commercialized [90]. However, in long-term field applications, these inoculants have to compete for their survival with native soil nematodes and microbes. This technology is still in its infancy and suitable approaches for the solubilization of mineral-fixed P are being explored. The issues with the application of this technology include (1) inoculum survival and colonization, (2) host plant inoculum specificity, and the (3) limited commercialization value and insufficient effects on plant growth and developmental processes [91]. PSM inoculants are applied to the soil and fixed P becomes solubilized; however, plant P uptake and biomass production are negatively affected by phosphorus fixation in the microbial biomass [92]. A previous study [93] reported that the application of phosphate-solubilizing bacteria along with nematodes efficiently enhanced plant P uptake and available P. Phosphorus solubilization is also influenced by soil texture. Aluminum (Al) and iron (Fe) chemical reactions with P in acidic or low-pH soil result in P reduction. However, tricalcium phosphate $(Ca_3PO_4)_3$ is formed in alkaline or high-pH soil, reducing plant-available P in soil [94]. Thus, soil alkalinity causes P unavailability—a process known as P fixation. Sorghum yield increased significantly after PSM inoculation [95]. To increase P availability, bacterial species, such as *Bacillus* spp., *Pseudomonas* spp., and *Agrobacterium* spp., are used as soil inoculants [96]. Phosphorus solubilization, which involves local acidification or alkalization, has been observed in some *Pseudomonas*, *Cyanobacteria*, and *Bacillus* species isolated from plant rhizospheres [97]. Organic phosphate is the most abundant source of soil phosphate, but its compounds are complex (nucleic acids, phospholipids, etc.) and must be transformed by microorganisms before they can be absorbed by plants [98].

4.5. Biosolid Application

In the 1990s, the term "biosolid" was first introduced for the selection of liquid, solid, and semi-solid materials produced from the treatment of domestic sewage sludge [99]. A satisfactory production process is needed for sludge to be applied to land [100]. For decades, biosolids have been recommended economically, are socially acceptable, and are part of traditional practices of land applications globally [101]. In addition, land application of biosolids is a viable technology for industrial-waste management, generally proposed by environmentalists. Biosolids contain large amounts of micro- and macroelements, such as copper (Cu), zinc (Zn), iron (Fe), manganese (Mn), potassium (K), nitrogen (N), calcium (Ca), sulphur (S), organic carbon (C), and phosphorus (P) [102,103]. These are essential elements for plant growth processes and faunal survival in soil. Usually, in biosolids, most P is in the form of aluminum phosphate, commonly adsorbed on the surfaces of calcium phosphate and iron phosphate [104]. The total inorganic P present in biosolids ranges from 70% to 90% [105]. Furthermore, biosolids contain smaller amounts of water-soluble P [80] as well as organic P in the form of phospholipids, orthophosphate monoesters, and orthophosphate diesters [106]. Wastewater-treatment-plant operations strongly affect the P concentrations in biosolids [107].

Phosphate fertilizer application affects the dynamics of P in biosolids; however, not all of the P in biosolids is available to plants [108]. Precipitation, dissolution, microbial decomposition, desorption, and sorption of P occurs when biosolids are land-applied [109]. The above-mentioned processes may be slow or fast, depending on the biological and physicochemical processes that occur in the applied soil. P availability in biosolid-treated soil increases with the rate of biosolid application [110]. Thus, elevated contents of inorganic P in biosolids may be linked to enhanced concentrations of P in the soils to which they are applied. However, in soils that have the highest capacities to retain P, or in P-concentrated soils, such variations were found to be less obvious [111]. The conversion of lower labile P to higher labile P species may also be linked to enhanced contents of bioavailable P in amended soils [112]. Another study revealed that calcium (Ca)-concentrated biosolids also contributed to enhanced levels of water-soluble P in amended soil [113]. Mineralization also increases with amendments of biosolids, which could unlock biosolid-borne organic P, thus contributing to enhanced contents of extractable P [114].

The relationship between plant-available P and the degradation of biosolids is more complicated. A previous study revealed that biosolid amendment not only changes soil physicochemical properties, such as pH, EC, dissolved organic matter, and biological properties, but also the adsorption capacity of amended soil [115]. The enzymes phosphodiesterase and phosphomonoesterase mainly hydrolyze organic phosphate esters into orthophosphate anions biochemically and secrete most of the assimilated P through soil microbes. Plant roots and soil microbes both passively and actively participate in the secretion of extracellular enzymes to make them available for plants through the mineralization of S, C, P, and N from complicated forms [116]. In the plant rhizosphere, phosphomonoesterase enzyme activity was identified as the major mechanism for the acquisition of P by plants, catalyzing a large number of anhydrides and orthophosphate minerals and thus releasing bound P [117]. In agricultural soils, enzyme phosphatase activity mainly contributes to agricultural activities, and, in forest soils, it responds to seasonal variations in moisture as well as temperature [118].

According to economic theory, in biosolid-applied soils, P is amended together with other nutrients, such as S, N, and S, thus stimulating the activity of phosphatase enzymes, and is utilized by soil microbes as an energy source, while, in soil amended with chemical fertilizers, phosphatase enzyme activity is suppressed [119], thereby revealing the clear difference between inorganic P amendment and fertilizer application [18]. Previously, in the agricultural sector, biosolid applications to soil enhanced acid phosphatase activity, nutrient concentrations, and microbial biomass [120]. In different soils, various persistence and production rates of enzyme activities have been studied. Soil pH-buffering capacity also varies with biosolid application, resulting in changes in phosphatase enzyme activities [121]. In biosolid-applied soil, the enzymes phosphomonoesterase and diesterase might be used as signals of P secretion from biosolids because, usually, sewage sludge includes different P forms [122]. The higher the microbial or plant origin of the enzyme, the greater the demand for mineral P [123]. Thus, competition or cooperation between plant roots and rhizosphere microbes is primarily determined by soil P status.

4.6. Carrier Application

China and India, as the largest consumers of P fertilizers [124], are facing the great challenge of gradually decreasing P resources [125]. Excessive and unbalanced P fertilization applications in most regions have been extensively reported [126]. To maintain high crop productivity, high rates of chemical P fertilizers (120 kg P_2O_5 ha^{-1}) have been applied by farmers, resulting in dramatic increases in P accumulation in agricultural soil [127], as well as P loss via runoff and leaching to the aquatic environment. Therefore, to achieve environmental, ecological, and economic goals, it is necessary to minimize the input of P fertilizers and improve the P status of agricultural soils through the feasible application of new materials in the modern agricultural sector. The application of biochar (a carbon-rich material) is an environmentally friendly and cost-effective approach to improve nutrient-deficient agricultural soil [128]. Biochar is produced from waste residues and is frequently recommended for the fertility of agricultural soil and carbon sequestration [129]. Using biochar as a carrier material for plant growth-promoting bacteria (PGPB) offers unique opportunities and benefits in the agricultural sector. Plant growth-promoting bacteria inoculation with biochar increases the value of biochar and enhances its commercialization as a biofertilizer [130]. Applications of biochar as a carrier have many benefits, as they ensure the survival of introduced PGPB in the treated soil and provide hot spots for microbial movements. Furthermore, biochar carrier amendment of soils alters many physicochemical properties, which enhances the survival efficiency of the introduced PGPB in treated soils. The positive results of biochar amendment of agricultural soil include increased soil pH, bulk density, fertility, water-holding capacity, nutrient retention, and aeration capacity [131].

In agricultural soils, the total concentration of P (Pt) ranges from 400 to 1200 mg kg^{-1}. However, in available forms, such as the orthophosphate ions $H_2PO_4^-$ and HPO_4^{-2}, only 1 mg kg^{-1} of Pt is present [132]. The non-soluble form of P is present in inorganic (Pi) and organic forms (Po). In soil, non-soluble Pi ranges from 20 to 50% in the form of PO_4^- ions [133]. These ions are adsorbed to various compounds of Ca, Fe, and Al—Ca-P, Fe-P, and Al-P—producing stable complexes. Due to the short and poor survival of PGPR in soil, post-soil inoculation is usually not recommended [134]. Furthermore, direct inoculation of liquid PGPR in soil becomes complex due to the adhesion of soil particles, which reduces their ability to colonize host root surfaces, as well as vertical transport [135]. Ordinary carriers, such as peat and vermiculate, have some drawbacks which limit their application at large scales [136]. However, compared to peat and vermiculate, biochar carriers seem to be environmentally friendly and cost-effective. Furthermore, biochar sterilization offers a reliable and premium-quality preparation for alternative carriers (PSMs loaded on biochar). Due to these benefits, biochar is considered an alternative, cost-effective carrier. The attachment of PSMs on biochar surfaces might provide defensive colonization as well as a safe zone [137]. One study [138] reported that when loaded on a biochar surface, *Azospirillum* had a 6-month long shelf life at room temperature. However, it is not clear whether PSMs loaded on biochar surfaces ensure survival; the matter needs further investigation. We applied different biochar carriers (PSMs loaded on biochar), such as rice-straw biochar (RSB), soybean-straw biochar (SSB), rice-husk biochar (RHB), peanut-shell biochar (PNB), wood biochar (WB), and corn-cob biochar (CCB), to agricultural soil collected from an agricultural field in Hailun City, Heilongjiang Province, China (Figure 3). The tested soil had a pH of 6.52 units and a total P concentration of 5.59 mg kg^{-1} [139] The incubation experiment was conducted at lab scale in a complete randomized block design (CRBD) in four replicates. Each carrier was amended separately at a rate of 3% to the tested soil. After one month of incubation experiment, the RSB and SSB amended-soil showed significantly higher concentrations of $NaHCO_3$ extractable Pi and Po, NaOH extractable Pi and Po, HCl Pi, residual P, and total P in the treated soils than RHB, PNB, WB, CCB, and CK-amended soil (Table 3). This increase in total P and P fractions in carrier-amended soils may have been due to the increased solubilization and mineralization of mineral-bound P.

Table 3. Sequentially extracted P fractionation (mg kg^{-1}) after application of different carrier materials into agricultural soil. Mean values are shown ±1 standard deviation (n = 4) (unpublished data).

Parameters	0.5 M $NaHCO_3$ Extractable		0.1 M NaOH Extractable		1 M HCl	P Residual	P Total	Extraction Efficiency (%)
	Pi	Po	Pi	Po	Pi			
CK	0.58 ± 0.03 a	0.32 ± 0.08 a	0.63 ± 0.09 a	1.72 ± 0.20 a	1.46 ± 0.14 a	0.88 ± 0.15 a	5.59 ± 0.51 a	99
RSB	0.69 ± 0.08 b	0.77 ± 0.15 b	0.67 ± 0.04 a	1.20 ± 0.15 b	2.49 ± 0.27 b	1.21 ± 0.23 b	6.98 ± 0.96 c	100
SSB	0.94 ± 0.09 c	0.64 ± 0.16 b	0.71 ± 0.12 a	0.96 ± 0.15 b	2.99 ± 0.36 c	1.41 ± 0.34 c	7.61 ± 1.06 c	100
RHB	0.66 ± 0.12 b	0.27 ± 0.08 a	0.64 ± 0.08 a	1.46 ± 0.23 a	2.08 ± 0.23 b	1.62 ± 0.35 c	6.83 ± 0.19 b	98
PNB	0.70 ± 0.06 b	1.03 ± 0.27 c	0.69 ± 0.14 a	0.73 ± 0.15 b	2.00 ± 0.29 b	1.53 ± 0.26 b	7.01 ± 0.98 bc	95
WB	0.65 ± 0.02 b	0.48 ± 0.29 b	0.62 ± 0.05 a	1.29 ± 0.23 a	2.20 ± 0.26 b	1.43 ± 0.28 b	7.11 ± 0.18 b	93
CCB	0.73 ± 0.05 b	0.35 ± 0.08 a	0.61 ± 0.09 a	0.96 ± 0.18 b	2.43 ± 0.34 b	1.42 ± 0.24 b	6.65 ± 0.19 b	97

Abbreviations: CK: Control, RSB: Rice-straw biochar carrier, SSB: Soybean-straw biochar carrier, RHB: Rice-husk biochar carrier, PNB: Peanut-shell biochar carrier, WB: Wood biochar carrier, CCB: Corn-cob biochar carrier. Extraction efficiency for each of the P fractions was calculated from the sum of P fractions divided by total P and multiplied by 100. Pi (organic P) and Po (organic P), P residual, and P total. Different letters in columns indicate significant differences ($p \leq 0.05$) between treatments, while similar letters indicate non-significant differences.

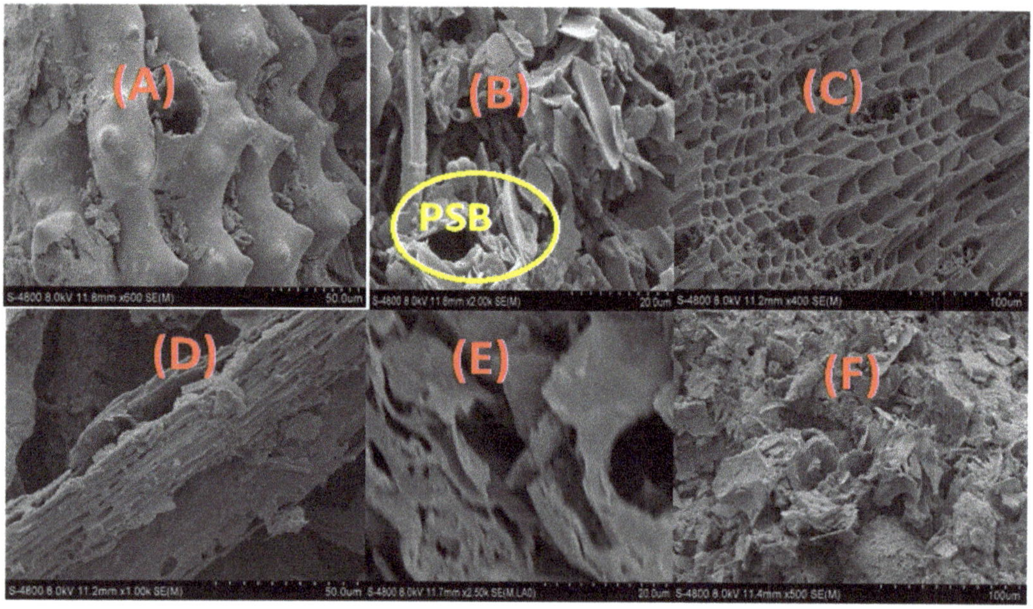

Figure 3. SEM images of carrier materials: (**A**) RHB, (**B**) SSB, (**C**) WB, (**D**) RSB, (**E**) PNB, and (**F**) CCB (unpublished data).

5. Conclusions

The current review has brief understanding of phosphorus mobilization and P dynamics between Pi and Po pools in agricultural soil. It has been discussed how, in P-deficient conditions, plants maintain P concentrations and utilize internal tissue-stored P. Progress in the development of the diffusive gradients in thin films technique to study P bioavailability was systematically reviewed. In addition, different inspired strategies for the improvement of plant-available P in soil were also reviewed. Finally, an update about the application of novel carriers was provided.

Author Contributions: Conceptualization, M.I. (Muhammad Ibrahim) and M.I. (Muhammad Iqbal); Methodology, S.K.; Data collection and analysis, G.L.; Writing—original draft preparation: M.I. (Muhammad Ibrahim); Data presentation, reviewing and editing, G.L., Y.-T.T. and D.-X.G. All authors have read and agreed to the published version of the manuscript.

Funding: This work was supported by Ningbo Bureau of Science and Technology (2022Z169, 2021Z101) and the National Natural Science Foundation of China (42177194).

Institutional Review Board Statement: Not applicable.

Informed Consent Statement: Not applicable.

Data Availability Statement: The data presented in this study are available on request from the corresponding author.

Conflicts of Interest: The authors declare that they have no known competing financial interests or personal relationships that could appear to have influenced the work reported in this review.

References

1. Wahid, F.; Fahad, S.; Danish, S.; Adnan, M.; Yue, Z.; Saud, S.; Siddiqui, M.H.; Brtnicky, M.; Hammerschmiedt, T.; Datta, R. Sustainable management with mycorrhizae and Phosphate solubilizing bacteria for enhanced Phosphorus uptake in calcareous soils. *Agriculture* **2020**, *10*, 334. [CrossRef]
2. Metson, G.S.; MacDonald, G.K.; Haberman, D.; Nesme, T.; Bennett, E.M. Feeding the Corn Belt: Opportunities for phosphorus recycling in U.S. agriculture. *Sci. Total. Environ.* **2016**, *542*, 1117–1126. [CrossRef] [PubMed]

3. Muhmood, A.; Lu, J.; Dong, R.; Wu, S. Formation of struvite from agricultural waste waters and its reuse on farmlands: Status and hindrances to closing the nutrient loop. *J. Environ. Manag.* **2019**, *230*, 1–13. [CrossRef] [PubMed]
4. Hummel, C.; Boitt, G.; Santner, J.; Lehto, N.J.; Condron, L.; Wenzel, W.W. Co-occurring increased phosphatase activity and labile P depletion in the rhizosphere of *Lupinus angustifolius* assessed with a novel, combined 2D-imaging approach. *Soil Biol. Biochem.* **2021**, *153*, 107963. [CrossRef]
5. Nesme, T.; Metson, G.S.; Bennett, E.M. Global phosphorus flows through agricultural trade. *Glob. Environ. Chang.* **2018**, *50*, 133–141. [CrossRef]
6. Wilson, H.; Elliott, J.; Macrae, M.; Glenn, A. Near surface soils as a source of Phosphorus in snow melt runoff from crop land. *J. Environ. Qual.* **2019**, *23*, 445–455.
7. Jiang, B.; Shen, J.; Sun, M.; Hu, Y.; Jiang, W.; Wang, J.; Wu, J. Soil phosphorus availability and rice phosphorus uptake in paddy fields under various agronomic practices. *Pedosphere* **2021**, *31*, 103–115. [CrossRef]
8. Helfenstein, J.; Tamburini, F.; von Sperber, C.; Massey, M.S.; Pistocchi, C.; Chadwick, O.A.; Frossard, E. Combining spectroscopic and isotopic techniques gives a dynamic view of phosphorus cycling in soil. *Nat. Commun.* **2018**, *9*, 312–320. [CrossRef]
9. Scholz, R.W.; Wellmer, F.W. Endangering the integrity of science by misusing unvalidated models and untested assumptions as facts: General considerations and the mineral and phosphorus scarcity fallacy. *Sustain. Sci.* **2021**, *16*, 2069–2086. [CrossRef]
10. Kreueder, A.; Santner, J.; Scharsching, V.; Oburger, E.; Hoefer, C.; Hann, S.; Wenzel, W.W. In situ observation of localized, sub-mm scale changes of phosphorus biogeochemistry in the rhizosphere. *Plant Soil* **2018**, *424*, 573–589. [CrossRef]
11. Tian, J.; Lu, X.; Chen, Q.; Kuang, X.; Liang, C.; Deng, L.; Lin, D.; Cai, K.; Tian, J. Phosphorus fertilization affects soybean rhizosphere phosphorus dynamics and the bacterial community in karst soils. *Plant Soil* **2022**, *475*, 137–152. [CrossRef]
12. De la Fuente Cantó, C.; Simonin, M.; King, E.; Moulin, L.; Bennett, M.J.; Castrillo, G.; Laplaze, L. An extended root phenotype: The rhizosphere, its formation and impacts on plant fitness. *Plant J.* **2020**, *103*, 951–964. [CrossRef]
13. Kruse, J.; Abraham, M.; Amelung, W.; Baum, C.; Bol, R.; Kühn, O.; Leinweber, P. Innovative methods in soil phosphorus research: A review. *J. Plant Nutr. Soil Sci.* **2015**, *178*, 43–88. [CrossRef]
14. Guan, D.X.; He, S.X.; Li, G.; Teng, H.H.; Ma, L.Q. Application of diffusive gradients in thin-films technique for speciation, bioavailability, modeling and mapping of nutrients and contaminants in soils. *Crit. Rev. Environ. Sci. Technol.* **2021**, *52*, 3035–3079. [CrossRef]
15. Santner, J.; Zhang, H.; Leitner, D.; Schnepf, A.; Prohaska, T.; Puschenreiter, M.; Wenzel, W.W. High-resolution chemical imaging of labile phosphorus in the rhizosphere of *Brassica napus* L. cultivars. *Environ. Exp. Bot.* **2012**, *77*, 219–226. [CrossRef]
16. Santner, J.; Larsen, M.; Kreuzeder, A.; Glud, R.N. Two decades of chemical imaging of Solutes in sediments and soils—A review. *Anal. Chim. Acta* **2015**, *878*, 9–42. [CrossRef]
17. Wang, Y.; Yuan, J.-H.; Chen, H.; Zhao, X.; Wang, D.; Wang, S.-Q.; Ding, S.-M. Small-scale interaction of iron and phosphorus in flooded soils with rice growth. *Sci. Total Environ.* **2019**, *660*, 911–919. [CrossRef]
18. Wang, Y.; Chen, H.; Wang, L.; Zhu, W.; Yuan, J.; Jaisi, D.P.; Zhao, X.; Wang, S. Using diffusive gradients in thin films technique for in-situ measurement of labile phosphorus around *Oryza sativa* L. roots in flooded paddy soils. *Pedosphere* **2021**, *31*, 76–82. [CrossRef]
19. Kreuzeder, A.; Santner, J.; Prohaska, T.; Wenzel, W.W. Gel for simultaneous chemical imaging of anionic and cationic solutes using diffusive gradients in thin films. *Anal. Chem.* **2013**, *85*, 12028–12036. [CrossRef]
20. Hong, C.; Su, Y.; Lu, S. Phosphorus availability changes in acidic soils amended with biochar, fly ash, and lime determined by diffusive gradients in thin films (DGT) technique. *Environ. Sci. Pollut. Res.* **2018**, *25*, 30547–30556. [CrossRef]
21. Tian, K.; Xing, Z.; Liu, G.; Wang, H.; Jia, M.; Hu, W.; Huang, B. Cadmium phytoavailability under greenhouse vegetable production system measured by diffusive gradients in thin films (DGT) and its implications for the soil threshold. *Environ. Pollut.* **2018**, *241*, 412–421. [CrossRef]
22. Wang, Y.; Lambers, H. Root-released organic anions in response to low phosphorus availability: Recent progress, challenges and future perspectives. *Plant Soil* **2020**, *447*, 135–156. [CrossRef]
23. Penn, C.J.; Camberato, J.J. A Critical Review on Soil Chemical Processes that Control How Soil pH Affects Phosphorus Availability to Plants. *Agriculture* **2019**, *9*, 120. [CrossRef]
24. Darch, T.; Giles, C.D.; Blackwell, M.S.A.; George, T.S.; Brown, L.K.; Menezes Blackburn, D.; Shand, C.A.; Stutter, M.I.; Lumsdon, D.G.; Mezeli, M.M.; et al. Inter and intra specific intercropping of barley cultivars and legume species as affected by soil phosphorus availability. *Plant Soil* **2018**, *427*, 125–138. [CrossRef]
25. Luo, X.; Mahdi, A.-K.; Han, X.-Z.; Chen, X.; Yan, J.; Biswas, A.; Zou, W.-X. Long-term application of fertilizer and manures affect P fractions in Mollisol. *Sci. Rep.* **2020**, *10*, 14793. [CrossRef]
26. Hill, J.O.; Simpson, R.J.; Moore, A.D.; Chapman, D.F. Morphology and response of roots of pasture species to phosphorus and nitrogen nutrition. *Plant Soil* **2006**, *286*, 7–19. [CrossRef]
27. Duque, L.O.; Villordon, A. Root Branching and Nutrient Efficiency: Status and Way Forward in Root and Tuber Crops. *Front. Plant Sci.* **2019**, *10*, 237. [CrossRef]
28. Lyu, Y.; Tang, H.; Li, H.; Zhang, F.; Rengel, Z.; Whalley, W.R.; Shen, J. Major crop species show differential balance between root morphological and physiological responses to variable Phosphorus supply. *Front. Plant Sci.* **2016**, *7*, 1939. [CrossRef]
29. Liu, J.; Yang, L.; Luan, M.; Wang, Y.; Zhang, C.; Zhang, B.; Luan, S. A vacuolar phosphate transporter essential for phosphate homeostasis in *Arabidopsis*. *Proc. Natl. Acad. Sci. USA* **2015**, *112*, E6571–E6578. [CrossRef]

30. Hassler, S.; Jung, B.; Lemke, L.; Novák, O.; Strnad, M.; Martinoia, E.; Neuhaus, H.E. Function of the Golgi-located phosphate transporter PHT4;6 is critical for senescence-associated processes in Arabidopsis. *J. Exp. Bot.* **2016**, *67*, 4671–4684. [CrossRef]
31. Hurley, B.A.; Tran, H.T.; Marty, N.J.; Park, J.; Sneddenn, W.A.; Mullen, R.T.; Plaxton, W.C. The dual-targeted purple acid phosphatase isozyme At-PAP26 is essential for efficient acclimation of Arabidopsis to nutritional phosphate deprivation. *Plant Physiol.* **2010**, *153*, 1112–1122. [CrossRef] [PubMed]
32. Dellero, Y.; Clouet, V.; Marnet, N.; Pellizzaro, A.; Dechaumet, S.; Niogret, M.-F.; Bouchereau, A. Leaf status and environmental signals jointly regulate proline metabolism in winter oilseed rape. *J. Exp. Bot.* **2020**, *71*, 2098–2111. [CrossRef] [PubMed]
33. Dissanayaka, D.M.S.B.; Ghahremani, M.; Siebers, M.; Wasaki, J.; Plaxton, W.C. Recent Insight into the Metabolic Adaptations of Phosphorus Deprived Plants. *J. Exp. Bot.* **2020**, *72*, 199–223. [CrossRef] [PubMed]
34. Li, M.; Welti, R.; Wang, X. Quantitative profiling of Arabidopsis polarglycero lipids in response to phosphorus starvation: Roles of phospholipases Dz1and Dz2 in Phosphatidyl choline hydrolysis and digalactosyl diacylglycerol accumulation in Phosphorus-starved plants. *Plant Physiol.* **2006**, *142*, 750–761. [CrossRef]
35. Wang, F.; Ding, D.; Li, J.; He, L.; Xu, X.; Zhao, Y.; Xu, J. Characterization of genes involved in galactolipids and sulfolipids metabolism in maize and Arabidopsis and their differential responses to phosphate deficiency. *Funct. Plant Biol.* **2020**, *47*, 279–292. [CrossRef]
36. Mehta, D.; Ghahremani, M.; Pérez-Fernández, M.; Tan, M.; Schläpfer, P.; Plaxton, W.C.; Uhrig, R.G. Phosphate and phosphite differentially impact the proteome and phosphoproteome of Arabidopsis suspension cell cultures. *Plant J.* **2020**, *105*, 924–941. [CrossRef]
37. Toscano, S.; Romano, D. Morphological, Physiological, and Biochemical Responses of Zinnia to Drought Stress. *Horticulturae* **2021**, *7*, 362. [CrossRef]
38. Rouached, H.; Arpat, A.B.; Poirier, Y. Regulation of phosphate starvation responses in plants: Signaling players and cross-talks. *Mol. Plant* **2010**, *3*, 288–299. [CrossRef]
39. Matsoukas, I.G. Interplay between sugar and hormone signaling pathways modulate floral signal transduction. *Front. Genet.* **2014**, *5*, 218. [CrossRef]
40. Menezes-Blackburn, D.; Inostroza, N.; Gianfreda, L.; Greiner, R.; Mora, M.L.; Jorquera, M.A. Phytase-producing *Bacillus* sp. inoculation increases phosphorus availability in Cattle manure. *J. Soil. Sci. Plant Nutr.* **2016**, *16*, 200–2010. [CrossRef]
41. Yuan, Z.; Jiang, S.; Sheng, H.; Liu, X.; Hua, H.; Liu, X.; Zhang, Y. Human Perturbation of the Global Phosphorus Cycle: Changes and Consequences. *Environ. Sci. Technol.* **2018**, *52*, 2438–2450. [CrossRef]
42. Irfan, M.; Aziz, T.; Maqsood, M.A.; Bilal, H.M.; Siddique, K.H.M.; Xu, M. Phosphorus (P) use efficiency in rice is linked to tissue-specific biomass and P allocation patterns. *Sci. Rep.* **2020**, *10*, 4278. [CrossRef]
43. Elgharably, A. Effects of rock phosphate added with farm yard manure or sugar juice residues on wheat growth and uptake of certain nutrients and heavy metals. *J. Soils Sediments* **2020**, *20*, 3931–3940. [CrossRef]
44. Alori, E.T.; Glick, B.R.; Babalola, O.O. Microbial Phosphorus Solubilization and Its Potential for Use in Sustainable Agriculture. *Front. Microbiol.* **2017**, *8*, 971. [CrossRef]
45. Lovio-Fragoso, J.P.; de Jesús-Campos, D.; López-Elías, J.A.; Medina-Juárez, L.Á.; Fimbres-Olivarría, D.; Hayano-Kanashiro, C. Biochemical and molecular aspects of Phosphorus limitation in diatoms and their relationship with biomolecule accumulation. *Biology* **2021**, *10*, 565. [CrossRef]
46. Prüter, J.; Leipe, T.; Michalik, D.; Klysubun, W.; Leinweber, P. Phosphorus speciation in sediments from the Baltic Sea, evaluated by a multi-method approach. *J. Soils Sediments* **2020**, *20*, 1676–1691. [CrossRef]
47. Laconi, A.; Mughini-Gras, L.; Tolosi, R.; Grilli, G.; Trocino, A.; Carraro, L.; Piccirillo, A. Microbial community composition and antimicrobial resistance in agricultural soils fertilized with livestock manure from conventional farming in northern Italy. *Sci. Total. Environ.* **2021**, *760*, 143404. [CrossRef]
48. Dou, Z.; Toth, J.D.; Galligan, D.T.; Ramberg, C.F., Jr.; Ferguson, J.D. Laboratory procedures for characterizing manure phosphorus. *J. Environ. Qual.* **2000**, *29*, 508–514. [CrossRef]
49. Neina, D. The role of soil pH in plant nutrition and soil remediation. *Appl. Environ. Soil Sci.* **2019**, *2019*, 5794869. [CrossRef]
50. Martins, M.A.; Santos, C.; Almeida, M.M.; Elisabete, M.; Costa, V. Hydroxy apatite micro and nanoparticles: Chleation and growth mechanisms in the presence of citrate species. *J. Colloid Interface Sci.* **2008**, *318*, 210–216. [CrossRef]
51. Omar, L.; Hasbullah, N.A. Phosphorus Transformation in Soils Following Co-Application of Charcoal and Wood Ash. *Agronomy* **2021**, *11*, 2010. [CrossRef]
52. Brundrett, M.C. Mycorrhizal associations and other means of nutrition of vascular plants: Understanding the global diversity of host plants by resolving conflicting information and developing reliable means of diagnosis. *Plant Soil* **2009**, *320*, 37–77. [CrossRef]
53. Borie, F.; Aguilera, P.; Castillo, C.; Valentine, A.; Alex Seguel, A.; José Miguel Barea, J.M.; Cornejo, P. Revisiting the Nature of Phosphorus Pools in Chilean Volcanic Soils as a Basis for Arbuscular Mycorrhizal Management in Plant P Acquisition. A review. *J. Soil. Sci. Plant Nutr.* **2019**, *19*, 390–401. [CrossRef]
54. Teste, F.P.; Jones, M.D.; Dickie, I.A. Dual mycorrhizal plants: Their ecology and relevance. *New Phytol.* **2019**, *225*, 1835–1851. [CrossRef] [PubMed]
55. Ghobadi, M.; Dehnavi, M.M.; Yadavi, A.R.; Parvizi, K.; Zafari, D.M. Reduced P fertilization improves Fe and Zn uptake in potato when inoculated with AMF in conditions of deficiency of phosphorus, iron and zinc in the soil. *Rhizosphere* **2020**, *15*, 100239. [CrossRef]

56. Tawaraya, K.; Naito, M.; Wagatsuma, T. Solubilization of insoluble inorganic phosphate by hyphal exudates of arbuscular mycorrhizal fungi. *J. Plant Nutr.* **2006**, *29*, 657–665. [CrossRef]
57. Richardson, A.E.; George, T.S.; Jakobsen, I.; Simpson, R.J. Plant utilization of inositol Phosphates. In *Inositol Phosphates: Linking Agriculture and the Environment*; Turner, B.L., Richardson, A.E., Mullaney, E.J., Eds.; CABI: Wallingford, UK, 2007; pp. 242–260.
58. Schnephf, A.; Roose, T.; Schweiger, P. Impact of growth and uptake patterns of arbuscular Mycorrhizal fungi on plant phosphorus uptake-a modelling study. *Plant Soil* **2008**, *312*, 85–99. [CrossRef]
59. Begum, N.; Qin, C.; Ahanger, M.A.; Raza, S.; Khan, M.I.; Ashraf, M.; Zhang, L. Role of Arbuscular Mycorrhizal Fungi in Plant Growth Regulation: Implications in Abiotic Stress Tolerance. *Front. Plant Sci.* **2019**, *10*, 1068. [CrossRef]
60. Saxena, J.; Rana, G.; Pandey, M. Impact of addition of biochar along with *Bacillus* sp. on growth and yield of French beans. *Sci. Hortic.* **2013**, *162*, 351–356. [CrossRef]
61. Etesami, H.; Adl, S.M. Plant growth-promoting rhizobacteria (PGPR) and their action mechanisms in availability of nutrients to plants. In *Phyto-Microbiome in Stress Regulation: Environmental and Microbial Biotechnology*; Kumar, M., Kumar, V., Prasad, R., Eds.; Springer: Singapore, 2020.
62. Glick, B.R.; Gamalero, E. Recent Developments in the Study of Plant Microbiomes. *Microorganisms* **2021**, *9*, 1533. [CrossRef]
63. Khatoon, Z.; Huang, S.; Rafique, M.; Fakhar, A.; Kamran, M.A.; Santoyo, G. Unlocking the potential of plant growth-promoting rhizobacteria on soil health and the sustainability of agricultural systems. *J. Environ. Manag.* **2020**, *273*, 111–118. [CrossRef]
64. Martinez, O.; Crowley, D.; Mora, M.L.; Jorquera, M.A. Short-term study shows that phytate-mineralizing rhizobacteria inoculation affects the biomass, phosphorus (P) uptake and rhizosphere properties of cereal plants. *J. Soil Sci. Plant Nutr.* **2015**, *15*, 153–166.
65. Hale, L.; Luth, M.; Kenney, R.; Crowley, D. Evaluation of pinewood biochar as a carrier of bacterial strain Enterobacter cloacae UW5 for soil inoculation. *Appl. Soil Ecol.* **2014**, *84*, 192–199. [CrossRef]
66. Barea, J.M.; Pozo, M.J.; Azcón, R.; Azcón-Aguilar, C. Microbialco-operation in the rhizosphere. *J. Exp. Bot.* **2005**, *56*, 1761–1778. [CrossRef]
67. Breitkreuz, C.; Buscot, F.; Tarkka, M.; Reitz, T. Shifts between and among Populations of Wheat Rhizosphere Pseudomonas, Streptomyces and Phyllo bacterium Suggest Consistent Phosphate Mobilization at Different Wheat Growth Stages under Abiotic Stress. *Front. Microbiol.* **2020**, *10*, 3109. [CrossRef]
68. Basu, S.; Kumar, G.; Chhabra, S.; Prasad, R. *New and Future Developments in Microbial Biotechnology and Bioengineering Phytomicrobiome for Sustainable Agriculture*; Elsevier: Amsterdam, The Netherlands, 2021; pp. 149–157.
69. Nahas, E. Factors determining rock phosphate solubilization by microorganisms isolated from soil. *World J. Microbiol. Biotechnol.* **2007**, *12*, 567–572. [CrossRef]
70. Zheng, B.X.; Ibrahim, M.; Zhang, D.P.; Bi, Q.-F.; Li, H.-Z.; Zhou, G.-W.; Yang, X.-R. Identification and characterization of inorganic-phosphate-solubilizing bacteria from agricultural fields with a rapid isolation method. *Appl. Microbiol. Biotechnol.* **2018**, *8*, 47. [CrossRef]
71. Guo, S.; Feng, B.; Xiao, C.; Wang, Q.; Chi, R. Phosphate-solubilizing microorganisms to enhance phytoremediation of excess phosphorus pollution in phosphate mining wasteland soil. *Biomed. J.* **2021**, *25*, 271–281. [CrossRef]
72. Amadou, I.; Houben, D.; Faucon, M.P. Unravelling the Role of Rhizosphere Microbiome and Root Traits in Organic Phosphorus Mobilization for Sustainable Phosphorus Fertilization. A Review. *Agronomy* **2021**, *11*, 2267. [CrossRef]
73. Waldrip, H.M.; He, Z.; Erich, M.S. Effects of poultry manure amendment on phosphorus uptake by ryegrass, soil phosphorus fractions and phosphatase activity. *Biol. Fertil. Soils* **2011**, *47*, 407–418. [CrossRef]
74. Rady, M.M.; Semida, W.M.; Hemida, K.A.; Abdelhamid, M.T. The effect of compost on growth and yield of *Phaseolus vulgaris* plants grown under saline soil. *Int. J. Recycl. Org. Waste Agric.* **2016**, *5*, 311–321. [CrossRef]
75. Silva-Leal, J.A.; Pérez-Vidal, A.; Torres-Lozada, P. Effect of biosolids on the nitrogen and phosphorus contents of soil used for sugarcane cultivation. *Heliyon* **2021**, *7*, 63360. [CrossRef]
76. Shen, Q.; Hedley, M.; Camps Arbestain, M.; Kirschbaum, M.U. Can biochar increase the bioavailability of phosphorus? *J. Soil Sci. Plant Nutr.* **2016**, *16*, 268–286. [CrossRef]
77. Adekiya, A.O.; Olaniran, A.F.; Adenusi, T.T.; Aremu, C.; Ejue, W.S.; Iranloye, Y.M.; Olayanju, A. Effects of cow dung and wood biochars and green manure on soil fertility and tiger nut (*Cyperus esculentus* L.) performance on a savanna Alfisol. *Sci. Rep.* **2020**, *10*, 21021. [CrossRef]
78. Zaidi, A.; Khan, S. Interactive effect of rhizotrophic microorganisms on growth, yield, and nutrient uptake of Wheat. *J. Plant Nutr.* **2005**, *28*, 2079–2092. [CrossRef]
79. Castanheira, N.L.; Dourado, A.C.; Pais, I.; Semedo, J.; Scotti-Campos, P.; Borges, N.; Carvalho, G.; Crespo, M.T.B.; Fareleira, P. Colonization and beneficial effects on annual ryegrass by mixed inoculation with plant growth promoting bacteria. *Microbial. Res.* **2017**, *198*, 47–55. [CrossRef]
80. El Maaloum, S.; Elabeda, A.; El Alaoui-Talibia, Z.; Meddicha, A.; Filali-Maltoufb, A.; Douirac, A.; Ibnsouda-Koraichid, S.; Amire, S.; El Modafar, C. Effect of arbuscular mycorrhizal fungi and Phosphate-solubilizing bacteria consortia associated with Phospho-compost on Phosphorus solubilization and growth of tomato seedlings (*Solanum lycopersicum* L.). *Commun. Soil Sci. Plant Anal.* **2019**, *51*, 622–634. [CrossRef]
81. Yasmeen, T.; Arif, M.S.; Shahzad, S.M.; Riaz, M.; Tufail, M.A.; Mubarik, M.S.; Ahmad, A.; Ali, S.; Albasher, G.; Shakoor, A. Abandoned agriculture soil can be recultivated by promoting biological phosphorus fertility when amended with nano-rock phosphate and suitable bacterial inoculant. *Ecotoxicol. Environ. Saf.* **2022**, *234*, 113385. [CrossRef]

82. Soumare, A.; Boubekri, K.; Lyamlouli, K.; Hafidi, M.; Ouhdouch, Y.; Kouisni, L. Efficacy of phosphate solubilizing Actinobacteria to improve rock phosphate agronomic effectiveness and plant growth promotion. *Rhizosphere* **2020**, *17*, 100284. [CrossRef]
83. Rosa, P.A.L.; Mortinho, E.S.; Jalal, A.; Galindo, F.S.; Buzetti, S.; Fernandes, G.C.; Teixeira Filho, M.C.M. Inoculation with growth-promoting bacteria associated with the reduction of phosphate fertilization in Sugarcane. *Front. Environ. Sci.* **2020**, *8*, 32. [CrossRef]
84. Richardson, A.E.; George, T.S.; Hens, M.; Simpson, R.J. Utilization of soil organic phosphorus by higher plants. In *Organic Phosphorus in the Environment*; Turner, B.L., Frossard, E., Baldwin, D.S., Eds.; CABI: Wallingford, UK, 2005; pp. 165–184.
85. Msimbira, L.A.; Smith, D.L. The Roles of Plant Growth Promoting Microbes in Enhancing Plant Tolerance to Acidity and Alkalinity Stresses. *Front. Sustain. Food Syst.* **2020**, *4*, 106. [CrossRef]
86. Raymond, N.S.; Gómez-Muñoz, B.; van der Bom, F.J.T.; Nybroe, O.; Stoumann Jensen, L.; Müller-Stöver, D.S.; Richardson, A.E. Phosphate solubilising microorganisms for improved crop productivity: A critical assessment. *New Phytol.* **2021**, *229*, 1268–1277. [CrossRef] [PubMed]
87. Oberson, A.; Joner, E.J. Microbial turnover of phosphorus in soil. In *Organic Phosphorus in the Environment*; Turner, B.L., Frossard, E., Baldwin, D.S., Eds.; CABI: Wallingford, UK, 2005; pp. 133–164.
88. Brimecombe, M.J.; De Leij, F.A.A.M.; Lynch, J.M. Rhizodeposition and microbial populations. In *The Rhizosphere Biochemistry and Organic Substances at the Soil-Plant Interface*; Pinton, R., Varanini, Z., Nannipieri, P., Eds.; CRC Press: Boca Raton, FL, USA, 2007; pp. 73–109.
89. Tian, J.; Ge, F.; Zhang, D.; Deng, S.; Liu, X. Roles of Phosphate Solubilizing Microorganisms from Managing Soil Phosphorus Deficiency to Mediating Biogeochemical P Cycle. *Biology* **2021**, *10*, 158. [CrossRef] [PubMed]
90. Owen, D.; Williams, A.P.; Griffith, G.W.; Withers, P.J.A. Use of commercial bio-inoculants to increase agricultural production through improved phosphorus acquisition. *Appl. Soil Ecol.* **2015**, *86*, 41–54. [CrossRef]
91. Maharana, R.; Basu, A.; Dhal, N.K.; Adak, T. Biosolubilization of rock phosphate by *Pleurotus ostreatus* with brewery sludge and its effect on the growth of maize (*Zea mays* L.). *J. Plant Nutr.* **2020**, *44*, 395–410. [CrossRef]
92. Martinez-Viveros, O.; Jorquera, M.; Crowley, D.; Gajardo, G.; Mora, M.L. Mechanisms and practicalconsiderations involved in plant growth promotion by rhizobacteria. *J. Soil Sci. Plant Nutr.* **2010**, *10*, 293–319. [CrossRef]
93. Menezes-Blackburn, D.; Jorquera, M.A.; Gianfreda, L.; Greiner, R.; Mora, M.L. A novel phosphorus biofertilization strategy using cattle manure treated with phytase-nanoclay complexes. *Biol. Fertil. Soils* **2014**, *50*, 583–592. [CrossRef]
94. Irshad, U.; Brauman, A.; Villenave, C.; Plassard, C. Phosphorus acquisition from phytate depends on efficient bacterial grazing, irrespective of the mycorrhizal status of *Pinus pinaster*. *Plant Soil* **2012**, *358*, 155–168. [CrossRef]
95. Islam, M.K.; Sano, A.; Majumder, M.S.; Hossain, M.A.; Sakagami, J.I. Isolation and molecular characterization of phosphate solubilizing filamentous fungi from subtropical soils in Okinawa. *Appl. Ecol. Environ. Res.* **2019**, *1*, 9145–9157. [CrossRef]
96. Joy, J.M.; Ravinder, J.; Rakesh, S.; Somashekar, G. A review article on integrated nutrient management in wheat crop. *Int. J. Chem. Stud.* **2018**, *6*, 697–700.
97. Babalola, O.O.; Glick, B.R. The use of microbial inoculants in African agriculture: Current practice and future prospects. *Agric. Environ.* **2012**, *10*, 540–549.
98. Billah, M.; Khan, M.; Bano, A.; Hassan, T.U.; Munir, A.; Gurmani, A.R. Phosphorus and phosphate solubilizing bacteria: Keys for sustainable agriculture. *Geomicrobiol. J.* **2019**, *36*, 904–916. [CrossRef]
99. Demirbas, A.; Edris, G.; Alalayah, W.M. Sludge production from municipal wastewater treatment in sewage treatment plant. *Energy Sources Part A Recovery Util. Environ. Eff.* **2017**, *39*, 999–1006. [CrossRef]
100. Lu, Q.; He, Z.L.; Stoffella, P.J. Land application of biosolids in the USA: A review. *Appl. Environ. Soil Sci.* **2012**, *2012*, 201462. [CrossRef]
101. Larney, F.J.; Angers, D.A. The role of organic amendments in soil reclamation: A review. *Can. J. Soil Sci.* **2012**, *92*, 19–38. [CrossRef]
102. Chambers, B.J.; Nicholson, F.A.; Aitken, M.; Cartmell, E.; Rowlands, C. Benefits of biosolids to soil quality and fertility. *Water Environ. J.* **2003**, *17*, 162–167. [CrossRef]
103. Tian, H.Q.; Lu, C.; Jerry, L.; Melillo, J.; Ren, W.; Huang, Y.; Xu, X.; Liu, M.; Zhang, C.; Chen, G.; et al. Food benefit and climate warming potential of nitrogen fertilizer uses in China. *Environ. Res. Lett.* **2012**, *7*, 044020. [CrossRef]
104. Ali, M.; Ahmed, T.; Abu-Dieyeh, M.; Al-Ghouti, M. Environmental Impacts of Using Municipal Biosolids on Soil, Plant and Groundwater Qualities. *Sustainability* **2021**, *13*, 8368. [CrossRef]
105. He, Z.; Zhang, H.; Toor, G.S.; Dou, Z.; Honeycutt, C.W.; Haggard, B.E.; Reiter, M.S. Phosphorus distribution in sequentially extracted fractions of biosolids, poultry litter, and granulated products. *Soil Sci.* **2010**, *175*, 154–161. [CrossRef]
106. Torri, S.I.; Alberti, C. Characterization of organic compounds from biosolids of Buenos Aires City. *J. Soil Sci. Plant Nutr.* **2012**, *12*, 143–152. [CrossRef]
107. White, J.W.; Coale, F.J.; Sims, J.T.; Shober, A.L. Phosphorus runoff from waste water treatment biosolids and poultry litter applied to agricultural soils. *J. Environ. Qual.* **2010**, *39*, 314–323. [CrossRef]
108. Codling, E.E. Long-term effects of biosolid-amended soils on phosphorus, copper, manganese and zinc uptake by wheat. *Soil Sci.* **2014**, *179*, 21–27. [CrossRef]
109. Shaheen, S.; Tsadilas, C. Phosphorus sorption and availability to canola grown in an Alfisol amended with various soil amendments. *Commun. Soil Sci. Plant Anal.* **2013**, *44*, 89–103. [CrossRef]

110. Ippolito, J.A.; Barbarick, K.A.; Norvell, K.L. Biosolids impact soil phosphorus accountability fractionation, and potential environmental risk. *J. Environ. Qual.* **2007**, *36*, 764–772. [CrossRef] [PubMed]
111. Haney, C.H.; Samuel, B.S.; Bush, J.; Ausubel, F.M. Associations with rhizosphere bacteria can confer an adaptive advantage to plants. *Nat. Plants* **2015**, *1*, 15051. [CrossRef] [PubMed]
112. Brandt, R.C.; Elliott, H.A.; O'Connor, G.A. Water-extractable phosphorus in biosolids: Implications for land-based recycling. *Water Environ. Res.* **2004**, *76*, 121–129. [CrossRef] [PubMed]
113. Moreira, J.G.; Delvaux, J.C.; Magela, M.L.M.; Pereira, V.J.; de Carmargo, R.; Lana, R.M.Q. Chemical changes in soil with use of pelletized organomineral fertilizer made from biosolids and sugarcane filter cake. *Aust. J. Crop Sci.* **2021**, *15*, 67–72. [CrossRef]
114. Scharenbroch, B.C.; Meza, E.N.; Catania, M.; Fite, K. Biochar and biosolids increase tree growth and improve soil quality for urban landscapes. *J. Environ. Qual.* **2013**, *42*, 1372–1385. [CrossRef]
115. Nannipieri, P.; Giagnoni, L.; Renella, G.; Puglisi, E.; Ceccanti, B.; Masciandaro, G.; Fornasier, F.; Marinari, S. Soil enzymology: Classical and molecular approaches. *Biol. Fertil. Soils* **2012**, *48*, 743–762. [CrossRef]
116. Arif, M.S.; Shahzad, S.M.; Yasmeen, T.; Riaz, M.; Ashraf, M.; Ashraf, M.A.; Kausar, R. Improving plant Phosphorus (P) acquisition by phosphate-solubilizing Bacteria. In *Essential Plant Nutrients*; Springer: Cham, Switzerland, 2017; pp. 513–556. [CrossRef]
117. Renella, G.; Szukics, U.; Landi, L.; Nannipieri, P. Quantitative assessment of hydrolase production andpersistence in soil. *Biol. Fertil. Soils* **2007**, *44*, 321–329. [CrossRef]
118. Margalef, O.; Sardans, J.; Maspons, J.; Molowny-Horas, R.; Fernández-Martínez, M.; Janssens, I.A.; Peñuelas, J. The effect of global change on soil phosphatase activity. *Glob. Chang. Biol.* **2021**, *27*, 5989–6003. [CrossRef]
119. Dodor, D.E.; Tabatabai, M.A. Effect of cropping systems on phosphatases in soils. *J. Plant Nutr. Soil Sci.* **2003**, *66*, 7–13. [CrossRef]
120. Asemaninejad, A.; Langley, S.; Mackinnon, T.; Spiers, G.; Beckett, P.; Mykytczuk, N.; Basiliko, N. Blended municipal compost and biosolids materials for mine reclamation: Long-term field studies to explore metal mobility, soil fertility and microbial communities. *Sci. Total Environ.* **2020**, *760*, 143393. [CrossRef]
121. Jiang, J.; Wang, Y.; Yu, D.; Yao, X.; Han, J.; Cheng, R.; Cui, H.G.; Zhang, X.; Zhu, G. Garbage enzymes effectively regulated the succession of enzymatic activities and the bacterial community during sewage sludge composting. *Bioresour. Technol.* **2021**, *327*, 124792. [CrossRef]
122. Moorhead, D.L.; Sinsabaugh, R.L. Simulated patterns of litter decay predict patterns of extracellular enzyme activities. *Appl. Soil Ecol.* **2000**, *14*, 71–79. [CrossRef]
123. IFA. IFASTAT. Consumption Database. International Fertilizer Association. 2021. Available online: https://www.ifastat.org/databases/plant-nutrition (accessed on 11 October 2021).
124. Lu, R.K. Effect of long-term fertilization on soil properties. In *Soil-Plant Nutrients Principles and Fertilization*; Lu, R.K., Xie, J.C., Cai, G.X., Eds.; Chemical Industry Press: Beijing, China, 1998; pp. 102–110. (In Chinese)
125. Wang, Y.; Zhao, X.; Wang, L.; Wang, Y.; Li, W.; Wang, X.; Xing, G. The regime and availability of omitting P fertilizer application for rice in rice/wheat rotation in the Taihu lake region of southern China. *J. Soils Sediments* **2015**, *15*, 844–853. [CrossRef]
126. Vanden Nest, T.; Amery, F.; Fryda, L.; Boogaerts, C.; Bilbao, J.; Vandecasteele, B. Renewable P sources: P use efficiency of digestate, processed animal manure, compost, biochar and struvite. *Sci. Total Environ.* **2020**, *750*, 141699. [CrossRef]
127. Ibrahim, M.; Li, G.; Chan, F.K.S.; Kay, P.; Liu, X.; Firebank, L.; Xu, Y. Biochar effects potentially toxic elements and antioxidant enzymes in *Lactuca sativa* L. grown in multi-metals contaminated soil. *Environ. Technol. Innov.* **2019**, *15*, 100427. [CrossRef]
128. Choudhary, T.K.; Khan, K.S.; Hussain, Q.; Ashfaq, M. Nutrient Availability to Maize Crop (*Zea mays* L.) in Biochar Amended Alkaline Subtropical Soil. *J. Soil Sci. Plant Nutr.* **2021**, *21*, 1293–1306. [CrossRef]
129. Melaku, T.; Ambaw, G.; Nigussie, A.; Woldekirstos, A.N.; Bekele, E.; Ahmed, M. Short-term application of biochar increases the amount of fertilizer required to obtain potential yield and reduces marginal agronomic efficiency in high phosphorus-fixing soils. *Biochar* **2020**, *2*, 503–511. [CrossRef]
130. Santos-Torres, M.; Romero-Perdomo, F.; Mendoza-Labrador, J.; Guti´errez, A.Y.; Vargas, C.; Castro-Rincon, E.; Caro-Quintero, A.; Uribe-Velez, D.; Estrada-Bonilla, G.A. Genomic and phenotypic analysis of rock phosphate-solubilizing rhizobacteria. *Rhizosphere* **2021**, *17*, 100290. [CrossRef]
131. Major, J.; Rondon, M.; Molina, D.; Riha, S.J.; Lehmann, J. Nutrient leaching in a Colombian savanna oxisol amended with biochar. *J. Environ. Qual.* **2012**, *41*, 1076. [CrossRef]
132. Balla Kovács, A.; Kremper, R.; Kátai, J.; Vágó, I.; Buzetzky, D.; Kovács, E.M.; Kónya, J.; Nagy, N.M. Characterisation of soil phosphorus forms in the soil-plant system using radio isotopic tracer method. *Plant Soil Environ.* **2021**, *67*, 367–375. [CrossRef]
133. Nunes, R.d.S.; de Sousa, D.M.G.; Goedert, W.J.; de Oliveira, L.E.Z.; Pavinato, P.S.; Pinheiro, T.D. Distribution of Soil Phosphorus Fractions as a Function of Long-Term Soil Tillage and Phosphate Fertilization Management. *Front. Earth Sci.* **2020**, *8*, 350. [CrossRef]
134. Pereira, S.I.A.; Abreu, D.; Moreira, H.; Vega, A.; Castro, P.M.L. Plant growth-promoting rhizobacteria (PGPR) improve the growth and nutrient use efficiency in maize (Zea mays L.) under water deficit conditions. *Heliyon* **2020**, *6*, e05106. [CrossRef]
135. Ajeng, A.A.; Abdullah, R.; Ling, T.C.; Ismail, S.; Lau, B.F.; Ong, H.C.; Chang, J.-S. Bioformulation of biochar as a potential inoculant carrier for sustainable agriculture. *Environ. Technol. Innov.* **2020**, *20*, 101168. [CrossRef]
136. Albareda, M.; Rodríguez-Navarro, D.N.; Camacho, M.; Temprano, F.J. Alternatives to peat as a carrier for rhizobia inoculants: Solid and liquid formulations. *Soil Biol. Biochem.* **2008**, *40*, 2771–2779. [CrossRef]

137. Pagnani, G.; Galieni, A.; Stagnari, F.; Pellegrini, M.; DelGallo, M.; Pisante, M. Open field inoculation with PGPR as a strategy to manage fertilization of ancient *Triticum aestivum* genotypes. *Biol. Fertil. Soils* **2020**, *56*, 111–124. [CrossRef]
138. Saranya, K.; Krishnan, P.S.; Kumutha, K.; French, J. Potential for biochar as an alternate carrier to lignite for the preparation of biofertilizers in India. *Int. J. Agric. Environ. Biotechnol.* **2011**, *4*, 167–172.
139. Murphy, J.; Riley, J.P. A modified single-solution method for the determination of phosphorus in natural waters. *Anal. Chim. Acta* **1962**, *27*, 31–36. [CrossRef]

agronomy

MDPI

Article

Development and Applications of an In Situ Probe for Multi-Element High-Resolution Measurement at Soil/Sediment-Water Interface and Rice Rhizosphere

Meng Zhao [1,2,3,†], Jiang Liu [1,2,†], Chuangchuang Zhang [1,2], Xuefeng Liang [2], Qian E [1,2], Rongle Liu [2], Yujie Zhao [1,2,*] and Xiaowei Liu [1,2,*]

1 Key Laboratory for Environmental Factors Control of Agro-Product Quality Safety, Ministry of Agriculture and Rural Affairs, Tianjin 300191, China; 82101181137@caas.cn (M.Z.); 15548810038@163.com (J.L.); chuangchuangzhang@126.com (C.Z.); eqian1993@126.com (Q.E.)
2 Agro-Environmental Protection Institute, Ministry of Agriculture and Rural Affairs, Tianjin 300191, China; liangfxuefeng@caas.cn (X.L.); liurongle@caas.cn (R.L.)
3 Institute of Plan Nutrition, Resources and Environment, Beijing Academy of Agriculture and Forestry Sciences, Beijing 100097, China
* Correspondence: zhaoyujie@caas.cn (Y.Z.); xwliu2006@163.com (X.L.)
† These two authors contributed equally to this work.

Abstract: The biogeochemistry of multi-elements, such as sulfur (S), phosphorus (P) and arsenic (As), is interlinked especially at interfaces of soil/sediment–water and plant rhizosphere. To explore the biogeochemical behavior of multi-elements such as S-P-As at interfaces, an in situ and high-resolution technology is required. In this study, we developed an in situ probe (LDHs-DGT) based on the diffusive gradients in thin-films technique using a single binding layer to realize the co-measurement of multi-elements including sulfide and oxyanions. Mg-Al layered double hydroxides (LDHs) were synthesized and incorporated into the probe's binding layer. Laboratorial characterization showed that the LDHs-DGT probe had a high capacity for sulfide, phosphate and arsenate and can effectively determine their levels across a wide range of solution conditions, i.e., pH from 5 to 8 and ionic strengths from 0.005–0.01 mol L^{-1} $NaNO_3$. The application potential of the LDH_S-DGT probe in capturing the concentration profiles of sulfide and oxyanions across the soil/sediment–water interface at a centimeter scale was demonstrated. The synchronous co-variations of labile sulfide and phosphate were observed along an intact river sediment core, demonstrating the redox driven behaviors of oxyanions at aerobic–anaerobic transition zones. Moreover, the LDH_S-DGT probe was further used to acquire the dynamic distributions of multi-elements in the plant rhizosphere at a two-dimensional millimeter scale. Compared to treatments of sodium sulfate and mercaptopygorskite fertilization, the addition of elementary S promoted the reduction of sulfate to sulfide along the whole growth stage and thus inhibited the activation of toxic metals in the rice rhizosphere. Collectively, this study provides a tool for convenient measurement of nutrients and metal(loid)s across soil–water/root interfaces at high resolution and thus, a broad application prospect of the tool in sustainable agriculture is expected.

Keywords: in situ sampling; nutrients; metal(loid)s; crop rhizosphere; sustainable agriculture

Citation: Zhao, M.; Liu, J.; Zhang, C.; Liang, X.; E, Q.; Liu, R.; Zhao, Y.; Liu, X. Development and Applications of an In Situ Probe for Multi-Element High-Resolution Measurement at Soil/Sediment-Water Interface and Rice Rhizosphere. *Agronomy* **2021**, *11*, 2383. https://doi.org/10.3390/agronomy11122383

Academic Editor: Stefano Cesco

Received: 14 October 2021
Accepted: 15 November 2021
Published: 24 November 2021

1. Introduction

For sustainable agriculture, it is important to keep the balance of mineral elements in soils, and to control the uptake and accumulation of toxic metal(loid)s in crops [1,2]. The biogeochemical cycling of mineral elements and toxic metal(loid)s in the environment is highly interlinked. For example, it has been found that sulfur (S) and phosphorus (P) affect the environmental behavior of arsenic (As) [1,3–5]. Under anoxic soil conditions, microorganisms promote the formation of sulfide and thus act as reducing agents for arsenic, promoting the release of As into the solution phase [6]. Alternatively, sulfide may immobilize As by inducing co-precipitation of iron sulfide minerals (e.g., pyrite) and arsenic sulfide (dioecious and realgar) [6,7]. Phosphates and arsenates are considered chemical analogues, which means they can be substituted for each other in chemical reactions, including adsorption/desorption reaction, precipitation/dissolution reaction [4]. Besides, the distribution of elements in agricultural soils, especially at the crop rhizosphere, are highly uneven at both spatial and temporal scales. Therefore, it is strongly necessary to develop in situ high-resolution tools for co-measurement of nutrients and metal(loid)s at the soil/sediment–water interface and crop rhizosphere.

Diffusive gradients in thin films (DGT) is a representative in situ measurement technique characterized by broad applicability, quantitative measurement, trace element identification and high spatial resolution [8,9]. It is widely used for measurement of over 100 analytes including nutrients, metal(loid)s, radionuclides and polar organic contaminants in water, sediment and soil [8,10–15]. Typically, DGT is consistent of a protective filter membrane, a well-defined diffusion gel and a functionalized binding gel [16]. For the determination of different types of analytes, specific binding gels with incorporated binding materials are required [14,17]. Ideally, the binding gel should (1) have a high capacity for long-term monitoring of the target analytes; (2) contain particular binding materials with even distribution and small particle size for high-resolution application at interfaces [11,18]. The development of new bindings agents has received increasing attention in recent years. Specially, for the co-measurement of sulfide and oxyanions (e.g., phosphate and arsenate) using DGT two strategies are currently employed. One strategy involves the simultaneous application of two DGT probes with different binding layers that separately measure sulfide and oxyanions, e.g., AgI-DGT for sulfide, and ferrihydrite, Zr-oxide, or Metsorb-DGT for oxyanions [8,12,19,20]. The second strategy involves the application of a single type of DGT with a mixed binding layer, in which two kinds of functionalized materials such as AgI and ferrihydrite, AgI and Zr-oxide, or AgI and Metsorb were incorporated [16,20–23]. Although significant progress has been made in the development of binding gels, they often have some defects, such as poor mechanical resistance, poor repeatability or unclear gel structure which limit their application [2]. In this study, we propose a third strategy to achieve the co-measurement of sulfide and oxyanions based on DGT. In this new strategy, a DGT probe with a single functionalized material in the binding layer is sufficient. This is achieved by exploiting the ability of layered double hydroxides (LDHs) to capture different types of ions, including sulfide and oxyanions, thus overcoming the current limitations of binding gels.

LDHs are a type of anion adsorption material that is widely used in diverse industries for catalysis, adsorption and ion exchange, and are also used as functionalized environmental materials [24]. LDHs are layered compounds that aggregate the positively charged main laminate and the interlaminar anions. Because the interlayer anions in LDHs can be replaced by other anions, the target anions are thus immobilized in the interlayer [24,25]. However, there are significant differences in the anion exchange rates for different types of LDHs. Mg-Al LDHs synthesized from the chloride salts of magnesium and aluminum show excellent sorption capacity towards sulfide and oxyanions (e.g., P and As), while the incorporated Cl^- helps reduce interference caused by other anions, e.g., nitrate [15,26,27]. In this study, Mg-Al LDHs nanoparticles were synthesized and used for as a single binding material in the LDHs-DGT probe. We hypothesized that the Mg-Al LDH_S nanocomposite binding could adsorb both sulfide and oxyanions. The objective of this study is to (1) construct an in situ probe and characterize its performance for accurate measurement of labile concentrations of multi-elements in waters; (2) demonstrate the application potential of the probe in high-resolution measurement of nutrients/metal(loid)s at the soil/sediment–water interface and rice rhizosphere. This study provides a useful tool for management of nutrients and metal(loid)s in sustainable agriculture.

2. Materials and Methods

2.1. Reagents, Materials, and Solutions

The glassware and plastic materials used for probe construction were rinsed with 10% HNO_3 followed by rinsing with ultra-pure water (18.2 Ω cm, Milli-Q, Millipore, USA) three times before use to avoid sample contamination. All the solutions were prepared using ultra-pure water and all reagents used were of reagent-grade purity. Stock solutions (1 g·L^{-1}) were prepared using KH_2PO_4, Na_2S, Na_2SO_4, $Na_2HAsO_4{\cdot}7H_2O$ and $NaAsO_2$. $MgCl_2$.6H2O, $AlCl_3$.6H2O, NH_4.OH, Diethylenetriaminepentaacetic acid (DTPA) were used to make LDH_S nanoparticles.

2.2. Laboratory Evaluation of DGT Performance

2.2.1. Synthesis and Characterization of LDHs

The LDHs were synthesized in accordance with a Chinese patented method [15]. An aqueous solution of magnesium chloride and aluminum chloride was prepared with a Mg^{2+} (0.5 mol L^{-1}):Al^{3+} (0.5 mol L^{-1}) ratio of 2:1 (in order to fully exchange anions in the solution to be tested), 0.5 mol L^{-1} DTPA was then added to the solution. The milli-Q water configuration was used for all solutions. Under a stirring frequency of 40–50 Hz/s, 20–25% ammonia solution was slowly added when necessary to keep the pH of the reaction solution between 9.5. The mixture was continuously stirred for 30 min, and the aging reaction was allowed to occur for 60–120 min (25 °C). To make the precipitation reaction complete, ultra-pure water was then continuously added to the reaction mixture. The resulting LDHs precipitate was cleaned using ultra-pure water to remove chloride ions. After being stored overnight at −60 °C in an ultra-low temperature freezer (ZX-LGJ-18A, Zhixin, CHN), the precipitate was finally freeze-dried to obtain the requisite Mg-Al LDHs nanomaterials.

The crystal phases of the LDHs samples were examined by X-ray powder diffraction (XRD, D/MAX2500, Rigaku Corporation, JPN). The size and morphology of the samples were characterized by field emission scanning electron microscopy (FESEM, LEO 1350VP, LEO Electron Microscopy Ltd., GER) and transmission electron microscopy (TEM, JEM 2010, JEOL Ltd., JPN).

2.2.2. Probe Gel Preparation

LDH_S gel strips (0.6 mm thickness, ~10 cm × 17 cm) were made according to an established procedure [15]. To synthesize the binding gel, LDHs nanomaterials (2 g) were subjected to ultrasonication for 2–3 min [28]. Thereafter, tetramethylethylenediamine (TEMED) catalyst (22.5 µL) and freshly prepared ammonium persulfate initiator (90 µL, 10% w/v) were added. After sufficient stirring, the mixture was immediately cast between two glass plates separated by plastic spacers (ensuring that the LDH_S gel thickness is 0.6 mm) and allowed to set at 42 °C for 1 h. Thereafter, the gel obtained was removed and placed in ultra-pure water, which was changed frequently within 24 h to remove the interference of nitrate ions. The diffusion gel was cross-linked with 15% acrylamide and 0.3% agarose derivatives according to an established procedure [12]. The diffusive gel and the binding gel were cut with a ceramic knife and assembled into DGT probes (Figure S1), which were then placed in a sealed pocket containing 0.01 M NaCl protective solution and stored in a refrigerator at 4 °C [15]. The gel strips were cut into circular discs (1 cm diameter) for use in LDH_S-DGT characterization experiments (Figure S2). For LDH_S-DGT deployment across the soil/sediment–water interface and rice rhizosphere, the gel strips were firstly cut into a size of 2.5 cm × 15 cm and 9 cm × 12 cm, respectively.

2.2.3. Chemical Analysis and Concentration Calculation

To avoid of oxidation of S(-II) and As(III), experiments were carried out under nitrogen. The solution temperature was monitored during probe deployment. Upon retrieval, the probes were rinsed with ultra-pure water and the binding gel was then eluted with HNO_3 (9 mL, 0.5 mol·L^{-1}) for 10 min. The concentrations of sulfur, phosphorus and arsenic in the $LDHs_S$-DGT eluates and solutions were determined via inductively coupled plasma mass spectrometry (ICP-MS, 7700×, Agilent, USA), high-performance liquid chromatography (HPLC, 1100, Agilent), and ion chromatography (IC, ICS3000, DIONEX, USA) calibrated using standard solutions of S(-II), phosphate (P), As(III) and As(V) and SO_4^{2-}. DGT measured concentrations were calculated based on Fick's first diffusion law [8]:

$$M = \frac{C_e\,(V_g + V_e)}{f_e} \tag{1}$$

$$C_{DGT} = \frac{M\Delta g}{DAt} \tag{2}$$

where M represents the mass of the analyte bound onto the gel (ng), C_e represents the concentration of the analyte in the eluent (μg L^{-1}), V_g represents the volume of the gel, V_e represents the eluent volume (mL), and f_e represents the elution efficiency (100 ± 5% for LDHs, which can be approximated as 1). Δg represents the sum of the thicknesses of the diffusive gel and the nitrocellulose film (0.089 cm), D represents the diffusion coefficient of the target analyte in the diffusive gel (cm^2 s^{-1}), A represents the area of the DGT window (3.15 cm^2), and t represents the DGT deployment time (86,400 s).

2.2.4. Diffusion Coefficients in the Probe's Diffusive Layer

The diffusion coefficient of each analyte in the diffusion gel was measured using the probe time-series deployment method [29]. Triplicate LDHs-DGT probes were placed in a solution containing S(-II), SO_4^{2-}, phosphate, As(V), or As(III) (3.5 L, pH = 5.5 ± 0.2) at 25.0 ± 0.5 °C, with an ionic strength of 0.01 M $NaNO_3$. The concentration of each ion in the test solution was approximately 700 μg L^{-1}. $LDHs_S$-DGT probes were retrieved after 2, 4, 6, 8, 10, 14 or 15 h, and the binding gels were taken out for analysis. An aliquot of the solution (2 mL) was collected each time and filtered (0.45 μm) for analysis. The effective diffusion coefficients of S(-II), SO_4^{2-}, phosphate, As(III) and As(V) were calculated using Equation (3) [12,29].

$$D = \frac{slope \times \Delta g}{c_{sol} \times A \times 60} \tag{3}$$

where "slope" refers to the linear regression slope of the mass (ng) of the ions accumulated in the gel with time (min); "c_{sol}" is the concentration of the substance to be measured in the preparation solution (μg L^{-1}).

2.2.5. Selectivity of LDHs-DGT Probe to Sulfide and Arsenate

To determine the selectivity of the probe toward different forms of S (i.e., S(-II) and SO_4^{2-}) and As (i.e., As(V) and As(III)), the LDHs-DGT probes were deployed in a mixed solution (3 L) containing S(-II) and SO_4^{2-}, both at a concentration of 50 μg L^{-1}. A similar experiment was conducted for As(III) and As(V), with both being at a concentration of 50 μg L^{-1}. The solution was stirred thoroughly, and the $LDHs_S$-DGT probe was retrieved at 2, 4, 6, 8, 12, 24, 48 and 72 h, and at each time point, aliquots (10 mL) of the solution were collected to determine the concentrations of the chemical species. The concentrations of S(-II) were determined via ICP-MS, whereas SO_4^{2-} was quantified by IC and As(III) and As(V) via HPLC-ICP-MS [30–32].

2.2.6. Effects of Ionic Strength and pH on Probe Measurement

To study the effect of pH on probe measurement, the LDHs-DGT probe was immersed in a multi-element solution containing three different ions (phosphate, S(-II) and As(V)) at a concentration of approximately 50 μg L^{-1} (KH_2PO_4, Na_2S, Na_2HAsO_4). HNO_3 and NaOH were used to adjust the solution pH in the range 4–9. The LDHs-DGT probe was deployed in the solution for 6 h. To test the performance of the probe under different ionic strengths, the $LDHs_S$-DGT probe was exposed to a multi-element solution containing three ions (S(-II), phosphate and As(V)) at a concentration of approximately 50 μg L^{-1} for 6 h.

The ionic strength in the solution was varied in the range 0.001–0.7 mol L^{-1} $NaNO_3$. The concentrations of phosphate, S(-II) and As(V) in the solutions and eluents of binding gel were analyzed.

2.2.7. Capacity of the Probe

To determine the capacity of the probe with respect to the S(-II), phosphate and As(V) ions, the LDH_S-DGT probe was immersed in a 1 L well-stirred solution with a single solution of 0.5–120 mg L^{-1} for 24 h. Before the deployment of the LDH_S-DGT probe, the test solution was stirred vigorously for three days, while ensuring that a constant solution pH at 5.5 using dilute solutions of HNO_3 and NaOH. After deployment of the LDH_S-DGT probes, DGT-measured S(-II), phosphate and As(V) concentrations were determined via ICP-MS and the capacity of the LDH_S-DGT probe thus obtained through comparison with the theoretical uptake concentrations.

2.2.8. Competition Effect among Different Elements

To test the possible interference of several kinds of ions on DGT measurement, the LDHs-DGT probe was deployed in a mixed solution (2 L) containing S(-II), phosphate and As(V). The ion concentration of the mixed solution was in the range of 4–90 μg L^{-1} at pH 5.5. The highest ion concentration of 90 μg L^{-1} was 9-fold to that of P and 1.8-fold to those of As and S, based on the Chinese Surface Water Environmental Quality Standard [33]. The probe deployed in each solution was retrieved after 8 h, and the concentrations of the ions in the solutions and eluants were both determined.

2.3. Co-Measurement of Multi-Elements Using LDH_S-DGT Probe in Water

Three water samples were collected from three different waterbodies (0~20 cm) in Tianjin, China. The first sample ("W1") is agricultural irrigation water from Tianjin, which is used to represent the baseline concentration of S(-II), phosphate and As(V). The other two sites were from the Jinhe River ("W2", represents high eutrophication levels) and the park fountain ("W3", represents low eutrophication levels) with different levels of S(-II), phosphate and As(V). These samples demonstrate some extreme scenarios challenging LDHs-DGT measurements. After sampling, all water samples were filtered through a 0.45 mm membrane. A pH meter (S8-FEID Kit, Mettler Toledo, GER) was used to determine the sample pH. The S(-II) concentration was determined by methylene blue spectrophotometry [5], whereas the concentration of phosphate was directly determined by ICP-MS after HNO_3 acidification and that of As(V) was determined by HPLC-ICP-MS. Triplicate LDH_S-DGT probes were deployed in the water sample (2 L) at 25 ± 1 °C [34]. The probes were retrieved after 24 h, rinsed with ultra-pure water. The binding gel was removed and eluted with 9 mL 0.5 mol L^{-1} HNO_3 eluent. The concentrations S(-II), phosphate and As(V) were determined in the eluants were then measured.

2.4. High-Resolution Measurement of Multi-Element Distribution at Soil/Sediment-Water Interface

To test the applicability of the LDH_S-DGT probe for high resolution measurement across the soil/sediment–water interface, an intact sediment core with overlying water was collected from the eutrophic Jinhe River in Tianjin, China in July 2020. Three deoxygenated LDH_S-DGT probes were inserted into the core for 48 h, and after retrieval, the binding gel was cut into 1 cm-long fragments. Each fragment was eluted with HNO_3 (9 mL, 0.5 mol L^{-1}), and the concentrations of S(-II), phosphate and As(V) were determined via ICP-MS and calculated according to Equation (2). By assigning DGT-measured concentrations to the ordinated vertical location information, 1D concentration profile of the three ions across the interface was obtained [35].

2.5. Application of LDHs-DGT Probe in Rice Rhizosphere for Imaging of the Dynamic Distributions of Mult-Elements

In order to study the spatial and temporal dynamic distribution characteristics of different forms of exogenous sulfur on S(-II) and toxic metals (e.g., Cd) in rice rhizosphere soil, a rhizo-box was built for efficient LDHs-DGT probe deployment. It can provide the technical means and theoretical basis for studying the temporal and spatial dynamic characteristics of S(-II) and the biogeochemical cycle of S(-II) and other elements in paddy fields. A paddy soil contaminated with toxic metal(loid)s including Cd was collected from Hunan, China and used for the rhizo-box experiment, which was carried out from April to October 2020. About 1.5 kg soil (dry weight) was filled into each rhizo-box with length × width × height of 15 cm × 5 cm × 25 cm. There were 4 treatments: A-control (CK), B-elementary S^0 (S^0: 100 mg/kg), C-sodium sulfate (Na_2SO_4: 100 mg/kg) and Dmercaptopygorskite (MP: 100 mg/kg), and each treatment has three replicates. Nutrients (N: urea 1.50 g/kg; P, K: potassium dihydrogen phosphate 1.10 g/kg) were mixed evenly with the soil samples and the overall soil layer was maintained with a same height. Then, a foil paper was placed around the transparent glass plate to protect it from light. Finally, deionized water was added to maintain a constant 2-cm flooded layer for 4 weeks.

The experimental setup is shown in Figures 1 and S3. Rice seedlings (Mingzhu silk-H variety) with 3 leaves were transplanted in the rhizo-box, which was then maintained at a 70° inclination to facilitate root growth along the lower front window (to ensure that rice roots grow along the glass plate). At rice growing stages of tillering, heading, maturing, the detachable front glass plate was removed, and the LDHs-DGT probe binding gel was placed at the rhizosphere zone and kept in close contact with the PVDF membrane (thickness of 100 μm and pore size of 0.45 μm). After overlying the gel with a nitrocellulose membrane protective filter, the removed glass plate was installed back. The probe gel was retrieved from the rhizo-box after deployment for four days. Afterwards, photos of the distribution of root system were taken using a digital camera. The distribution of S(-II) and a common toxic metal, Cd, in the gel was analyzed using ICP-MS. 2D images of S(-II) and Cd at rice rhizosphere were acquired after calculation using equation 2 and position assignation to the ordinated rhizosphere environment.

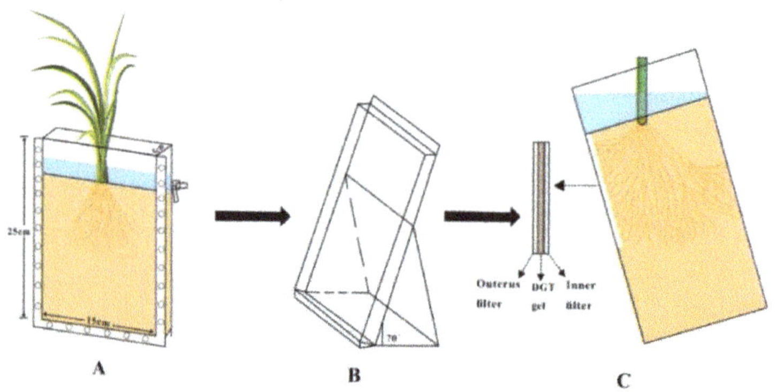

Figure 1. Experimental setup for imaging of the dynamic distributions of multi-elements in rice rhizosphere using LDHs-DGT probe. (**A**) The dimensions and arrangement of the rhizotron; (**B**) the rhizotron was tilted to anangle of 70° throughout the whole experiment; (**C**) cross-section showing the deployment sequence of innerfilte. Modified from He [36].

3. Results and Discussion

3.1. Laboratorial Characterization of the Performance of the LDHs-DGT Probe

3.1.1. Morphology and Structural Characterization of LDH_S

The morphology and particle size of the synthesized LDH_S were identified by FESEM and TEM. The FESEM micrographs in Figure S4A show that the LDHs exhibited a plate-like morphology, which is typical for layered double hydroxides, and the platelet size was approximately 200 nm. TEM characterization confirmed the hexagonal shape of the particles (Figure S4B). The existence of LDHs was confirmed by the X-ray diffraction pattern (Figure S4C), in which the major diffraction peaks matched well with the simulated pattern of LDH_S. The d-spacing of LDH_S samples was approximately 1.42 nm, and the channel height was approximately 0.94 nm. This result indicates that LDHs exist at an angle relative to the brucite layers. The molecular structure of the LDH_S is shown in Figure S4D.

3.1.2. Diffusion Coefficient of Multi-Elements in LDH_S-DGT Diffusive Layer

The diffusion coefficients of phosphate, SO_4^{2-}, S(-II), As(III) and As(V) were determined by the DGT time-series deployment method [19,37]. Linear relationships were also observed for the accumulated mass of phosphate, SO_4^{2-}, S(-II), As(III), and As(V) with their concentrations (Figure S5). The diffusion coefficients of S(II), SO_4^{2-}, phosphate, As(III) and As(V) at 25 °C (Table 1) in the LDHs-DGT diffusion gel were in the order of 10^{-6} [24,38], but were somewhat lower than those of the APA gel and the agarose gel. This is because the diffusion gel of LDHs-DGT is made of acrylamide, which has a small pore size (<5 nm). However, the APA and agarose gels had larger pore sizes of >5 nm and >20 nm, respectively [34]. A larger pore size will produce a lower tortuosity within the gel and promote ion diffusion [34]. Additionally, APA and agarose gels have a small positive charge, and thus, Donnan partitioning at the gel–solution interface is possible, thereby increasing the diffusion coefficients of the anionic species [39].

Table 1. Diffusion coefficient (E^{-6} cm^2 s^{-1}) measured by LDH_S-DGT devices at 25 °C. Data are mean ± SD (n = 3).

Analyte	S(-II)	SO₄(II)	Phosphate	As(III)	As(V)
D_{DGT}	5.99 ± 0.12	7.50 ± 0.28	4.57 ± 0.14	3.73 ± 0.15	3.41 ± 0.10

3.1.3. Selectivity of LDH_S-DGT Probe for Sulfide and Arsenate

Realizing the in situ measurement of S(-II) and As(V) is important in determining the selectivity of the LDHs-DGT probe. The amount of As(V) captured by LDHs-DGT from the mixed solution increased linearly with time, while the amount of retained As(III) was below 4% even after 72 h (Figure S6). Similarly, when SO_4^{2-} and S(-II) ions co-existed in the solution, the mass of S(-II) by LDHs-DGT increased linearly with time as expected, but the capture of SO_4^{2-} was highly inefficient, i.e., below 4% after 72 h (Figure S6). Therefore, LDH_S-DGT can achieve in situ selective adsorption of S(-II) and As(V).

3.1.4. Effects of pH and Ionic Strength on the Measurement of LDH_S-DGT Probe

The effects of environmental conditions on the performance of the LDHs-DGT were investigated by deploying the LDHs-DGT in a mixed solution containing phosphate, S(-II), and As(V) under different pH conditions (pH 4–9) and ionic strengths (0.001–0.7 mol L^{-1} $NaNO_3$) (Figure 2). The ratio of DGT-measured concentrations to those in solutions (C_{DGT}/C_{sol}) is widely used to judge the performance of the DGT probe and 0.9–1.1 indicates a good performance [12]. In the solution with pH in the range of 5–8, the ratios were all in the range of 0.9–1.1. However, at pH < 5 or pH > 8, the ratio was below 0.9. This can be explained by that the presence of excess hydrogen ions at pH < 5 that inhibit the adsorption of the analytes onto the binding gel [12,28]. Further, under low pH conditions, the LDHs structure is less stable, thereby reducing the adsorption of the analytes [40]. Furthermore, under higher pH conditions (pH > 8), the hydroxide precipitation that occurs on the surface of the LDH_S may affect the gel's adsorption performance [40]. LDHs-DGT performed well when ionic strength was at 5–100 mmol L^{-1} $NaNO_3$ (Figure 2B). The ratio of C_{DGT} to C_{sol} decreases with an increase in ionic strength due to the fact that the ionic strength of the solution affects the activity coefficient of the ions [40]. At extremely low ionic strength (1 mmol L^{-1} $NaNO_3$), the ratio for S(-II) and As(V) was slightly higher than

1.1. This is probably caused by the formation of an electric double layer complex within the LDHs materials which improves adsorption [24] or electrostatic interactions due to particle aggregation in LDHs [41]. At high ionic strength (700 mmol L^{-1} $NaNO_3$), the solvation shell around the ions becomes larger and the activity coefficient increases [40], causing a lower ratio than 0.9. Therefore, the LDHs-DGT is suitable for measuring multi-elements at pH of 4–8 and ionic strength at 5–100 mmol L^{-1} $NaNO_3$. For sole measurement of phosphate using LDHs-DGT, the environmental conditions can be slightly extended.

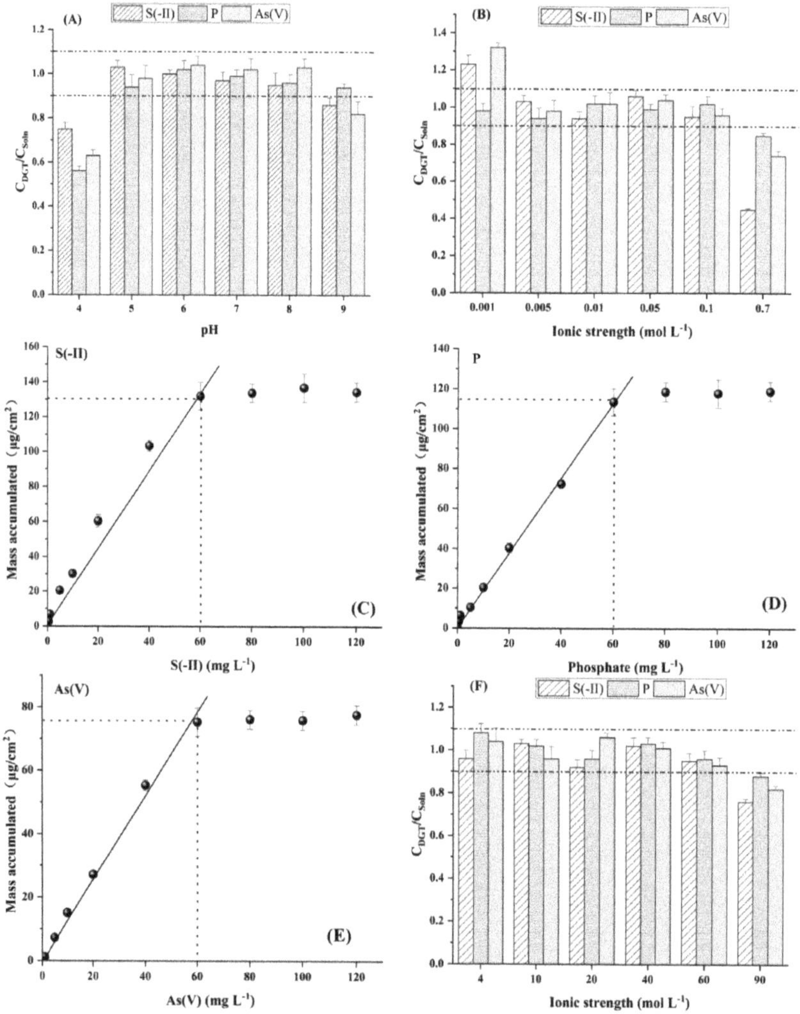

Figure 2. Effect of pH (**A**) and ionic strength (**B**) on measurement of S(-II), P and As(V) by LDH_S-DGT, time-depended mass accumulation of S(-II) (**C**), P (**D**) and As(V) € onto LDHS-DGT binding gel, and competition effect among S(-II), P and As(V) for their LDH_S-DGT measurement (**F**). In (**A**,**B**), the solutions contained 50 μg L−1 mixed ions of S(-II), P and As(V). The data are presented as average ± standard deviations (n = 3). The lower and higher dotted lines in (**A**,**B**,**F**) represent the values of 0.9 and 1.1, respectively, whereas the dotted liens in (**C**–**E**) indicate the capacity for S(-II), P and As(V) measured by LDH_S-DGT.

3.1.5. Capacity of LDHs-DGT Probe for Measurement of Sulfide, Phosphate and Arsenate and Their Competition Effects

The capacity of LDH_S-DGT was investigated by deploying the probes in solutions containing S(-II), phosphate and As(V) with concentrations in the ranges ~0.5–120 mg L^{-1} (pH = 5.50, IS = 0.01 mol L^{-1} $NaNO_3$). The mass of the ions measured in the LDH_S-DGT binding gel increased linearly with their concentration in the solution (Figure 2C–E) until the capacity was reached. Specifically, the capacities for S(-II), phosphate and As(V) were as high as 131.98, 113.56 and 75.31 μg cm^2, respectively. The capacity for S(-II) of LDH_S-DGT, was slightly lower than those of AgI and ZrO-AgI [37,38], but the capacity of LDHs-DGT for phosphate was higher than those of Chelex-100 (102 μg cm^2) and ZrO-Chelex (90 μg cm^2). Even though the capacity of LDH_S-DGT for As was lower than those of ZrO-Chelex and ZrO-CA (87.4 and 90.0 μg cm^2, respectively), it was still much higher than those of precipitated zirconia, Metsorb, precipitated/slurry ferrihydrite, and ZrO-CA (s) (42.6, 26.1, 27.7/10 and 47.2 μg cm^2, respectively) [18,29,42,43]. This is because LDHs have a large surface area and a flexible intermediate region, which enables them to absorb large amounts of oxyanions [24]. Therefore, LDHs-DGT has a high capacity for multi-elements and is suitable for long-term deployment in diverse environments.

The possible competition effect among the tested analytes was investigated and the results are shown in Figure 2F. When the ion concentration of the mixed solution was in the range of 4–60 μg L^{-1}, the ratio (C_{DGT}/C_{soln}) was within the expected range of values (0.9–1.1) indicating that there is no significant competition between unstable species at the binding sites [19]. Only a slight competition effect was observed at high ion concentration at 90 μg L^{-1}. Generally, tor LDH_S-DGT, the competition effect among ions was negligible.

3.2. Measurement of Labile Concentrations of Multi-Elements in Field Waters Using LDH_S-DGT Probe

The chemistries and hydrological conditions of the water body affect the mobility and species of redox sensitive elements, and thus affect the water quality [44,45]. The concentrations of S(-II), phosphate and As(V) in sample W1 were 15.50, 24.62 and 8.53 μg L^{-1}, respectively (Table 2), which represent the characteristics of a common water body. Among them, the concentration of phosphate exceeds the pollution threshold for surface water (0.02 mg L^{-1}) [46]. The concentration of phosphate in sample W2, representing an eutrophicated water body, is 74.52 μg L^{-1}, which is 3.8 times higher than the threshold. The concentration of S(-II), phosphate and As(V)) in sample W3, which was taken from an artificial lake, were 10.26, 19.45 and 12.22 μg L^{-1}, respectively, which indicates that the surface water is uncontaminated. This study is the first time that the LDH_S-DGT technique has been applied in the simultaneous determination of S(-II), phosphate and As(V) levels in water. The R value (ratio of C_{DGT} to C_{sol}) at the range of 0.9–1.1 (Table 2) reflects the reliability of LDHs-DGT measurement in real water samples.

Table 2. The pH of the water samples, concentrations of S(-II), phosphate, and As(V) in water (C_{water}) and those measured by LDH_S-DGT (C_{DGT}), and the ratios (R) of C_{water}/C_{DGT}. Data are mean ± SD (n = 3).

Water	W1	W2	W3
pH	6.81 ± 0.07	5.93 ± 0.06	7.26 ± 0.07
S(-II)-C_{water} (μg L^{-1})	15.50 ± 0.62	58.46 ± 4.09	10.26 ± 0.51
Phosphate-C_{water} (μg L^{-1})	24.62 ± 0.49	74.52 ± 2.24	19.45 ± 0.58
As(V)-C_{waterl} (μg L^{-1})	8.53 ± 0.17	4.34 ± 0.09	12.22 ± 0.49
S(-II)-C_{DGT} (μg L^{-1})	14.11 ± 0.42	54.92 ± 1.65	9.23 ± 0.28
Phosphate-C_{DGT} (μg L^{-1})	22.90 ± 0.69	72.28 ± 2.89	17.89 ± 0.54

Table 2. *Cont.*

Water	W1	W2	W3
As(V)-C_{DGT} (μg L^{-1})	8.10 ± 0.32	4.04 ± 0.08	11.36 ± 0.34
[a] $R_{S(-II)}$	0.91	0.94	0.90
[a] $R_{phosphate}$	0.93	0.97	0.92
[a] $R_{As(V)}$	0.95	0.94	0.90

[a] Obtained by averaging the ratios of C_{DGT} divided by C_{water}.

3.3. Application of LDHs-DGT Probe in Soil/Sediment-Water Interface for Measurement of Multi-Elements at 1D Centimeter Scale

The vertical distributions of S(-II), phosphate and As(V) in sediments were measured using the LDH_S-DGT (Figure 3). The average DGT concentrations of S(-II) and phosphate were higher in the vertical direction (51 and 56.3 μg L^{-1}), and the average DGT concentration of As(V) (5 μg L^{-1}) was significantly lower than those of phosphate and S(-II) ($p < 0.05$). Additionally, the changes in the vertical concentrations of S(-II), phosphate and As(V) were larger, and that of S(-II) was largest. This confirms that the redox reaction, which mainly involves Fe and Mn cycling, is the main factor leading to the heterogeneity of the chemical distribution [34].

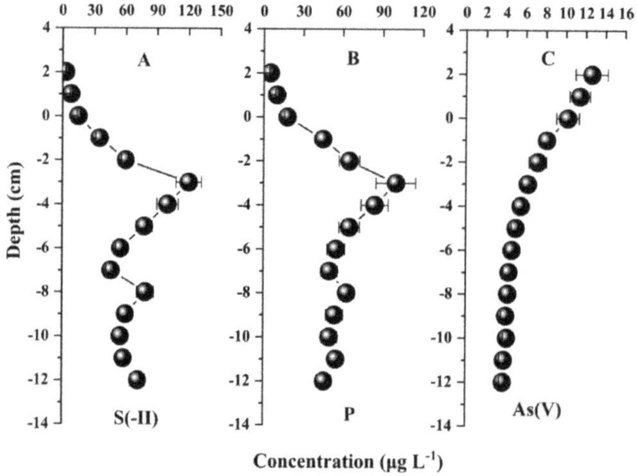

Figure 3. Vertical distributions of labile concentrations of S(-II) (**A**), phosphate (**B**) and As(V) (**C**) across sediment–water interface. The sediment core was collected from Jinhe River, Tianjin, China. Data were presented as the average values and standard deviations (n = 3). The resolution of the concentration profile is 1 cm.

Even though the chemical distribution was uneven in the vertical direction, the vertical distributions of S(-II), phosphate and As(V) were similar. Their concentration increased rapidly from the overlying water to a maximum value at a depth of approximately 3 cm below the sediment–water interface. However, the concentration of S(-II) first decreased to a depth of 8 cm below the interface, then increased, and then decreased again; the concentration of phosphate decreased rapidly, and then stabilized to the bottom of the profile, whereas the concentration of As(V) dropped directly to the bottom of the profile. The unstable changes in the concentrations of S(-II), phosphate and As(V) in the DGT further demonstrate that they are released during the reductive dissolution of Fe and Mn oxides [25,44,47]. This generally shows that LDH_S-DGT is highly useful for acquiring centimeter scale resolution concentration profiles of multi-elements and this function can be further played for exploring the processes in the soil–water interfaces.

3.4. Measurement of the Dynamic Distribution of Multi-Elements at Rice Rhizosphere at 2D Millimeter Scale

Imaging of element distribution at the rice rhizosphere tells previously hardly reached information at microniches. The images of S(-II) and Cd at three typical stages of rice growth are shown in Figures 4 and S7, with the resolution being at a millimeter scale (the data were interpolated to some extend). The uneven and dynamic distributions of the two elements were clearly shown. This highlights the necessity of 2D high spatio-resolution measurements.

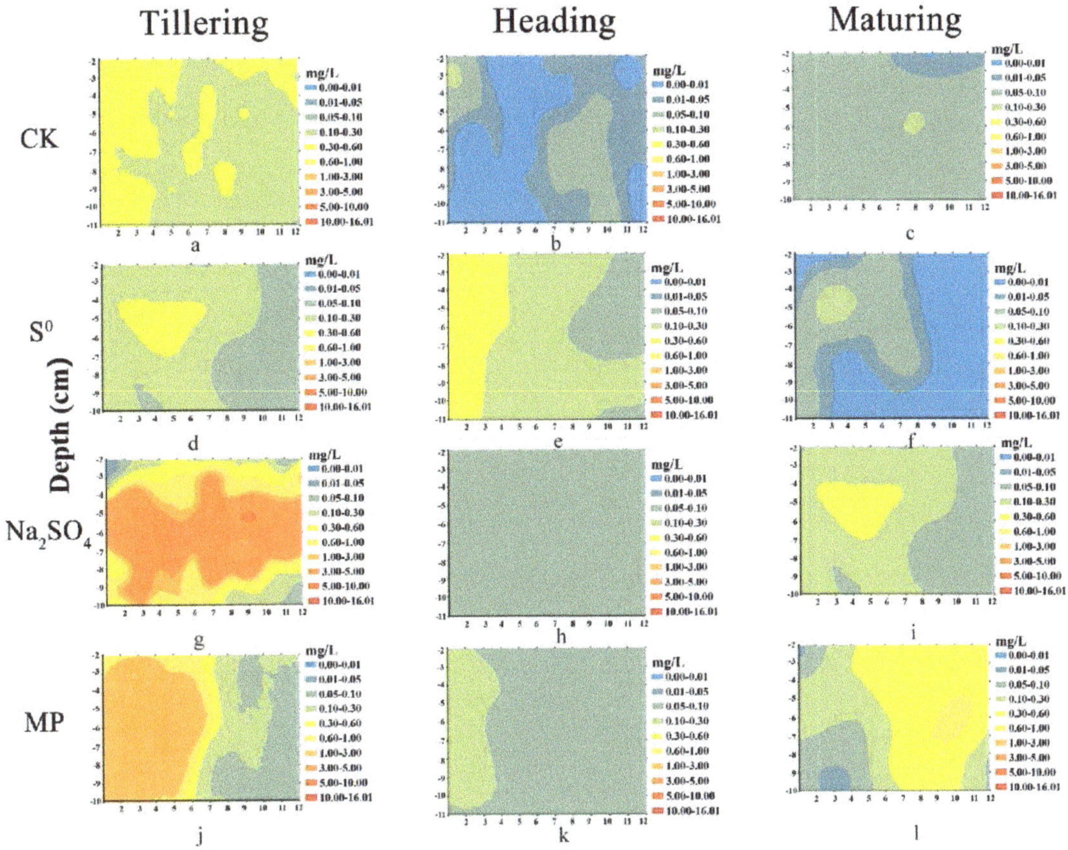

Figure 4. Two-dimensional distribution of S(-II) in the rhizosphere of rice at different growth stages (Tillering: **a**,**d**,**g**,**j**; Heading: **b**,**e**,**h**,**k**; Maturing: **c**,**f**,**i**,**l**). MP refers to mercaptopygorskite. Arc GIS 10.8 software was used to draw this figure.

The application of sulfur promoted the increase of S(-II) in rice root zone soil (Figure 4). This is mainly because flooding conditions can promote the increase of sulfate reducing bacteria and enhance sulfate reduction [48,49], thus promoting the increase of S(-II). In order to adapt to long-term flooding conditions, rice roots transport oxygen from aboveground leaves to the root system, and the root system secretes oxygen to the surrounding environment [50,51]. Therefore, the concentration of S(-II) in rhizosphere soil increased firstly, then decreased and then increased in different growth stages of rice. At heading stage, due to the formation of rhizome nodes, the process of oxygen transport from the ground to the root system is blocked [50]. In addition, S(-II) and Fe will form FeS precipitation [50], so the concentration of S(-II) in rhizosphere soil is the lowest. Thus, the formation of CdS is reduced, and Cd has the highest activity.

Sulfate reduction is an important process in controlling the dynamics of toxic metal(loid)s (e.g., Cd) in paddy soils [36,52,53]. The experimental results showed that the addition of S promoted the reduction of sulfate and inhibited the activation of Cd (Figures 4 and S4). Sulfate reduction can promote the formation of organic bound Cd and residual Cd, thus reducing labile Cd [15]. Therefore, the activity of Cd increased first and then decreased during the growth period of rice (Figure S4). Different S treatment groups resulted in distinct sulfuric acid reduction degree. Na_2SO_4 had a stronger effect than that of mercaptopygorskite, and that of S^0 is the lowest, which highlights the importance of valence state of S. The results show that $LDHs_S$-DGT is able to capture the dynamic processes of multi-elements in the rice rhizosphere at a millimeter-scale high spatio-resolution. Further, $LDHs_S$-DGT can be combined with other imaging tools such as planar optode for pH and O_2 and soil zymography for enzyme activities to provide a holistic understanding of the highly complex processes in crop rhizosphere.

4. Conclusions

This work provides a useful tool based on DGT technique for co-measurement of multi-elements, i.e., sulfides (S(-II)) and oxyanions (phosphate, As(V)). This is realized by taking advantage of the superior ion exchange properties of Mg-Al LDHs, which were used to prepare a novel binding gel in the $LDHs_S$-DGT probe. $LDHs_S$-DGT can be applied in a wide range of pH and ionic strength conditions. $LDHs_S$-DGT is able to capture the dynamic processes of multi-elements at soil/sediment–water interfaces and rice rhizosphere at high spatio-resolution (centimeter to millimeter scale). The co-measurement of 1D/2D processes in heterogeneous soils can significantly improve our understanding of the micro-niche biogeochemistry of inter-linked elements (e.g., sulfides and oxyanions) and prompt the establishment of fine management measures in sustainable agriculture.

Supplementary Materials: The following are available online at https://www.mdpi.com/article/10.3390/agronomy11122383/s1, Figure S1: $LDHs_S$-DGT probe device, Figure S2: $LDHs_S$-DGT test stand, Figure S3: Experimental setup for imaging of the dynamic distributions of mult-elements in rice rhizosphere using LDHs-DGT probe, Figure S4: SEM (A) and TEM images (B) of LDHs samples; XRD patterns of the standard and as-prepared LDHs (C); Schematic representation of LDHs structure (D), Figure S5: The accumulated mass of S(-II) (A), SO_4^{2-} (B), phosphate (C), As(III) (D), As(V) (E) in the binding gel with the time of LDHs-DGT deployment. Values are mean ± SD (n = 3), Figure S6: LDHs-DGT uptake curve in mixed solution of S(-II) and SO_4^{2-} (A), and As(III) and As(V) (B). The value is the average ± SD (pH = 5.50, n = 3), Figure S7: Two-dimensional distribution of labile Cd concentrations in rhizosphere at different growth stages of rice. MP referes to mercaptopygorskite. Arc GIS 10.8 (Esri, USA) software was used to draw this figure.

Author Contributions: Data curation, formal analysis, investigation, methodology, writing—original draft, writing—review & editing, M.Z. and J.L.; investigation, resources, validation, C.Z., X.L. (Xuefeng Liang), Q.E. and R.L.; funding acquisition, methodology, project administration, supervision, writing—review & editing, Y.Z. and X.L. (Xiaowei Liu). All authors have read and agreed to the published version of the manuscript.

Funding: This research was funded by the National Key Research and Development Program of China (No. 2016YFD0800307-4, and No. 2016YFD0800301-01).

Data Availability Statement: Data sharing not applicable.

Acknowledgments: We would like to thank Xiangyan Wang for the sample collection and preparation.

Conflicts of Interest: The authors declare no conflict of interest.

References

1. Deng, X.Z.; Guang, H.D.; Weng, J.Z.; Ya, G.; Hong, B.J.; Wei, R.; Ying, J.; Dai, C.W. Remediation of arsenic-contaminated paddy soil: Effects of elemental sulfur and gypsum fertilizer application. *Ecotoxicol. Environ. Saf.* **2021**, *223*, 112606.
2. Qin, W.; Gu, Y.; Wang, G.; Wu, T.; Zhang, H.; Tang, X.; Zhang, Y.; Zhao, H. Zirconium metal organic frameworks-based DGT technique for in situ measurement of dissolved reactive phosphorus in waters. *Water Res.* **2018**, *147*, 223–232. [CrossRef]

3. Pi, K.; Wang, Y.; Xie, X.; Ma, T.; Su, C.; Liu, Y. Role of sulfur redox cycling on arsenic mobilization in aquifers of Datong Basin, northern China. *Appl. Geochem.* **2017**, *77*, 31–43. [CrossRef]
4. Strawn, D.G. Review of interactions between phosphorus and arsenic in soils from four case studies. *Geochem. Transactions* **2018**, *19*, 10. [CrossRef] [PubMed]
5. Silva,, M.; Da, S.; Abate,, G.; Masini,, J. Spectrophotometric determination of acid volatile sulfide in river sediments by sequential injection analysis exploiting the methylene blue reaction. *Talanta* **2001**, *53*, 843–850. [CrossRef]
6. Kirk, M.F.; Roden, E.E.; Crossey, L.J.; Brealey, A.J.; Spilde, M.N. Experimental analysis of arsenic precipitation during microbial sulfate and iron reduction in model aquifer sediment reactors. *Geochim. Cosmochim. Acta* **2010**, *74*, 2538–2555. [CrossRef]
7. Kocar, B.; Borch, T.; Fendorf, S. Arsenic repartitioning during biogenic sulfidization and transformation of ferrihydrite. *Geochim. Cosmochim. Acta* **2010**, *74*, 980–994. [CrossRef]
8. Davlson, W.; Zhang, H. In situspeciation measurements of trace components in natural waters using thin-film gels. *Nature* **1994**, *367*, 546–548. [CrossRef]
9. Kreuzeder, A.; Santner, J.; Prohaska, T.; Wenzel, W.W. Gel for simultaneous chemical imaging of anionic and cationic solutes using diffusive gradients in thin films. *Anal. Chem.* **2013**, *85*, 12028–12036. [CrossRef]
10. Anonymous, J.; Broney, C. *Environmental Chemistry Study Findings on Environmental Chemistry Are Outlined in Reports from A*; Stockdale and Colleagues: Atlanta, GA, USA, 2008; p. 371.
11. Guan, D.X. Diffusive Gradients in Thin-Films (DGT): An Effective and Simple Tool for Assessing Contaminant Bioavailability in Waters, Soils and Sediments. *Encycl. Environ. Health* **2019**, *1025*, 111–124.
12. Zhang, H.; Davison, W. Performance characteristics of diffusion gradients in thin films for the in situ measurement of trace metals in aqueous solution. *Anal. Chem.* **1995**, *67*, 3391. [CrossRef]
13. Zhang, H.; Davison, W.; Knight, B.; McGrath, S. In situ measurements of solution concentrations and fluxes of trace metals in soils using DGT. *Environ. Sci. Technol.* **1998**, *32*, 704–710. [CrossRef]
14. Zhang, H.; Zhao, F.J.; Sun, B.; Davison, W.; McGrath, S.P. A new method to measure effective soil solution concentration predicts copper availability to plants. *Environ. Sci. Technol.* **2001**, *35*, 2602–2607. [CrossRef]
15. Zhao, M.; Qian, E.; Zhang, F.; Liu, R.; Liu, X.; Zhao, Y.; Liang, X. Spatiotemporal dynamics of labile Cd in soil during rice growth. *Sci. Total Environ.* **2020**, *738*, 139832. [CrossRef] [PubMed]
16. Luo, J.; Zhang, H.; Santner, J.; Davison, W. Performance characteristics of diffusive gradients in thin films equipped with a binding gel layer containing precipitated ferrihydrite for measuring arsenic (V), selenium (VI), vanadium (V), and antimony (V). *Anal. Chem.* **2010**, *82*, 8903–8909. [CrossRef] [PubMed]
17. Knutsson, J.; Rauch, S.; Morrison, G. Estimation of Measurement Uncertainties for the DGT Passive Sampler Used for Determination of Copper in Water. *Int. J. Anal. Chem.* **2014**, *389125*, 1–7. [CrossRef]
18. Guan, D.X.; Williams, P.N.; Luo, J.; Zheng, J.L.; Xu, H.C.; Cai, C.; Ma, L.Q. Novel precipitated zirconia-based DGT technique for high-resolution imaging of oxyanions in waters and sediments. *Environ. Sci. Technol.* **2015**, *49*, 3653–3661. [CrossRef]
19. Ding, S.; Xu, D.; Wang, Y.; Wang, Y.; Li, Y.; Gong, M.; Zhang, C. Simultaneous Measurements of Eight Oxyanions Using High-Capacity Diffusive Gradients in Thin Films (Zr-Oxide DGT) with a High-Efficiency Elution Procedure. *Environ. Sci. Technol.* **2016**, *50*, 7572–7580. [CrossRef]
20. Österlund, H.; Chlot, S.; Faarinen, M.; Widerlund, A.; Rodushkin, I.; Ingri, J.; Baxter, D.C. Simultaneous measurements of As, Mo, Sb, V and W using a ferrihydrite diffusive gradients in thin films (DGT) device. *Anal. Chim. Acta* **2010**, *682*, 59–65. [CrossRef]
21. Fan, X.; Ding, S.; Chen, M.; Gao, S.; Fu, Z.; Gong, M.; Tsang, D.; Wang, Y.; Zhang, C. Peak Chromium Pollution in Summer and Winter Caused by High Mobility of Chromium in Sediment of a Eutrophic Lake: In Situ Evidence from High Spatiotemporal Sampling. *Environ. Sci. Technol.* **2019**, *53*, 4755–4764. [CrossRef]
22. Fan, X.; Ding, S.; Chen, M.; Gao, S.; Fu, Z.; Gong, M.; Wang, Y.; Zhang, C. Mobility of chromium in sediments dominated by macrophytes and cyanobacteria in different zones of Lake Taihu. *Sci. Total Environ.* **2019**, *666*, 994–1002. [CrossRef] [PubMed]
23. Panther, J.G.; Stewart, R.R.; Teasdale, P.R.; Bennett, W.W.; Welsh, D.T.; Zhao, H. Titanium dioxide-based DGT for measuring dissolved As (V), V (V), Sb (V), Mo (VI) and W (VI) in water. *Talanta* **2013**, *105*, 80–86. [CrossRef]
24. Wu, P. *Anionic Clay Intercalation Construction and Environmental Remediation Technology*; Science Press: Beijing, China, 2016.
25. Chen, T.; Zhan, X.; Fan, M.; Chen, G.; Sun, J. Intercalation of pyrophosphate in the structure of anionic clay for instant synthesis. *Miner. Rock* **2005**, *76*, 105–108.
26. Türk, T.; Alp, İ.; Deveci, H. Adsorption of As (V) from water using Mg–Fe-based hydrotalcite (FeHT). *J. Hazard. Mater.* **2009**, *171*, 665–670. [CrossRef] [PubMed]
27. Yang, K.; Yan, L.G.; Yang, Y.M.; Yu, S.J.; Shan, R.R.; Yu, H.Q.; Zhu, B.C.; Du, B. Adsorptive removal of phosphate by Mg–Al and Zn–Al layered double hydroxides: Kinetics, isotherms and mechanisms. *Sep. Purif. Technol.* **2014**, *124*, 36–42. [CrossRef]
28. Liang, X.F.; Hou, W.G.; Xu, Y.M.; Sun, G.H.; Wang, L.; Sun, Y.; Qin, X. Sorption of lead ion by layered double hydroxide intercalated with diethylenetriaminepentaacetic acid. *Colloids Surf. a-Physicochem. Eng. Asp.* **2010**, *366*, 50–57. [CrossRef]
29. Wang, Y.; Ding, S.; Ren, M.; Li, C.; Xu, S.; Sun, Q.; Xu, L. Enhanced DGT capability for measurements of multiple types of analytes using synergistic effects among different binding agents. *Sci. Total Environ.* **2019**, *657*, 446–456. [CrossRef]
30. Hou, H.; Cui, W.; Xu, Q.; Tao, Z.; Guo, Y.; Deng, T. Arsenic Species Analysis at Trace Level by High Performance Liquid Chromatography with Inductively Coupled Plasma Mass Spectrometry. *Int. J. Anal. Chem.* **2019**, *4*, 1–6. [CrossRef]

31. Komorowicz, I.; Sajnóg, A.; Barałkiewicz, D. Total arsenic and arsenic species determination in freshwater fish by ICP-DRC-MS and HPLC/ICP-DRC-MS techniques. *Molecules* **2019**, *24*, 3. [CrossRef]
32. Morales, J.A.; de Graterol, L.S.; Mesa, J. Determination of chloride, sulfate and nitrate in groundwater samples by ion chromatography. *J. Chromatogr. A* **2000**, *884*, 185–190. [CrossRef]
33. Zhao, X.; Wang, H.; Tang, Z.; Zhao, T.; Qin, N.; Li, H.; Wu, F.; Giesy, J. Amendment of water quality standards in China: Viewpoint on strategic considerations. *Environ. Sci. Pollut. Res.* **2018**, *25*, 3078–3092. [CrossRef]
34. Wang, Y.; Ding, S.; Gong, M.; Xu, S.; Xu, W.; Zhang, C. Diffusion characteristics of agarose hydrogel used in diffusive gradients in thin films for measurements of cations and anions. *Anal. Chim. Acta* **2016**, *945*, 47–56. [CrossRef]
35. Guan, D.X.; He, S.X.; Li, G.; Teng, H.H.; Ma, L.Q. Application of diffusive gradients in thin-films technique for speciation, bioavailability, modeling and mapping of nutrients and contaminants in soils. *Crit. Rev. Environ. Sci. Technol.* **2021**, *1*, 1–45. [CrossRef]
36. He, S.; Wang, X.; Wu, X.; Yin, Y.; Ma, L.Q. Using rice as a remediating plant to deplete bioavailable arsenic from paddy soils. *Environ. Int.* **2020**, *141*, 105799. [CrossRef] [PubMed]
37. Ding, S.; Sun, Q.; Xu, D.; Jia, F.; He, X.; Zhang, C. High-resolution simultaneous measurements of dissolved reactive phosphorus and dissolved sulfide: The first observation of their simultaneous release in sediments. *Environ. Sci. Technol.* **2012**, *46*, 8297–8304. [CrossRef] [PubMed]
38. Teasdale, P.R.; Hayward, S.; Davison, W. In situ, High-Resolution Measurement of Dissolved Sulfide Using Diffusive Gradients in Thin Films with Computer-Imaging Densitometry. *Anal. Chem.* **1999**, *71*, 2186–2191. [CrossRef] [PubMed]
39. Guan, D.X.; Williams, P.N.; Xu, H.C.; Li, G.; Luo, J.; Ma, L.Q. High-resolution measurement and mapping of tungstate in waters, soils and sediments using the low-disturbance DGT sampling technique. *J. Hazard. Mater.* **2016**, *316*, 69–76. [CrossRef]
40. Liang, X.; Xu, Y.; Wang, L.; Sun, Y.; Lin, D.; Sun, Y.; Qin, X.; Wan, Q. Sorption of Pb^{2+} on mercapto functionalized sepiolite. *Chemosphere* **2013**, *90*, 548–555. [CrossRef]
41. Criscenti, L.; Sverjensky, D. The role of electrolyte anions (ClO_4^-, NO_3^-, and Cl^-) in divalent metal (M^{2+}) adsorption on oxide and hydroxide surfaces in salt solutions. *Am. J. Sci.* **1999**, *299*, 828–899. [CrossRef]
42. Bennett, W.W.; Teasdale, P.R.; Panther, J.G.; Welsh, D.T.; Jolley, D.F. New diffusive gradients in a thin film technique for measuring inorganic arsenic and selenium(IV) using a titanium dioxide based adsorbent. *Anal. Chem.* **2010**, *82*, 7401–7407. [CrossRef]
43. Wang, Y.; Ding, S.; Shi, L.; Gong, M.; Xu, S.; Zhang, C. Simultaneous measurements of cations and anions using diffusive gradients in thin films with a ZrO-Chelex mixed binding layer. *Anal. Chim. Acta* **2017**, *972*, 1–11. [CrossRef] [PubMed]
44. Coyte, R.M.; Vengosh, A. Factors controlling the risks of co-occurrence of the redox-Sensitive elements of arsenic, chromium, vanadium, and uranium in groundwater from the eastern united states. *Environ. Sci. Technol.* **2020**, *54*, 4367–4375. [CrossRef] [PubMed]
45. Puca, C.; Moldovan, M.; Silaghi-Dumitrescu, L.; Ungureanu, L.; Silaghi-Dumitrescu, R. On the apparent redox reactivity of "oxygen-enriched water". *Biol. Trace Elem. Res.* **2020**, *198*, 350–358.
46. Yan, Y.; Qian, Y.; Wang, Z.Y.; Yang, X.Y.; Wang, H.W. Ecological risk assessment from the viewpoint of surface water pollution in Xiamen City, China. *Int. J. Sustain. Dev. World Ecol.* **2018**, *25*, 403–410. [CrossRef]
47. Coffin, M.; Courtenay, S.C.; Pater, C.C.; Heuvel, M. An empirical model using dissolved oxygen as an indicator for eutrophication at a regional scale. *Mar. Pollut. Bull.* **2018**, *133*, 261–270. [CrossRef]
48. Jiang, M.; Sheng, Y.; Liu, Q.; Wang, W.; Liu, X. Conversion mechanisms between organic sulfur and inorganic sulfur in surface sediments in coastal rivers. *Sci. Total Environ.* **2020**, *752*, 141829. [CrossRef]
49. Wu, B.; Liu, F.; Fang, W.; Yang, T.; Wang, S. Microbial Sulfur metabolism and environmental implications. *Sci. Total Environ.* **2021**, *778*, 146085. [CrossRef]
50. Armstrong, J.; Armstrong, W. Rice: Sulfide-induced barriers to root radial oxygen loss, fesup2plus/sup and water uptake, and lateral root emergence. *Ann. Bot.* **2005**, *96*, 625–638. [CrossRef]
51. Xu, C.; Chen, L.; Chen, S.; Chun, G.; Wang, D.; Zhang, X. Effects of rhizosphere oxygen concentration on root physiological characteristics and anatomical structure at the tillering stage of rice. *Ann. Appl. Biol.* **2020**, *177*, 61–73. [CrossRef]
52. Wu, Z.Y.; Zhang, C.H.; Yan, J.L.; Yue, Q.; Ge, Y. Effects of sulfur supply and hydrogen peroxide pretreatment on the responses by rice under cadmium stress. *Plant Growth Regul.* **2015**, *77*, 299–306. [CrossRef]
53. Zhang, D.; Du, G.; Chen, D.; Shi, G.; Rao, W.; Li, X.; Jiang, Y.; Liu, S.; Wang, D. Effect of elemental sulfur and gypsum application on the bioavailability and redistribution of cadmium during rice growth. *Sci. Total Environ.* **2019**, *657*, 1460–1467. [CrossRef] [PubMed]

MDPI
St. Alban-Anlage 66
4052 Basel
Switzerland
www.mdpi.com

Agronomy Editorial Office
E-mail: agronomy@mdpi.com
www.mdpi.com/journal/agronomy

www.ingramcontent.com/pod-product-compliance
Lightning Source LLC
Chambersburg PA
CBHW061627210926
51850CB00009B/2467